AF335102

Multi-Carrier Techniques for Broadband Wireless Communications

A Signal Processing Perspective

Communications and Signal Processing

Editors: Prof. A. Manikas & Prof. A. G. Constantinides
(Imperial College London, UK)

Communications and Signal Processing – Vol. 3

Multi-Carrier Techniques for Broadband Wireless Communications

A Signal Processing Perspective

Man-On Pun

Princeton University, USA

Michele Morelli

University of Pisa, Italy

C-C Jay Kuo

University of Southern California, USA

Imperial College Press

Published by

Imperial College Press
57 Shelton Street
Covent Garden
London WC2H 9HE

Distributed by

World Scientific Publishing Co. Pte. Ltd.

5 Toh Tuck Link, Singapore 596224

USA office: 27 Warren Street, Suite 401-402, Hackensack, NJ 07601

UK office: 57 Shelton Street, Covent Garden, London WC2H 9HE

British Library Cataloguing-in-Publication Data
A catalogue record for this book is available from the British Library.

Communications and Signal Processing — Vol. 3
MULTI-CARRIER TECHNIQUES FOR BROADBAND WIRELESS COMMUNICATIONS
A Signal Processing Perspective

ISBN-13 978-1-86094-946-3
ISBN-10 1-86094-946-0

Desk editor: Tjan Kwang Wei

Printed in Singapore by Mainland Press Pte Ltd

To my wife Ying and my mother.
(Man-On Pun)

To my wife Monica and my son Tommaso.
(Michele Morelli)

To my parents, my wife Terri and my daughter Allison.
(C.-C. Jay Kuo)

Preface

The demand for multimedia wireless communications is growing today at an explosive pace. One common feature of many current wireless standards for high-rate multimedia transmission is the adoption of a multicarrier air interface based on either orthogonal frequency-division multiplexing (OFDM) or orthogonal frequency-division multiple-access (OFDMA). The latest examples of this trend are represented by the IEEE 802.11 and IEEE 802.16 families of standards for wireless local area networks (WLANs) and wireless metropolitan area networks (WMANs). Although the basic principle of OFDM/OFDMA is well established among researchers and communication engineers, its practical implementation is far from being trivial as it requires rather sophisticated signal processing techniques in order to fully achieve the attainable system performance.

This book is intended to provide an accessible introduction to OFDM-based systems from a signal processing perspective. The first part provides a concise treatment of some fundamental concepts related to wireless communications and multicarrier systems, whereas the second part offers a comprehensive survey of recent developments on a variety of critical design issues including synchronization techniques, channel estimation methods, adaptive resource allocation and practical schemes for reducing the peak-to-average power ratio of the transmitted waveform. The selection and treatment of topics makes this book quite different from other texts in digital communication engineering. In most books devoted to multicarrier transmissions the issue of resource assignment is not discussed at all while synchronization and channel estimation are only superficially addressed. This may give the reader the erroneous impression that these tasks are rather trivial and the system can always operate close to the limiting case of ideal synchronization and channel estimation. However, as discussed in this book, special design attentions are required for successfully accomplishing these tasks. In many cases, the trade-off between performance and system complexity has to be carefully taken into consideration in the practical implementation of multicarrier systems.

Most of the presented material originates from several projects and research activities conducted by the authors in the field of multicarrier transmissions. In order to keep the book concise, we do not cover advanced topics in multiple-input multiple-output (MIMO) OFDM systems as well as latest results in the field of resource assignment based on game theory. Also, we do not include a description of current wireless standards employing OFDM or OFDMA which are available in many other texts and journal papers.

The book is written for graduate students, design engineers in telecommunications industry as well as researchers in academia. Readers are assumed to be familiar with the basic concepts of digital communication theory and to have a working knowledge of Fourier transforms, stochastic processes and estimation theory. Whenever possible, we have attempted to keep the presentation as simple as possible without sacrificing accuracy. We hope that the book will contribute to a better understanding of most critical issues encountered in the design of a multicarrier communication system and may motivate further investigation in this exciting research area.

The authors acknowledge contributions of several people to the writing of this book. Many thanks go to Prof. Umberto Mengali who reviewed several portions of the manuscript and suggested valuable improvements to its original version. Without his advice and encouragement, this book would never have seen the light of day. We would also like to express appreciation to our co-workers and friends Antonio D'Amico, Marco Moretti and Luca Sanguinetti who reviewed the manuscript in detail and offered corrections and insightful comments. To all of them we owe a debt of gratitude. Special thanks go to Ivan Cosovic from NTT-DoCoMo, who critically read a first draft of the manuscript and provided invaluable suggestions.

M. Pun would like to thank his former colleagues at the SONY corporation, particularly Takahiro Okada, Yasunari Ikeda, Naohiko Iwakiri and Tamotsu Ikeda for first teaching him about the principle of OFDM. M. Pun would also like to acknowledge the Sir Edward Youde Foundation and the Croucher Foundation for supporting him in his research activity. M. Morelli would like to thank his wife Monica and son Tommaso for their support and understanding during the time he devoted to writing this book, and to his parents for their endless sacrifices. C.-C. J. Kuo would like to thank his parents, his wife Terri and daughter Allison for their encouragement and support for years.

Man-On Pun
Michele Morelli
C.-C. Jay Kuo

Contents

Chapter 1

Introduction

1.1   Aim of this book

The pervasive use of wireless communications is more and more conditioning lifestyle and working habits in many developed countries. Examples of this trend are the ever increasing number of users that demand Internet connection when they are traveling, the use of cellular phones to check bank accounts and make remote payments, or the possibility of sharing moments in our lives with distant friends by sending them images and video clips. In the last few years, the proliferation of laptop computers has led to the development of wireless local area networks (WLANs), which are rapidly supplanting wired systems in many residential homes and business offices. More recently, wireless metropolitan area networks (WMANs) have been standardized to provide rural locations with broadband Internet access without the costly infrastructure required for deploying cables. A new generation of wireless systems wherein multimedia services like speech, audio, video and data will converge into a common and integrated platform is currently under study and is expected to become a reality in the near future.

The promise of portability is clearly one of the main advantages of the wireless technology over cabled networks. Nevertheless, the design of a wireless communication system that may reliably support emerging multimedia applications must deal with several technological challenges that have motivated an intense research in the field. One of this challenge is the harsh nature of the communication channel. In wireless applications, the radiated electromagnetic wave arrives at the receiving antenna after being scattered, reflected and diffracted by surrounding objects. As a result, the receiver observes the superposition of several differently attenuated and

1

delayed copies of the transmitted signal. The constructive or destructive combination of these copies induces large fluctuations in the received signal strength with a corresponding degradation of the link quality. In addition, the characteristics of the channel may randomly change in time due to unpredictable variations of the propagation environment or as a consequence of the relative motion between the transmitter and receiver. A second challenge is represented by the limited amount of available radio spectrum, which is a very scarce and expensive resource. It suffices to recall that European telecommunication companies spent over 100 billion dollars to get licenses for third-generation cellular services. To obtain a reasonable return from this investment, the purchased spectrum must be used as efficiently as possible. A further impairment of wireless transmissions is the relatively high level of interference arising from channel reuse. Although advanced signal processing techniques based on multiuser detection have recently been devised for interference mitigation, it is a fact that mobile wireless communications will never be able to approach the high degree of stability, security and reliability afforded by cabled systems. Nevertheless, it seems that customers are ready to pay the price of a lower data throughput and worse link quality in order to get rid of wires.

The interest of the communication industry in wireless technology is witnessed by the multitude of heterogeneous standards and applications that have emerged in the last decade. In the meantime, the research community has worked (and is still working) toward the development of new broadband wireless systems that are expected to deliver much higher data rates and much richer multimedia contents than up-to-date commercial products. The ability to provide users with a broad range of applications with different constraints in terms of admissible delay (latency), quality of service and data throughput, demands future systems to exhibit high robustness against interference and channel impairments, as well as large flexibility in radio resource management. The selection of a proper air-interface reveals crucial for achieving all these features. The multicarrier technology in the form of orthogonal frequency-division multiplexing (OFDM) is widely recognized as one of the most promising access scheme for next generation wireless networks. This technique is already being adopted in many applications, including the terrestrial digital video broadcasting (DVB-T) and some commercial wireless LANs. The main idea behind OFDM is to split a high-rate data stream into a number of substreams with lower rate. These substreams are then transmitted in parallel over orthogonal subchannels characterized by partially overlapping spectra. Compared to single-carrier transmissions, this approach provides the system with increased resistance against narrowband interference and channel distortions. Furthermore, it ensures a high level of flexibility since modulation parameters like constellation size and coding rate can independently be selected over each subchannel. OFDM can also be combined with conventional multiple-access techniques for operation in a multiuser scenario. The most prominent scheme in this area is represented by orthogonal frequency-division multiple-access (OFDMA), which has become part of the emerging standards for wireless MANs.

Even though the concept of multicarrier transmission is simple in its basic principle, the design of practical OFDM and OFDMA systems is far from being a trivial task. Synchronization, channel estimation and radio resource management are only a few examples of the numerous challenges related to multicarrier technology. As a result of continuous efforts of many researchers, most of these challenging issues have been studied and several solutions are currently available in the open literature. Nevertheless, they are scattered around in form of various conference and journal publications, often concentrating on specific performance and implementation issues. As a consequence, they are hardly useful to give a unified view of an otherwise seemingly heterogeneous field. The task of this book is to provide the reader with a harmonized and comprehensive overview of new results in the rapidly growing field of multicarrier broadband wireless communications. Our main goal is to discuss in some detail several problems related to the physical layer design of OFDM and OFDMA systems. In doing so we shall pay close attention to different trade-offs that can be achieved in terms of performance and complexity.

1.2 Evolution of wireless communications

Before proceeding to a systematic study of OFDM and OFDMA, we think it useful to review some basic applications of such schemes and highlight the historical reasons that led to their development. The current section is devoted to this purpose, and illustrates the evolution of wireless communication systems starting from the theoretical works of Maxwell in the nineteenth century till the most recent studies on broadband wireless networks. Some historical notes on multicarrier transmissions are next provided in the last section of this introductory chapter.

1.2.1 *Pioneering era of wireless communications*

The modern era of wireless communications began with the mathematical theory of electromagnetic waves formulated by James Clerk Maxwell in 1873. The existence of these waves was later demonstrated by Heinrich Hertz in 1887, when for the first time a radio transmitter generated a spark in a receiver placed several meters away. Although Nikola Tesla was the first researcher who showed the ability of electromagnetic waves to convey information, Guglielmo Marconi is widely recognized as the inventor of wireless transmissions. His first publicized radio experiment took place in 1898 from a boat in the English Channel to the Isle of Wight, while in 1901 his radio telegraph system sent the first radio signal across the Atlantic Ocean from Cornwall to Newfoundland. Since then, the wireless communication idea was constantly investigated for practical implementation, but until the 1920s mobile radio systems only made use of the Morse code. In 1918 Edwin Armstrong invented the superheterodyne receiver, thereby opening the way to the first broadcast radio transmission that took place at Pittsburgh in 1920. In the subsequent years the radio became widespread all over the world, but in the meantime the research community was studying the possibility of transmitting real-time moving images through the air. These efforts culminated in 1929 with the first experiment of TV transmission made by Vladimir Zworykin. Seven years later the British Broadcasting Corporation (BBC) started its TV services.

Although radio and TV broadcasting were the first widespread wireless services, an intense research activity was devoted to develop practical schemes for bi-directional mobile communications, which were clearly appealing for military applications and for police and fire departments. The first mobile radio telephones were employed in 1921 by the Detroit Police Department's radio bureau, that began experimentation for vehicular mobile services. In subsequent years, these early experiments were followed by many others. In the 1940s, radio equipments called "carphones" occupied most of the police cars. These systems were powered by car batteries and allowed communications among closed group of users due to lack of interconnection with the public switched telephone network (PSTN). In 1946, mobile telephone networks interconnected with the PSTN made their first appearance in several cities across the United States. The main shortcoming of these systems was the use of a single access point to serve an entire metropolitan area, which limited the number of active users to the number of allocated frequency channels. This drawback motivated investigations as how to enlarge the number of users for a given allocated frequency band. A solution was found in 1947 by the AT&T's Bell Labs with the advent of the cellular concept [131], which represented a fundamental contribution in the development of wireless communications. In cellular communication systems, the served area is divided into smaller regions called *cells*. Due to its reduced dimension, each cell requires a relatively low power to be covered. Since the power of the transmitted signal falls off with distance, users belonging to adequately distant cells can operate over the same frequency band with minimal interference. This means that the same frequency band can be reused in other (most often non adjacent) cells, thereby leading to a more efficient use of the radio spectrum.

In 1957, the Union Soviet launched its first satellite Sputnik I and the United States soon followed in 1958 with Explorer I. The era of space exploration and satellite communications had begun. Besides being used for TV services, modern satellite networks provide radio coverage to wide sparsely populated areas where a landline infrastructure is absent. Typical applications are communications from ships, offshore oil drilling platforms and war or disaster areas.

1.2.2 *First generation (1G) cellular systems*

Despite its theoretical relevance, the cellular concept was not widely adopted during the 1960s and 1970s. To make an example, in 1976 the Bell Mobile Phone had only 543 paying customers in the New York City area, and mobile communications were mainly supported by heavy terminals mounted on cars. Although the first patent describing a portable mobile telephone was granted to Motorola in 1975 [25], mobile cellular systems were not introduced for commercial use until the early 1980s, when the so-called first generation (1G) of cellular networks were deployed in most developed countries. The common feature of 1G systems was the adoption of an analog transmission technology. Frequency modulation (FM) was used for speech transmission over the 800-900 MHz band and frequency-division multiple-access (FDMA) was adopted to separate users' signals in the frequency domain. In practice, a fraction of the available spectrum (subchannel) was exclusively allocated to a given user during the call set-up and retained for the entire call.

In the early 1980s, 1G cellular networks experienced a rapid growth in Europe, particularly in Scandinavia where the Nordic Mobile Telephony (NMT) appeared in 1981, and in United Kingdom where the Total Access

Communication System (TACS) started service in 1985. The Advanced Mobile Phone Service (AMPS) was deployed in Japan in 1979, while in the United States it appeared later in 1983. These analog systems created a critical mass of customers. Their main limitations were the large dimensions of cellphones and the reduced traffic capacity due to a highly inefficient use of the radio spectrum.

At the end of the 1980s, progress in semiconductor technology and device miniaturization allowed the production of small and light-weight hand-held phones with good speech quality and acceptable battery lifetime. This marked the beginning of the wireless cellular revolution that took almost everyone by surprise since in the meantime many important companies had stopped business activities in cellular communications, convinced that mobile telephony would have been limited to rich people and would have never attracted a significant number of subscribers.

1.2.3　*Second generation (2G) cellular systems*

The limitations of analog radio technology in terms of traffic capacity became evident in the late 1980s, when 1G systems saturated in many big cities due to the rapid growth of the cellular market. Network operators realized that time was ripe for a second generation (2G) of cellular systems that would have marked the transition from analog to digital radio technology. This transition was not only motivated by the need for higher network capacity, but also by the lower cost and improved performance of digital hardware as compared to analog circuitry.

Driven by the success of NMT, in 1982 the Conference of European Posts and Telecommunications (CEPT) formed the Group Spècial Mobile (GSM) in order to develop a pan-European standard for mobile cellular radio services with good speech quality, high spectral efficiency and the ability for secure communications. The specifications of the new standard were approved in 1989 while its commercial use began in 1993. Unlike 1G systems, the GSM was developed as a digital standard where users' analog signals are converted into sequences of bits and transmitted on a frame-by-frame basis. Within each frame, users transmit their bits only during specified time intervals (slots) that are exclusively assigned at the call setup according to a time-division multiple-access (TDMA) approach. Actually, the GSM is based on a hybrid combination of FDMA and TDMA, where FDMA is employed to divide the available spectrum into 200 kHz-wide subchannels while TDMA is used to separate up to a maximum of eight users allocated over the same subchannel. In Europe the operating frequency band is 900 MHz, even though in many big cities the 1800 MHz band is also being adopted to accommodate a larger number of users. Many modern European GSM phones operate in a "dual-band" mode by selecting either of the two recommended frequencies. In the United States, the 1900 MHz frequency band is reserved to the GSM service.

In addition to circuit-switched applications like voice, the adoption of a digital technology enabled 2G cellular systems to offer low-rate data services including e.mail and short messaging up to 14.4 kbps. The success of GSM was such that by June 2001 there were more than 500 millions GSM subscribers all over the world while in 2004 the market penetration exceeded 80% in Western Europe. The reasons for this success can be found in the larger capacity and many more services that the new digital standard offered as compared to previous 1G analog systems. Unfortunately, the explosive market of digital cellphones led to a proliferation of incompatible 2G standards that sometimes prevent the possibility of roaming among different countries. Examples of this proliferation are the Digital Advanced Mobile Phone Services (D-AMPS) which was introduced in the United States in 1991 and the Japanese Pacific Digital Cellular (PDS) [67]. The Interim Standard 95 (IS-95) became operative in the United States starting from 1995 and was the first commercial system to employ the code-division multiple-access (CDMA) technology as an air interface.

1.2.4 *Third generation (3G) cellular systems*

At the end of the 1990s it became clear that GSM was not sufficient to indefinitely support the explosive number of users and the ever-increasing data rates requested by emerging multimedia services. There was the need for a new generation of cellular systems capable of supporting higher transmission rates with improved quality of service as compared to GSM. After long deliberations, two prominent standards emerged: the Japanese-European Universal Mobile Telecommunication System (UMTS) [160] and the American CDMA-2000 [161]. Both systems operate around the 2 GHz frequency band and adopt a hybrid FDMA/CDMA approach. In practice, groups of users are allocated over disjoint frequency subbands, with users sharing a common subband being distinguished by quasi-orthogonal spreading codes. The CDMA technology has several advantages over TDMA and FDMA, including higher spectral efficiency and increased flexibility in radio resource management. In practical applications, however, channel distortions may destroy orthogonality among users' codes, thereby resulting in multiple-access interference (MAI). In the early 1990s, problems related to MAI mitigation spurred an intense research activity on CDMA and other spread-spectrum techniques. This led to the development of a large number of multiuser detection (MUD) techniques [164], where the inherent structure of interfering signals is exploited to assist the data detection process.

The introduction of 3G systems offered a wide range of new multimedia applications with the possibility of speech, audio, images and video transmissions at data rates of 144-384 kbps for fast moving users up to 2 Mbps for stationary or slowly moving terminals. In addition to the increased data rate, other advantages over 2G systems are the improved spectral efficiency, the ability to multiplex several applications with different quality of service requirements, the use of variable bit rates to offer bandwidth on demand and the possibility of supporting asymmetric services in the uplink and downlink directions, which is particularly useful for web browsing and high-speed downloading operations. Unfortunately, the impressive costs paid by telecom providers to get 3G cellular licenses slackened the deployment of the 3G infrastructure all over the world and led to a spectacular crash of the telecom stock market during the years 2000/2001. As a result, many startup companies went bankrupt while others decreased or stopped at all their investments in the wireless communication area. This also produced a significant reduction of public funding for academic research.

1.2.5   *Wireless local and personal area networks*

In the first years of the new millennium, the development of personal area networks (PANs) and wireless local area networks (WLANs) has suscitated a renewed interest in the wireless technology. These products provide wireless connectivity among portable devices like laptop computers, cordless phones, personal digital assistants (PDAs) and computer peripherals. Compared to wired networks they promise portability, allow simple and fast installation and save the costs for deploying cables. Because of their relatively limited coverage range, both technologies are mainly intended for indoor applications.

Several standards for PAN products have been developed by the IEEE 802.15 working group [62]. Among them, Bluetooth is perhaps the most popular scheme. The first release of Bluetooth appeared in 1999 while the first headset was produced by Ericsson in the year 2000. This technology enables low-powered transmissions with short operating ranges up to 10 meters. It provides wireless connection among closely spaced portable devices with limited battery power and must primarily be considered as a substitute for data transfer cables. Typical applications are the interconnection between a hands-free headset and a cellular phone, a DVD player and a television set, a desktop computer and some peripheral devices like a printer, keyboard or mouse. Bluetooth operates over the unlicensed Industrial, Scientific and Medical (ISM) frequency band, which is centered around 2.4 GHz. The allocated spectrum is divided into 79 adjacent subchannels which are accessed by means of a frequency-hopping spread-spectrum (FHSS) technique. Each subchannel has a bandwidth of 1 MHz for a data rate approaching 1 Mbps [44].

WLANs have a wider coverage area as compared to PANs and are mainly used to distribute the Internet access to a bunch of portable devices (typically laptop computers) dislocated in private homes or office buildings. A typical application is represented by a user who needs to be able to carry out a laptop into a conference room without losing network connection. WLANs are also being used in hotels, airports or coffee shops to create "hotspots" for public access to the Internet. The number of users that can simultaneously be served is usually limited to about 10, even though in principle more users could be supported by lowering the individual data rates. The typical network topology of commercial WLANs is based on a cellular architecture with cell radii up to 100 meters. In this case, several user terminals (UTs) establish a wireless link with a fixed access point (AP) which is connected to the backbone network as illustrated in Fig. 1.1. An alternative configuration is represented in Fig. 1.2, where an ad-hoc network is set up for peer-to-peer communications without involving any AP.

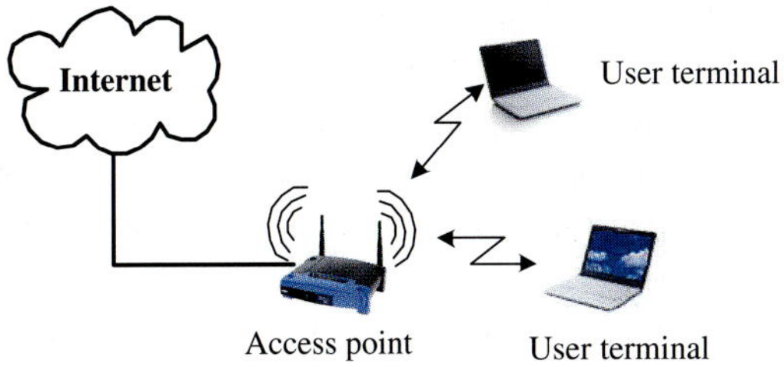

Fig. 1.1   Illustration of a WLAN with fixed access point.

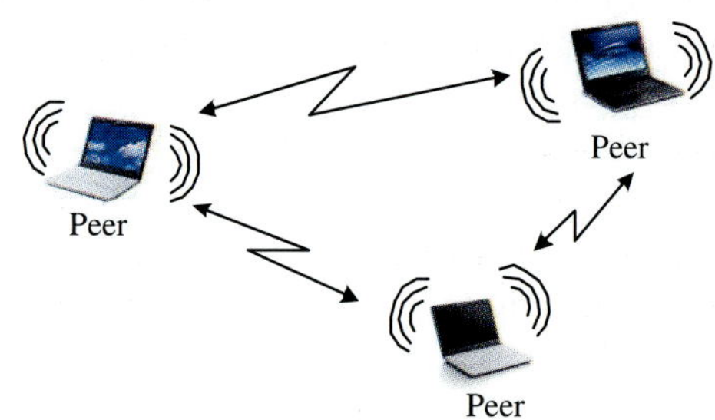

Fig. 1.2   Illustration of a WLAN for peer-to-peer communications.

The most successful class of WLAN products is based on the IEEE 802.11 family of standards. The first 802.11 release appeared in 1997 [58] and was intended to provide data rates of 1 and 2 Mbps. Three different physical layer architectures were recommended. The first two operate over the 2.4 GHz band and employ either a direct-sequence spread-spectrum or frequency-hopping technology. The third operational mode is based on infrared light and has rarely been used in commercial products. A first amendment called 802.11b was ratified in 1999 to improve the data rate up to 11 Mbps [60]. This product was adopted by an industry group called WiFi (Wireless Fidelity) and became soon very popular. In the same year a new amendment called 802.11a recommended the use of OFDM to further increase the data rate up to 54 Mbps [59]. This standard operates over the 5 GHz band, which is unlicensed in the US but not in most other countries. A TDMA approach is used to distinguish users within a cell while FDMA is employed for cell separation. A further evolution of the 802.11 family was approved in 2003 and is called 802.11g [61]. This standard is similar to 802.11a, except that it operates over the ISM band, which is license-exempt in Europe, United States and Japan.

Other examples of WLAN standards include the Japanese multimedia mobile access communication (MMAC) and the European high performance LAN (HiperLAN2) [41]. The physical layers of these systems are based on OFDM and only present minor modifications with respect to IEEE 802.11a. The major differences lie in the MAC layer protocols. Actually, HiperLAN2 employs a reservation based access scheme where each UT sends a request to the AP before transmitting a data packet, while 802.11 adopts Carrier-Sense Multiple-Access with Collision Avoidance (CSMA-CA), where each

UT determines whether the channel is currently available and only in that case it starts transmitting data. As for MMAC, it supports both of the aforementioned protocols.

The current generation of WLANs offers data rates of tens of Mbps and is characterized by low mobility and relatively limited coverage areas. The challenge for future WLANs is to extend the radio coverage and support new services like real-time video applications that are highly demanding in terms of data rate and latency.

1.2.6 *Wireless metropolitan area networks*

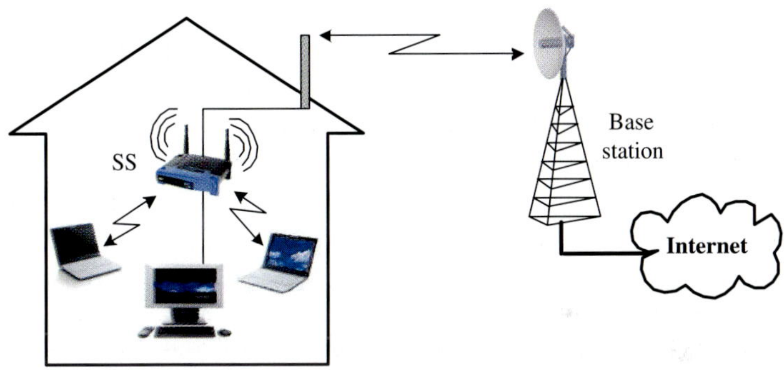

Fig. 1.3   Illustration of a WMAN providing wireless Internet access to a remote SS.

Wireless metropolitan area networks (WMANs) represent the natural evolution of WLANs. The purpose of these systems is to provide network access to residential or enterprise buildings through roof-top antennas communicating with a central radio base station, thereby replacing the wired "last mile" connection by a wireless link. This offers an appealing alternative to cabled access networks or digital subscriber line (DSL) links, and promises ubiquitous broadband access to rural or developing areas where broadband is currently unavailable for lack of a cabled infrastructure. Figure 1.3 depicts a typical scenario where the WMAN provides wireless Internet access to a Subscriber Station (SS) placed within a building. A WLAN or a backbone local network is used inside the building to connect the SS to the user terminals. In a more challenging application, the SS is mounted on a moving vehicle like a car or a train to provide passengers with continuous

Internet connectivity.

Several options for the WMAN air interface and MAC protocols are specified by the IEEE 802.16 Working Group, who started its activity in 1998. The goal was to deliver high data rates up to 50 Mbps over metropolitan areas with cell radii up to 50 kilometers. At the beginning, the interest of the Group focused on the 10-66 GHz band where a large amount of unlicensed spectrum is available worldwide. The first 802.16 release appeared in 2002 [63] and was specifically intended for line-of-sight (LOS) applications due to the severe attenuations experienced by short wavelengths when passing through walls or other obstructions. This standard adopts single-carrier (SC) modulation in conjunction with a TDMA access scheme. Transmission parameters like modulation and coding rates are adaptively adjusted on a frame-by-frame basis depending on the actual interference level and channel quality. The LOS requirement was the main limitation of this first release since rooftop antennas mounted on residential buildings are typically too low for a clear sight line to the base station antenna. For this reason, in the same year 2002 a first amendment called 802.16a was approved to support non line-of-sight (NLOS) operations over the 2-11 GHz band [112]. This novel standard defines three different air interfaces and a common MAC protocol with a reservation based access. The first air interface relies on SC transmission, the second employs OFDM-TDMA while the third operates according to the OFDMA principle in which users' separation is achieved at subcarrier level. Among the three recommended air interfaces, those based on OFDM and OFDMA seem to be favored by the vendor community due to their superior performance in NLOS applications. The last evolution of the 802.16 family is represented by the 802.16e specifications, whose standardization process began in the year 2004 [113]. This emerging standard adopts a scalable OFDMA physical layer and promises mobility at speeds up to 120 km/h by using adaptive antenna arrays and improved inter-cell handover. Its main objective is to provide continuous Internet connection to mobile users moving at vehicular speed.

In order to ensure interoperability among all 802.16-based devices and rapidly converge to a worldwide WMAN standard, an industry consortium called WiMax (Worldwide Interoperability for Microwave Access) Forum has been created. However, due to the large variety of data rates, coverage ranges and potential options specified in the standards, it is currently difficult to predict what type of performance WiMax-certified devices will reasonably provide in the near future.

1.2.7 *Next generation wireless broadband systems*

The demand for novel high-rate wireless communication services is growing today at an extremely rapid pace and is expected to further increase in the next years. This trend has motivated a significant number of research and development projects all over the world to define a fourth generation (4G) of wireless broadband systems that may offer increased data rates and better quality of service than current 3G products. The new wireless technology will support multimedia applications with extremely different requirements in terms of reliability, bit rates and latency. The integration of the existing multitude of standards into a common platform represents one of the major goals of 4G systems, which can only be achieved through the adoption of a flexible air interface with high scalability and interoperability [57, 138].

Software Defined Radio (SDR) represents a viable solution to provide 4G systems with the necessary level of flexibility and reconfigurability [4, 159, 170]. The main concept behind SDR is that different transceiver functions are executed as software programs running on suitable processors. Once the software corresponding to existing standards has been pre-loaded on the system, the SDR platform guarantees full compatibility among different wireless technologies. In addition, SDR can easily incorporate new standards and protocols by simply loading the specific application software.

A second challenge for next generation systems is the conflict between the increasing demand for higher data rates and the scarcity of the radio spectrum. This calls for an air interface characterized by an extremely high spectral efficiency. Recent advances in information theory has shown that large gains in terms of capacity and coverage range are promised by multiple-input multiple-output (MIMO) systems, where multiple antennas are deployed at both ends of the wireless link [46]. Based on these results, it is likely that the MIMO technology will be widely adopted in 4G networks. An alternative way for improving the spectral efficiency is the use of flexible modulation and coding schemes, where system resources are adaptively assigned to users according to their requested data rates and channel quality. As mentioned previously, the multicarrier technique is recognized as a potential candidate for next generation broadband wireless systems thanks to its attractive features in terms of robustness against channel distortions and narrowband interference, high spectral efficiency, high flexibility in resource management and ability to support adaptive modulation schemes. Furthermore, multicarrier transmissions can easily be combined with MIMO technology as witnessed by recent advances on

MIMO-OFDM [149] and MIMO-OFDMA.

1.3 Historical notes on multicarrier transmissions

The first examples of multicarrier (MC) modems operating in the High-Frequency (HF) band date back to the 1950s. In these early experiments, the signal bandwidth was divided into several non-overlapping frequency subchannels, each modulated by a distinct stream of data coming from a common source. On one hand, the absence of any spectral overlap between adjacent subchannels helped to eliminate interference among different data streams (interchannel interference). On the other, it resulted into a very inefficient use of the available spectrum. The idea of orthogonal MC transmission with partially overlapping spectra was introduced by Chang in 1966 with his pioneering paper on parallel data transmission over dispersive channels [15]. In the late 1960s, the MC concept was adopted in some military applications such as KATHRYN [184] and ANDEFT [120]. These systems involved a large hardware complexity since parallel data transmission was essentially implemented through a bank of oscillators, each tuned on a specific subcarrier. As a consequence, in that period much of the research effort was devoted to find efficient modulation and demodulation schemes for MC digital communications [121, 139]. A breakthrough in this sense came in 1971, when Weinstein and Ebert eliminated the need for a bank of oscillators and proposed the use of the Fast Fourier Transform (FFT) for baseband processing. They also introduced the *guard band* concept to eliminate interference among adjacent blocks of data. The new FFT-based technique was called orthogonal frequency-division multiplexing (OFDM). Despite its reduced complexity with respect to previously developed MC schemes, practical implementation of OFDM was still difficult at that time because of the limited signal processing capabilities of the electronic hardware. For this reason, OFDM did not attract much attention until 1985, when was suggested by Cimini for high-speed wireless applications [21].

Advances in digital and hardware technology in the early 1990s enabled the practical implementation of FFTs of large size, thereby making OFDM a realistic option for both wired and wireless transmissions. The ability to support adaptive modulation and to mitigate channel distortions without the need for adaptive time-domain equalizers made OFDM the selected access scheme for asymmetric digital subscriber loop (ADSL) applications in the USA [19]. In Europe, Digital Audio Broadcasting (DAB) standardized by ETSI was the first commercial wireless system to use OFDM as an air interface in 1995 [39]. This success continued in 1997 with the adoption of OFDM for terrestrial Digital Video Broadcasting (DVB-T) [40] and in 1999 with the release of the WLAN standards HiperLAN2 [41] and IEEE 802.11a [59], both based on OFDM-TDMA. More recently, OFDM has been used in the interactive terrestrial return channel (DVB-RCT) [129] and in the IEEE 802.11g WLAN products [61]. In 1998 a combination of OFDM and FDMA called orthogonal frequency-division multiple-access (OFDMA) was proposed by Sari and Karam for cable TV (CATV) networks [140]. The main advantages of this scheme over OFDM-TDMA are the increased flexibility in resource management and the ability for dynamic channel assignment. Compared to ordinary FDMA, OFDMA offers higher spectral efficiency by avoiding the need for large guard bands between users' signals. A hybrid combination of OFDMA and TDMA has been adopted in the uplink of the DVB-RCT system while both OFDM-TDMA and OFDMA are recommended by the IEEE 802.16a standard for WMANs [112]. An intense research activity is currently devoted to study MIMO-OFDM and MIMO-OFDMA as promising candidates for 4G wireless broadband systems.

1.4 Outline of this book

The remaining chapters of this book are organized in the following way.

Chapter 2 lays the groundwork material for further developments and is divided into three parts. The first is concerned with the statistical characterization of the wireless channel. Here, some relevant parameters are introduced ranging from the channel coherence bandwidth and Doppler spread to the concept of frequency-selective and time-selective fading. The second part illustrates the basic idea of OFDM and how this kind of modulation can be implemented by means of FFT-based signal processing. The OFDMA principle is described in the third part of the chapter, along with some other popular multiple-access schemes based on OFDM.

Chapter 3 provides a comprehensive overview of synchronization methods for OFDMA applications. A distinction is made between downlink and uplink transmissions, with a special attention to the uplink situation which is particularly challenging due to the presence of many unknown synchronization parameters. Several timing and frequency recovery schemes are presented, and comparisons are made in terms of system complexity and estimation accuracy. Some methods for compensating the synchronization errors in an uplink scenario are illustrated in the last part of this chapter.

Chapter 4 deals with channel estimation and equalization in OFDM systems. After illustrating how channel distortions can be compensated for through a bank of one-tap complex-valued multipliers, we present a large variety of methods for estimating the channel frequency response over each subcarrier. A number of these schemes are based on suitable interpolation of pilot symbols which are inserted in the transmitted frame following some specified grid patterns. Other methods exploit the inherent redundancy introduced in the OFDM waveform by the use of the cyclic prefix and/or virtual carriers. The chapter concludes by illustrating recent advances in the context of joint channel estimation and data detection based on the expectation-maximization (EM) algorithm.

Chapter 5 extends the discussions of the previous two chapters and presents a sophisticated receiver structure for uplink OFDMA transmissions where the tasks of synchronization, channel estimation and data detection are jointly performed by means of advanced iterative signal processing techniques. At each iteration, tentative data decisions are exploited to improve the synchronization and channel estimation accuracy which, in turn, produces more reliable data decisions in the next iteration. Numerical results demonstrate the effectiveness of this iterative architecture.

Chapter 6 covers the topic of dynamic resource allocation in multicarrier systems, where power levels and/or data rates are adaptively adjusted over each subcarrier according to the corresponding channel quality. We begin by reviewing the rate-maximization and margin-maximization concepts and discuss several bit and power loading techniques for single-user OFDM. The second part of the chapter presents a survey of state-of-the-art allocation techniques for OFDMA applications. In this case, the dynamic assignment of subcarriers to the active users provides the system with some form of multiuser diversity which can be exploited to improve the overall data throughput.

Finally, Chapter 7 provides a thorough discussion of the peak-to-average power ratio (PAPR) problem, which is considered as one of the main obstacles to the practical implementation of OFDM/OFDMA. After providing a detailed statistical characterization of the PAPR, a large number of PAPR reduction schemes are presented, starting from the conventional clipping technique till some sophisticated encoding approaches based on Reed-Muller codes and Golay complementary sequences.

Chapter 2

Fundamentals of OFDM/OFDMA Systems

This chapter lays the groundwork for the material in the book and addresses several basic issues. Section 2.1 describes the main features of the wireless communication channel and introduces the concept of frequency-selective and time-selective fading. In Sec. 2.2 we review conventional approaches to mitigate the distortions induced by the wireless channel on the information-bearing signal. Section 2.3 introduces the principle of Orthogonal Frequency-Division Multiplexing (OFDM) as an effective means for high-speed digital transmission over frequency-selective fading channels. We conclude this chapter by illustrating how OFDM can be combined with conventional multiple-access techniques to provide high-rate services to several simultaneously active users. In particular, we introduce the concept of Orthogonal Frequency-Division Multiple-Access (OFDMA), where each user transmits its own data by modulating an exclusive set of orthogonal subcarriers. The advantages of OFDMA are highlighted through comparisons with other popular multiplexing techniques.

2.1 Mobile channel modeling

In a mobile radio communication system, information is conveyed by a digitally modulated band-pass signal which is transmitted through the air. The band-pass signal occupies an assigned portion of the radio frequency (RF) spectrum and is mathematically expressed as

$$s_{RF}(t) = \Re e \left\{ s(t) e^{j2\pi f_c t} \right\}, \qquad (2.1)$$

where $\Re e \left\{ \cdot \right\}$ denotes the real part of the enclosed quantity, $s(t)$ is the complex envelope of $s_{RF}(t)$ and f_c is the carrier frequency. Since only the amplitude and phase of $s(t)$ are modulated by the information symbols, in

17

the ensuing discussion we can restrict our attention to $s(t)$ without any loss of generality. Furthermore, in order to highlight the performance degradation caused by channel impairments, we temporarily neglect the effect of thermal noise and other disturbance sources. This enables a better understanding of the OFDM ability to cope with severe channel distortions.

2.1.1   *Parameters of wireless channels*

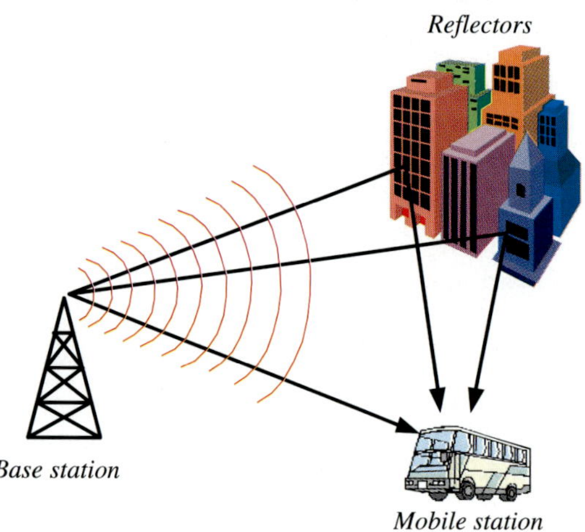

Fig. 2.1   The basic principle of multipath propagation.

Figure 2.1 depicts a typical wireless communication environment where radio waves are scattered, reflected and diffracted from surrounding objects like buildings, trees or hills. In such a scenario, the transmitted waveform arrives at the receiving antenna after traveling through several distinct paths, each characterized by a specific attenuation, phase and propagation delay. The received signal is thus the superposition of a possibly large number of attenuated, phase-shifted and delayed versions of the transmitted waveform known as *multipath components*. This results into a linear (and possibly time-varying) distortion of the information-bearing signal while it propagates through the transmission medium. A schematic situation is depicted in Fig. 2.2, where a narrow pulse is spread over a relatively large time interval as a consequence of multipath propagation.

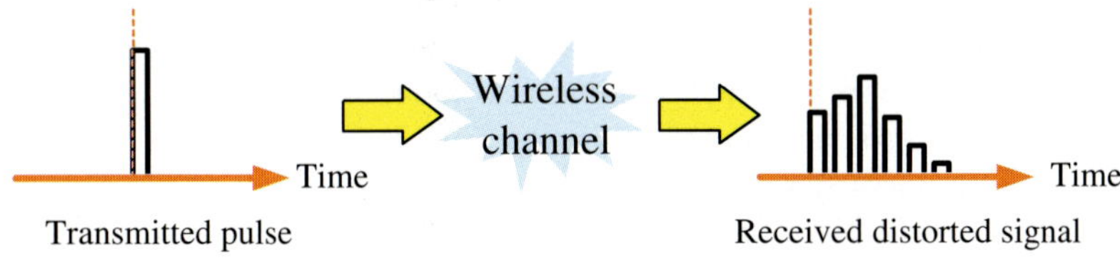

Fig. 2.2   Distortion introduced by multipath propagation.

At the receiving antenna, the multipath components may overlap in a constructive or destructive fashion depending on their relative phase shifts. Therefore, the received signal strength is subject to unpredictable fluctuations due to random variations of the propagation scenario or in consequence of the relative motion between the transmitter and receiver. Since each multipath component undergoes a phase shift of 2π over a travel distance as short as one wavelength, power fluctuations induced by multipath propagation occur over a very small time-scale and, for this reason, they are normally referred to as *small-scale fading*. In addition, the mean received power (averaged over small-scale fading) may still randomly fluctuate because of several obstructions (walls, foliage or other obstacles) encountered by radio waves along their way. These fluctuations occur over distances up to a few hundreds of wavelengths (tens of meters), and result in *large-scale fading*.

From the ongoing discussion it should be clear that wireless propagation is mostly governed by a large number of unpredictable factors which can hardly be described in a rigorous fashion. For this reason, it is often preferable to characterize the wireless channel from a statistical viewpoint using some fundamental parameters that are now introduced.

2.1.1.1   *Path loss*

The path loss is a statistical measure of the attenuation incurred by the transmitted signal while it propagates through the channel. Assume that the transmitter and the receiver are separated by a distance d and let P_T and P_R be the average transmitted and received powers, respectively. Then, in the absence of any shadowing effect, it has been empirically found that

$$P_R = \beta d^{-n} P_T \qquad (2.2)$$

where n is the path-loss exponent and β is a parameter that depends on the employed carrier frequency, antenna gains and other environmental factors. For free-space propagation the path-loss exponent is 2, while in urban environment it takes values between 4 and 6.

The path loss $L_{path}(d)$ at a specified distance d is defined as the ratio P_R/P_T expressed in decibel (dB). From Eq. (2.2) it follows that

$$L_{path}(d) = L_{path}(d_0) + 10n \log_{10}\left(\frac{d}{d_0}\right), \qquad (2.3)$$

where d_0 is an arbitrarily chosen reference distance. It is worth noting that power fluctuations induced by large-scale fading are not contemplated in Eq. (2.3). The common approach to take these fluctuations into account is to assume a Gaussian distribution of the received power around the value in Eq. (2.3). This amounts to setting

$$L_{path}(d) = L_{path}(d_0) + 10n \log_{10}\left(\frac{d}{d_0}\right) + Z, \qquad (2.4)$$

where Z is a Gaussian random variable with zero-mean and standard deviation σ_Z (measured in dB). Since the path loss expressed in logarithmic dB scale follows a normal distribution, the model Eq. (2.4) is usually referred to as log-normal shadowing. Typical values of σ_Z lie between 5 and 12 dB.

2.1.1.2   *Excess delay*

The wireless channel is fully described by its channel impulse response (CIR) $h(\tau, t)$. This represents the response of the channel at time t to a Dirac delta function applied at time $t - \tau$, *i.e.*, τ seconds before. Denoting N_p the number of resolvable multipath components, we may write

$$h(\tau, t) = \sum_{\ell=1}^{N_p} \alpha_\ell(t) e^{j\theta_\ell(t)} \delta\left(\tau - \tau_\ell(t)\right), \qquad (2.5)$$

where $\alpha_\ell(t)$, $\theta_\ell(t)$ and $\tau_\ell(t)$ are the time-varying attenuation, phase shift and propagation delay of the ℓth path, respectively. Without loss of generality, we assume that the path delays are arranged in an increasing order of magnitude and define the ℓth *excess* delay $\Delta\tau_\ell(t)$ as the difference between $\tau_\ell(t)$ and the delay $\tau_1(t)$ of the first arriving multipath component, *i.e.*, $\Delta\tau_\ell(t) = \tau_\ell(t) - \tau_1(t)$. At the receiver side, it is a common practice to use a time scale such that $\tau_1(t) = 0$. In this case, the excess delays reduce to $\Delta\tau_\ell(t) = \tau_\ell(t)$ for $\ell > 1$.

If a signal $s_{RF}(t)$ is transmitted over a wireless channel characterized by the CIR given in Eq. (2.5), the complex envelope of the received waveform takes the form

$$r(t) = \sum_{\ell=1}^{N_p} \alpha_\ell(t)e^{j\theta_\ell(t)}s\left(t - \tau_\ell(t)\right). \tag{2.6}$$

2.1.1.3 *Power delay profile*

The power delay profile (PDP) is a statistical parameter indicating how the power of a Dirac delta function is dispersed in the time-domain as a consequence of multipath propagation. The PDP is usually given as a table where the average power associated with each multipath component is provided along with the corresponding delay. In particular, the average power $p(\tau_\ell)$ of the ℓth path is defined as

$$p(\tau_\ell) = \mathrm{E}\{|\alpha_\ell(t)|^2\}, \tag{2.7}$$

where $|\cdot|$ is the magnitude of the enclosed complex-valued quantity while $\mathrm{E}\{\cdot\}$ denotes statistical expectation. Clearly, summing all quantities $p(\tau_\ell)$ provides the total average received power P_R. In practice, however, the PDP is normalized so that the sum of $p(\tau_\ell)$ is unity, *i.e.*,

$$\sum_{\ell=1}^{N_p} p(\tau_\ell) = 1. \tag{2.8}$$

In this case, the CIR $h(\tau, t)$ in Eq. (2.5) must be multiplied by a factor $\sqrt{A}$, where A is a log-normal random variable which takes into account the combined effect of path loss and large-scale fading.

Table 2.1   The PDP of a typical urban (TU) channel

Path number ℓ	Typical Urban Channel	
	Delay τ_ℓ (μs)	Average power $p(\tau_\ell)$
0	0.0	0.1897
1	0.2	0.3785
2	0.5	0.2388
3	1.6	0.0951
4	2.3	0.0600
5	5.0	0.0379

Table 2.1 provides the PDP of a typical urban (TU) wireless channel [89]. A pictorial illustration of the same PDP is given in Fig. 2.3.

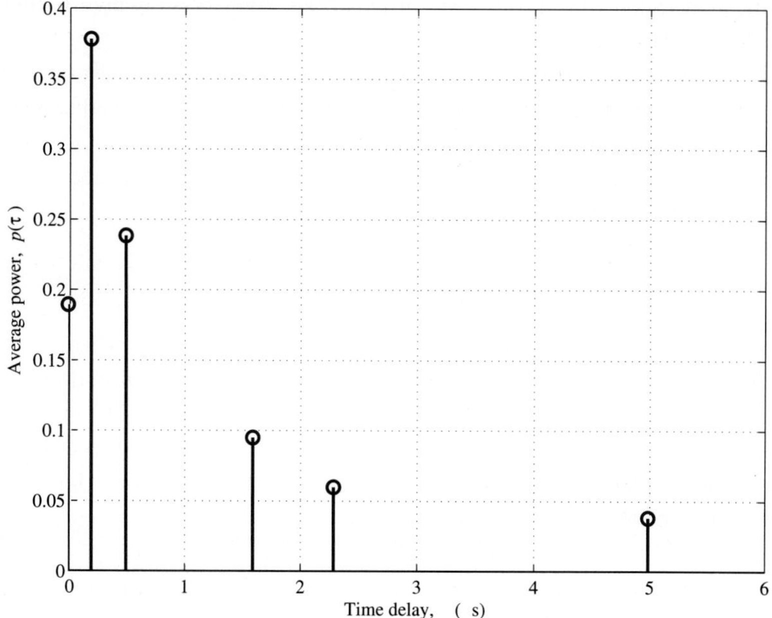

Fig. 2.3   PDP of the TU channel in Table 2.1.

2.1.1.4   *Root-mean-squared (RMS) delay spread*

The root-mean-squared (RMS) delay spread provides a measure of the time dispersiveness of a multipath channel. This parameter is defined as

$$\tau_{rms} = \sqrt{\overline{\tau^2} - (\bar{\tau})^2},\tag{2.9}$$

where $\bar{\tau}$ and $\overline{\tau^2}$ are obtained from the PDP of the channel in the form

$$\bar{\tau} = \sum_{\ell=1}^{N_p} \tau_\ell p(\tau_\ell)\tag{2.10}$$

and

$$\overline{\tau^2} = \sum_{\ell=1}^{N_p} \tau_\ell^2 p(\tau_\ell).\tag{2.11}$$

From the normalization condition Eq. (2.8), it appears evident that the quantities $p(\tau_\ell)$ for $\ell = 1, 2, \ldots, N_p$ can be interpreted as a probability mass function. In this respect, τ_{rms} represents the standard deviation of the path delays τ_ℓ.

Typical values of τ_{rms} are in the order of nanoseconds for indoor applications and of microseconds for outdoor environments. For example, using the PDP in Table 2.1 it is found that $\tau_{rms} = 1.0620\ \mu s$ for the TU channel. This statistical parameter is an important indicator for evaluating the impact of multipath distortion on the received signal. Actually, the distortion is negligible if the symbol duration T_s is adequately larger than τ_{rms}, say $T_s > 10\tau_{rms}$. Otherwise, appropriate techniques must be employed to compensate for the disabling effects of multipath distortion on the system performance. For example, in the IEEE 802.11a/g standards for wireless local area networks (WLANs) the symbol duration is $T_{s,WLAN} = 50\ ns$. Since in a typical urban channel we have $\tau_{rms} = 1.0620\ \mu s$, it follows that $T_{s,WLAN} \ll \tau_{rms}$. As a result, some compensation procedures are required at the receiver to avoid severe performance degradations.

2.1.1.5 *Coherence bandwidth*

The channel frequency response at time t is defined as the Fourier transform of $h(\tau,t)$ with respect to τ, *i.e.*,

$$H(f,t) = \int_{-\infty}^{\infty} h(\tau,t)e^{-j2\pi f\tau}\, d\tau. \tag{2.12}$$

To characterize the variations of $H(f,t)$ with f at a given time instant t, we introduce the concept of coherence bandwidth B_c as a measure of the "flatness" of the channel frequency response. More precisely, two samples of $H(f,t)$ that are separated in frequency by less than B_c can be assumed as highly correlated. It is well-known that B_c is inversely proportional to τ_{rms}. In particular, for a 0.5-correlation factor it is found that

$$B_c \approx \frac{1}{5\tau_{rms}}. \tag{2.13}$$

If the bandwidth B_s of the transmitted signal is smaller than B_c, the channel frequency response can be considered as approximately flat over the whole signal spectrum. In this case the spectral characteristics of the transmitted signal are preserved at the receiver. Vice versa, if B_s is much larger than B_c, the signal spectrum will be severely distorted and the channel is said to be frequency-selective. From the above discussion it turns out that it is not meaningful to say that a given channel is flat or frequency-selective without having any information about the transmitted signal. Recalling that the signal bandwidth is strictly related to the speed at which information is transmitted, a given channel may appear as flat or frequency-selective depending on the actual transmission rate.

Example 2.1 The RMS delay spread of the TU channel in Table 2.1 has been found to be 1.0620 μs. Hence, the 0.5-correlation coherence bandwidth is given by

$$B_c \approx \frac{1}{5 \times 1.0620 \ \mu s} = 0.2 \ \text{MHz}. \tag{2.14}$$

This means that the frequency response of the TU channel can be considered as nearly *flat* over frequency intervals not larger than 0.2 MHz. This fact can also be inferred by inspecting Fig. 2.4, which illustrates the amplitude $|H(f)|$ of the frequency response as a function of f.

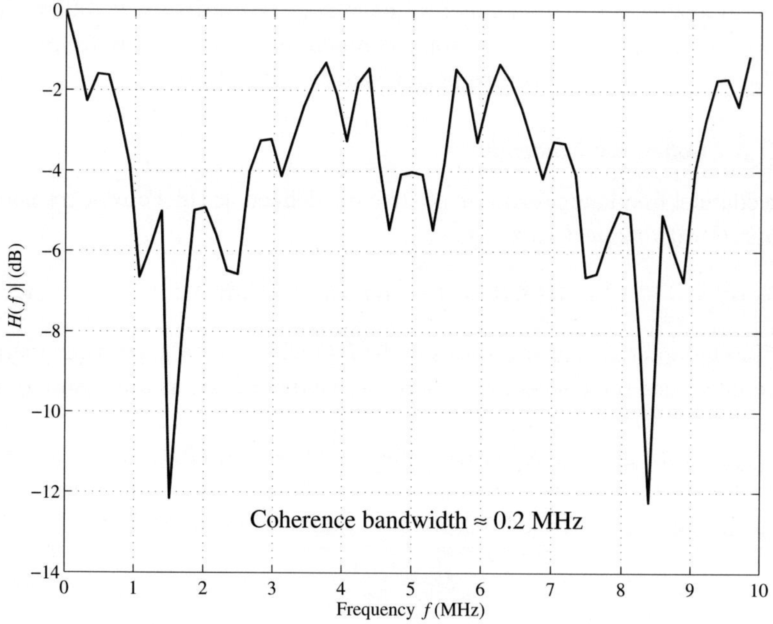

Fig. 2.4   Frequency response of the TU channel in Table 2.1.

2.1.1.6 *Doppler spread*

In a mobile communication environment, the physical motion of the transmitter, receiver and surrounding objects induces a Doppler shift in each multipath component. To fix the ideas, assume that a pure sinusoid of frequency f_c is transmitted over the channel and received by a mobile antenna traveling at a speed of v m/s. Defining ψ_ℓ the angle between the direction of the receiver motion and the direction of arrival of the ℓth multipath component, the corresponding Doppler shift is given by

$$f_{D,\ell} = \frac{f_c v}{c} \cos(\psi_\ell), \qquad (2.15)$$

where $c = 3 \times 10^8$ m/s is the speed of light in the free space. In the presence of several multipath components, the received signal is a superposition of many sinusoidal waveforms, each affected by an unpredictable frequency shift due to the random nature of the angles $\{\psi_\ell\}$. This phenomenon results into a spectral broadening of the received spectrum known as *Doppler spread*. The maximum Doppler shift is obtained from Eq. (2.15) by setting the cosine function to unity and reads

$$f_{D,\mathrm{max}} = \frac{f_c v}{c}. \qquad (2.16)$$

In practice, $f_{D,\mathrm{max}}$ provides information about the frequency interval over which a pure sinusoid is received after propagating through the channel. Specifically, if f_c is the transmitted frequency, the received Doppler spectrum will be confined in the range $[f_c - f_{D,\mathrm{max}}, f_c + f_{D,\mathrm{max}}]$.

Example 2.2 Assume that a laptop computer is moving at a speed of 20 km/h in a IEEE 802.11g local area network operating around the 2.2 GHz frequency band. From Eq. (2.16) it follows that the maximum Doppler shift is given by

$$f_{D,\mathrm{max}} = \frac{2.2 \times 10^9 \cdot (20 \times 10^3/3600)}{3 \times 10^8} \approx 40.7 \ \mathrm{Hz}. \qquad (2.17)$$

Figure 2.5 illustrates the power of the received signal $r(t)$ as a function of t when $f_{D,\mathrm{max}} = 40.7$ Hz. We see that the power occasionally drops far below its expected value. This is a manifestation of the small-scale fading, which is caused by non-coherent superposition of the multipath components at the receiving antenna. Inspection of Fig. 2.5 indicates that in the presence of destructive superposition the received power may drops dramatically. When this happens, we say that the channel is experiencing a *deep fade*.

The rate of occurrence of fade events is measured by the so-called level crossing rate (LCR). This parameter is defined as the expected rate at which the received power goes beyond a preassigned threshold level κ. The frequency of threshold crossings is a function of κ and is expressed by [64]

$$N_\kappa = f_{D,\mathrm{max}} \frac{\kappa}{\sqrt{\sigma_r^2/\pi}} e^{-\frac{\kappa^2}{2\sigma_r^2}}, \qquad (2.18)$$

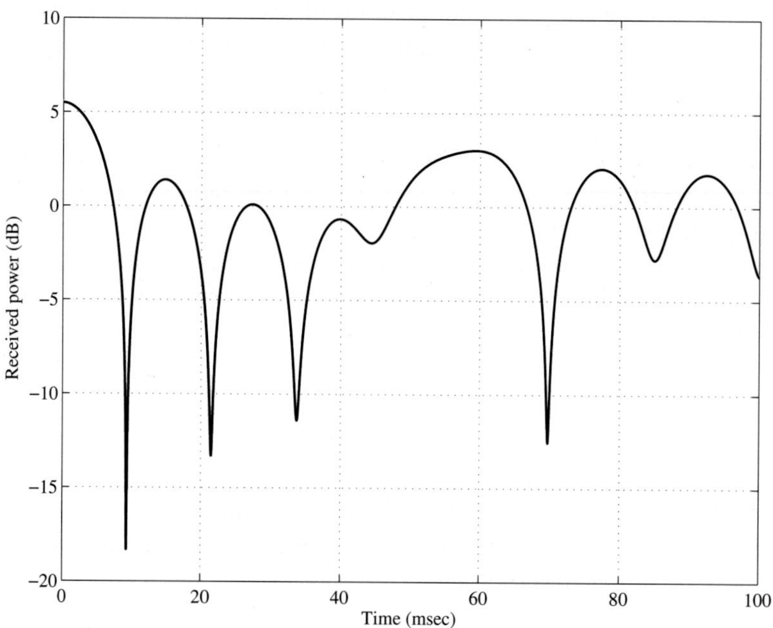

Fig. 2.5 Fluctuations of the received signal power with $f_{D,\mathrm{max}} = 40.7$ Hz.

where $\sigma_r^2 = \mathrm{E}\{|r(t)|^2\}$. The maximum of N_κ is found by computing the derivative of Eq. (2.18) with respect to κ and setting it to zero. This yields $N_{\kappa,\mathrm{max}} = f_{D,\mathrm{max}}e^{-1/2}\sqrt{\pi} \simeq 1.07 \cdot f_{D,\mathrm{max}}$, meaning that the expected number of fade events is approximately equal to the maximum Doppler shift $f_{D,\mathrm{max}}$. This result is validated by computer simulations shown in Fig. 2.5, where four deep fades are observed over a time interval of 0.1 s when $f_{D,\mathrm{max}} = 40.7$ Hz.

2.1.1.7 *Coherence time*

The coherence time T_c is a measure of how fast the channel characteristics vary in time. From a theoretical viewpoint, this parameter is defined as the maximum time lag between two highly correlated channel snapshots. In a more practical sense, T_c can be regarded as the time interval over which the CIR is time-invariant.

The coherence time is proportional to the inverse of the maximum

Doppler shift. For a correlation threshold of 0.5, it is well approximated by

$$T_c = \frac{9}{16\pi f_{D,\max}}.$$ (2.19)

If the signaling period T_s is smaller than T_c, each data symbol is subject to stationary propagation conditions. In such a case we say that the channel is slowly fading. Vice versa, if $T_s > T_c$ the propagation environment may significantly vary over a symbol period and the channel is thus affected by fast fading. We conclude that the same channel can appear as slowly or fast fading depending on the actual signaling rate.

Example 2.3 Assuming a maximum Doppler shift of 40.7 Hz as in Example 2.2, from Eq. (2.19) we find

$$T_c = \frac{9}{16\pi \cdot 40.7} \; s \approx 4.4 \;\; ms.$$ (2.20)

Since the duration of each data block in the IEEE 802.11a/g standards is about 4.0 μs, the TU channel can be considered as time invariant over one block.

2.1.2 *Categorization of fading channels*

As discussed earlier, the impact of multipath propagation on the reliability of a wireless link is strictly related to the characteristics of the transmitted signal. In general, we can distinguish four distinct types of channels. The latter are summarized in Fig. 2.6 and are now discussed in some detail.

	$B_c > B_s$	$B_c < B_s$
$T_c > T_s$	Frequency-nonselective slowly-fading	Frequency-selective fading
$T_c < T_s$	Time-selective fading	Frequency and time-selective fading

Fig. 2.6   Categorization of fading channels.

2.1.2.1 *Frequency-nonselective and slowly-fading channels*

In many practical applications such as *fixed* communications within local areas, the coherence time T_c is much greater than the symbol duration T_s. In this case, the channel is affected by slowly-fading and the multipath parameters in Eq. (2.5) may be regarded as approximately *invariant* over many signaling intervals. As a result, the CIR becomes independent of t and can be rewritten as

$$h(\tau) = \sum_{\ell=1}^{N_p} \alpha_\ell e^{j\theta_\ell} \delta\left(\tau - \tau_\ell\right), \qquad (2.21)$$

while the corresponding channel frequency response is given by

$$H(f) = \sum_{\ell=1}^{N_p} \alpha_\ell e^{j\theta_\ell} e^{-j2\pi f \tau_\ell}. \qquad (2.22)$$

If the path delays are much smaller than the symbol duration, then we may reasonably set $\tau_\ell \approx 0$ into Eqs. (2.21) and (2.22). This yields

$$h(\tau) \approx \rho e^{j\varphi} \delta(\tau) \qquad (2.23)$$

and

$$H(f) \approx \rho e^{j\varphi}, \qquad (2.24)$$

where we have defined

$$\rho e^{j\varphi} = \sum_{\ell=1}^{N_p} \alpha_\ell e^{j\theta_\ell}. \qquad (2.25)$$

Inspection of Eq. (2.24) reveals that $H(f)$ is practically constant over the whole signal bandwidth, and the channel is therefore frequency-nonselective or flat. In this case the complex envelope of the received signal takes the form

$$r(t) = \rho e^{j\varphi} s(t) \qquad (2.26)$$

and is simply an attenuated and phase-rotated version of $s(t)$.

As indicated in Eq. (2.25), the multiplicative factor $\rho e^{j\varphi}$ is the sum of N_p statistically independent contributions, each associated with a distinct multipath component. Thus, invoking the central limit theorem [2], the real and imaginary parts of $\rho e^{j\varphi}$ can reasonably be approximated as two statistically independent Gaussian random variables with the same variance σ^2 and expected values η_R and η_I, respectively. In the absence of any line-of-sight (LOS) path between the transmitter and receiver, no dominant multipath component is present and we have $\eta_R = \eta_I = 0$. In such a case the phase term φ is found to be uniformly distributed over $[-\pi, \pi)$, while the amplitude ρ follows a Rayleigh distribution with probability density function (pdf)

$$p(\rho) = \frac{\rho}{\sigma^2} \exp\left(-\frac{\rho^2}{2\sigma^2}\right), \quad \rho \geq 0. \tag{2.27}$$

In some applications including satellite or microcellular mobile radio systems, a LOS is normally present in addition to a scattered component. In this case ρ has a Rician distribution and its pdf is given by

$$p(\rho) = \frac{2\rho(K+1)}{P_\rho} \exp\left\{-\left[K + \frac{(K+1)\rho^2}{P_\rho}\right]\right\} I_0\left(2\rho\sqrt{\frac{K(K+1)}{P_\rho}}\right), \tag{2.28}$$

where $\rho \geq 0$ and $P_\rho = \mathrm{E}\{\rho^2\} = 2\sigma^2 + \eta_R^2 + \eta_I^2$ while $K = (\eta_R^2 + \eta_I^2)/(2\sigma^2)$ is the Rician factor, which is defined as the ratio between the power of the LOS path and the average power of the scattered component. Moreover, $I_0(x)$ is the zeroth-order modified Bessel function of the first kind, which reads

$$I_0(x) = \frac{1}{2\pi} \int_0^{2\pi} e^{x \cos \alpha} d\alpha. \tag{2.29}$$

Note that in the absence of any LOS component ($K = 0$) the Rician distribution in Eq. (2.28) boils down to the Rayleigh pdf in Eq. (2.27) because of the identities $P_\rho = 2\sigma^2$ and $I_0(0) = 1$.

2.1.2.2 *Frequency-selective fading channels*

Assume for simplicity that the channel is slowly-fading and consider its frequency response as given in Eq. (2.22). If the transmitted signal has a bandwidth B_s larger than the channel coherence bandwidth, its spectral components will undergo different attenuations while propagating from the transmitter to the receiver. In this case the channel is frequency-selective and the received waveform is a linearly distorted version of the transmitted signal. The frequency selectivity of a channel can also be checked in the time-domain. Bearing in mind that B_s and B_c are inversely proportional to T_s and τ_{rms}, respectively, the channel appears as frequency-selective if $T_s < \tau_{rms}$ and frequency-nonselective (or flat) otherwise. The most prominent impairment caused by frequency-selective fading is the insurgence of intersymbol interference (ISI) in the received signal. A schematic illustration of the ISI phenomenon is shown in Fig. 2.7, where a train of pulses separated by T_s seconds is transmitted over a frequency-selective channel. If T_s is shorter than the channel delay spread, each received pulse overlaps with neighboring pulses, thereby producing ISI.

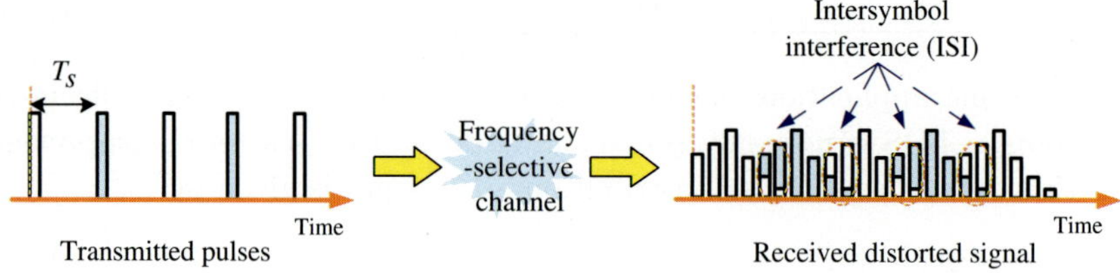

Fig. 2.7   Illustration of the intersymbol interference (ISI) phenomenon.

Figure 2.8 depicts a frequency-selective and slowly-fading channel where the channel frequency response keeps approximately constant over each symbol interval, but slowly varies from one interval to another.

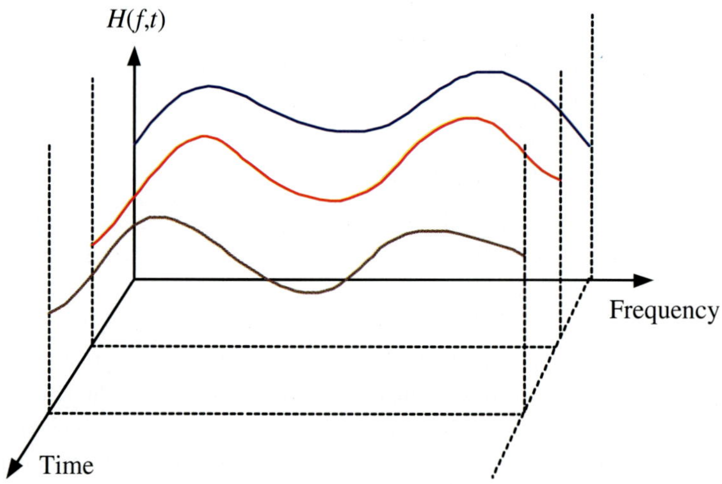

Fig. 2.8   Illustration of a frequency-selective and slowly-fading channel.

2.1.2.3 *Time-selective fading channels*

The concept of time-selective fading is typically introduced by considering a frequency-flat channel in which the delay spread is much smaller than the symbol duration. As discussed previously, in this case we may reasonably substitute $\tau_\ell = 0$ into Eq. (2.5) to obtain

$$h(\tau, t) = \rho(t)e^{j\varphi(t)}\delta(\tau) \tag{2.30}$$

with

$$\rho(t)e^{j\varphi(t)} = \sum_{\ell=1}^{N_p} \alpha_\ell(t)e^{j\theta_\ell(t)}. \tag{2.31}$$

The corresponding channel frequency response is given by

$$H(f, t) = \rho(t)e^{j\varphi(t)} \tag{2.32}$$

and its amplitude is schematically depicted in Fig. 2.9 at some different time instants t.

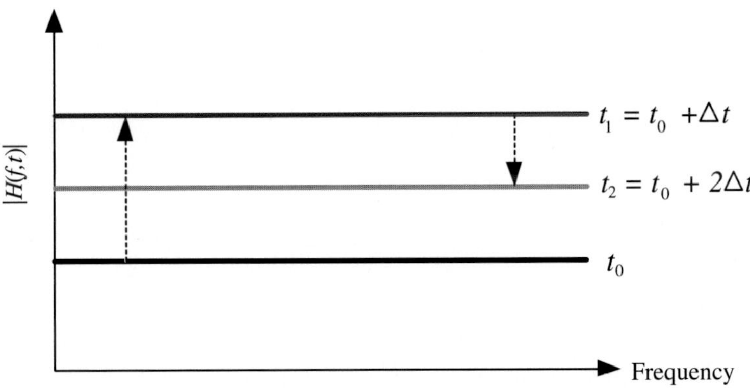

Fig. 2.9   Illustration of a time-selective fading channel.

Substituting $\tau_\ell = 0$ into Eq. (2.6) and using Eq. (2.31), yields

$$r(t) = \rho(t)e^{j\varphi(t)}s(t), \tag{2.33}$$

from which it follows that the received signal is a replica of the transmitted waveform $s(t)$ except for a time-varying multiplicative distortion.

If the symbol period is greater than the channel coherence time, the multiplicative factor $\rho(t)e^{j\varphi(t)}$ may significantly vary over a signaling interval. In such a case the channel is said to be time-selective and produces a Doppler spread of the received signal spectrum. A classical model to statistically characterize the multiplicative distortion induced by time-selective fading is due to Jakes [64]. This model applies to a scenario similar to that illustrated in Fig. 2.10, where an omni-directional antenna receives a large number of multipath contributions in the horizontal plane from uniformly distributed scatterers.

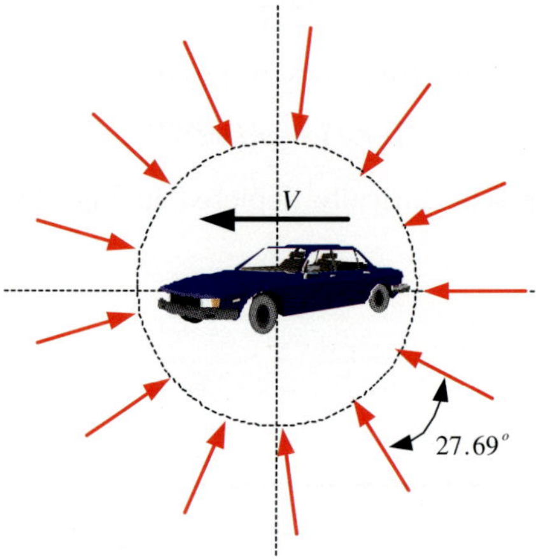

Fig. 2.10   A typical scenario for application of the Jakes model.

In the above hypothesis, the quadrature components of $\rho(t)e^{j\varphi(t)}$ are statistically independent zero-mean Gaussian processes with power σ^2 and autocorrelation function

$$R(\tau) = \sigma^2 J_0(2\pi f_{D,\max}\tau), \tag{2.34}$$

where $J_0(x)$ is the zeroth-order Bessel function of the first kind while $f_{D,\max}$ denotes the maximum Doppler shift. In this case $\rho(t)$ follows a Rayleigh distribution and the corresponding Doppler power spectrum (which is defined as the Fourier transform of $2R(\tau)$) is given by

$$P(f) = \begin{cases} \dfrac{2\sigma^2}{\pi\sqrt{f_{D,\max}^2 - f^2}} & |f| \le f_{D,\max} \\ 0 & \text{otherwise.} \end{cases} \tag{2.35}$$

Function $P(f)$ exhibits the classical "bowl-shaped" form depicted in Fig. 2.11. However, in many practical situations the Doppler power spectrum can considerably deviate from the Jakes model.

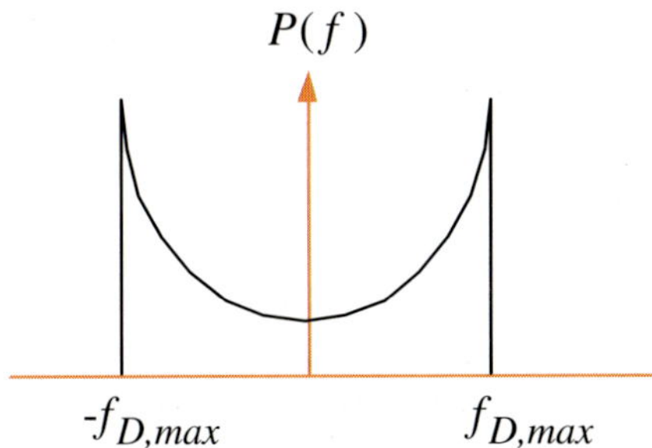

Fig. 2.11   The "bowl-shaped" Doppler power spectrum of the Jakes model.

The main impairment of a time-selective Rayleigh fading channel is that $\rho(t)$ may occasionally drop to very low values (deep fades). When this happens, the signal-to-noise ratio (SNR) becomes poor and the communication system is thus vulnerable to the additive noise.

2.1.2.4   *Frequency- and time-selective fading channels*

In some applications it may happen that the symbol period and transmission bandwidth of the information-bearing signal are larger than the channel coherence time and coherence bandwidth, respectively. In this case the transmitted signal undergoes frequency-selective as well as time-selective fading (often referred to as doubly-selective fading), and the received waveform is the superposition of several time-varying multipath components, each characterized by a non-negligible path delay as indicated in Eq. (2.6). In general, compensating the distortions induced by doubly-selective fading is a rather difficult task.

2.2   Conventional methods for channel fading mitigation

Channel fading represents a major drawback in digital wireless communications. Numerous research efforts have been devoted to combating its detrimental effects and different solutions have been devised depending on whether the channel can be categorized as time- or frequency-selective.

2.2.1   *Time-selective fading*

As mentioned previously, signals experiencing time-selective fading are occasionally plagued by deep fades which lead to severe attenuation of the received signal power. In this case data symbols are highly vulnerable to the additive noise and "bursts" of errors are likely to occur. Channel coding can be used to cope with the drop of SNR associated with deep fades. The main idea is to introduce some redundancy in the transmitted data stream so as to protect the information symbols against additive noise [26]. Since channel coding is more effective in the presence of sparse errors, time interleaving is typically employed to break up error bursts. In addition to interleaving and channel coding, diversity techniques have been proposed to combat time-selective fading.

2.2.2   *Frequency-selective fading*

The main impairment induced by frequency-selective fading is the occurrence of ISI in the received signal. A classical approach to compensate for ISI is to pass the received signal into a properly designed linear filter called *channel equalizer*. Several approaches have been proposed for the filter design. Figure 2.12 illustrates the zero-forcing (ZF) solution, where the frequency response of the equalizer is taken as the inverse of the channel frequency response $H(f)$. In this case ISI is completely removed at the expense of some noise enhancement. Better results are obtained with the classical minimum mean-square error (MMSE) solution, which aims at minimizing the mean-square error (MSE) between the received samples and the transmitted data symbols. In this way the equalizer can reduce the ISI with much lower noise enhancement as compared to the ZF equalizer.

Example 2.4 We consider a wireless channel with three multipath components and the following frequency response

$$H(f) = 0.815 - 0.495e^{-j2\pi f T_s} - 0.3e^{-j4\pi f T_s}. \qquad (2.36)$$

If we neglect the contribution of thermal noise, the nth received sample is given by

$$r(n) = 0.815 \cdot c(n) - 0.495 \cdot c(n-1) - 0.3 \cdot c(n-2), \qquad (2.37)$$

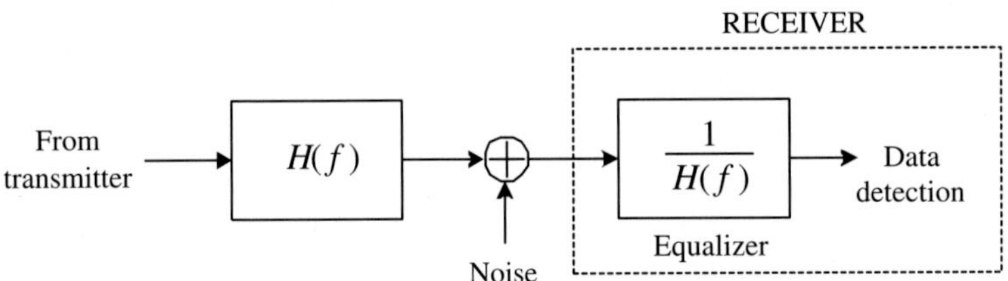

Fig. 2.12　Structure of a conventional zero-forcing (ZF) equalizer.

where $c(n)$ is the nth transmitted symbol. A ZF equalizer is used to compensate for the linear distortion produced by $H(f)$. As shown in Fig. 2.13, the equalizer is implemented as a finite impulse response (FIR) filter of length M and with weighting coefficients

$$p_m = 1.143 \cdot (0.981)^m - 0.631 \cdot (-0.542)^m, \qquad m = 0, 1, \ldots, M-1. \quad (2.38)$$

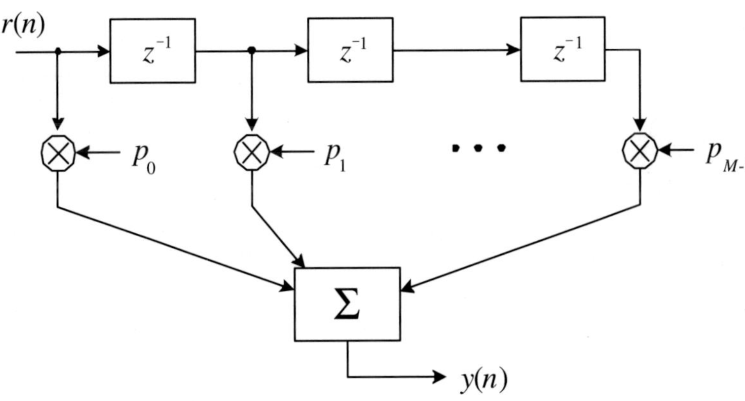

Fig. 2.13　FIR implementation of the ZF equalizer in Example 2.4.

The performance of the equalizer is usually given in terms of the output MSE. This parameter is defined as

$$MSE = \mathrm{E}\{|y(n) - c(n)|^2\}, \qquad (2.39)$$

where $y(n)$ is the equalizer output and represents a soft estimate of $c(n)$

Figure 2.14 illustrates the impact of the equalizer length M on the output MSE as obtained through Monte-Carlo simulations. These results indicate that efficient ISI compensation requires an equalizer with at least 70 weighting coefficients. A longer filter is necessary if the propagation channel comprises more multipath components with larger path delays, thereby increasing the complexity of the receiving terminal. This is clearly undesirable since mobile receivers have usually limited computational resources and strict power constraints. A straightforward solution to reduce the ISI is to make the symbol duration adequately longer than the maximum channel delay spread. However, since τ_{rms} is only determined by the physical characteristics of the propagation channel, this approach translates into a suitable enlargement of the symbol period with a corresponding reduction of the achievable throughput. All these facts indicate that frequency-selective fading is in general a serious obstacle for broadband wireless communications.

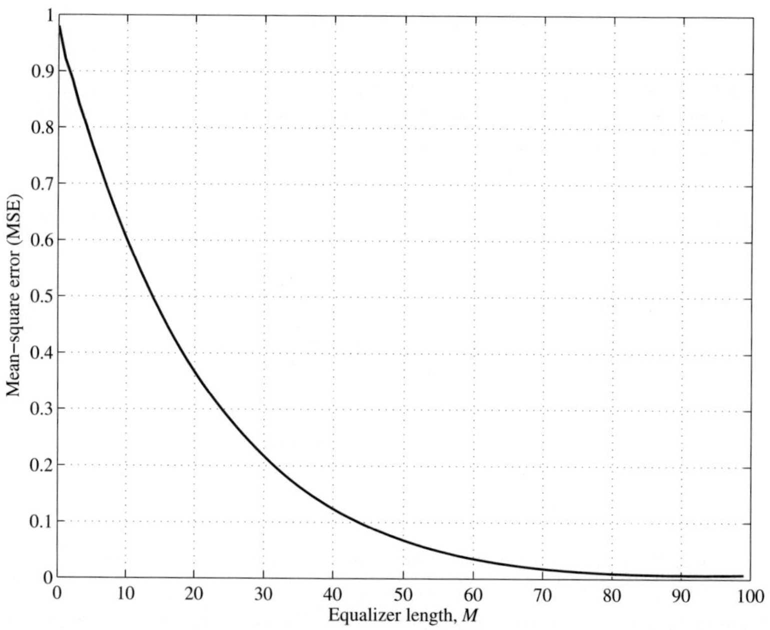

Fig. 2.14   Output MSE as a function of the equalizer length M.

2.3 OFDM systems

2.3.1 *System architecture*

Orthogonal frequency-division multiplexing (OFDM) is a signaling technique that is widely adopted in many recently standardized broadband communication systems due to its ability to cope with frequency-selective fading. Figure 2.15 shows the block diagram of a typical OFDM system.

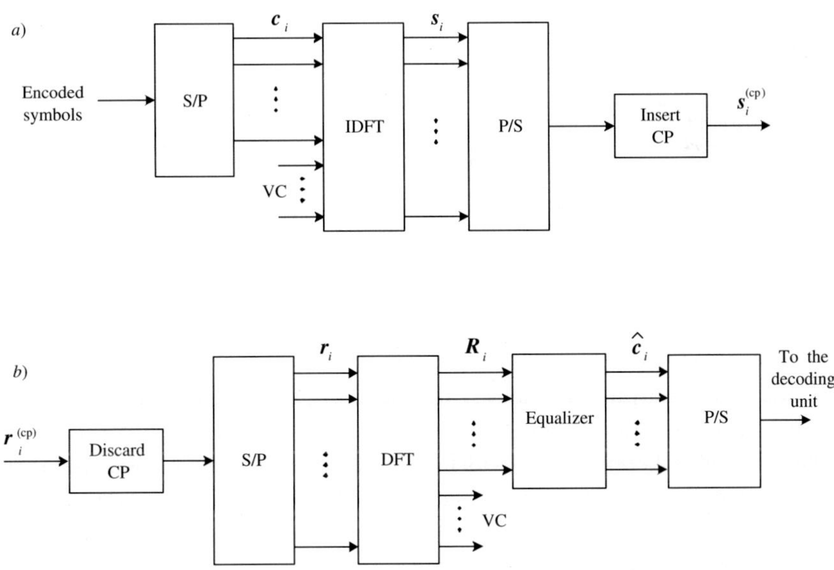

Fig. 2.15   Block diagram of a typical OFDM system: *a*) transmitter; *b*) receiver.

The main idea behind OFDM is to divide a high-rate encoded data stream into N_u parallel substreams that are modulated onto N_u orthogonal carriers (referred to as *subcarriers*). This operation is easily implemented in the discrete-time domain through an N-point inverse discrete Fourier transform (IDFT) unit with $N \geqslant N_u$. The $N - N_u$ unused inputs of the IDFT are set to zero and, in consequence, they are called *virtual carriers* (VCs). In practice, VCs are employed as guard bands to prevent the transmitted power from leaking into neighboring channels. By modulating the original data onto N subcarriers, OFDM increases the symbol duration by a factor of N, thereby making the transmitted signal more robust against frequency-selective fading. The essence of this process is illustrated in Fig. 2.16 through a simple example where the symbol duration is doubled by dividing the original data stream into two parallel substreams. A comparison with Fig. 2.7 reveals that lengthening the symbol duration provides an effective means to cope with ISI.

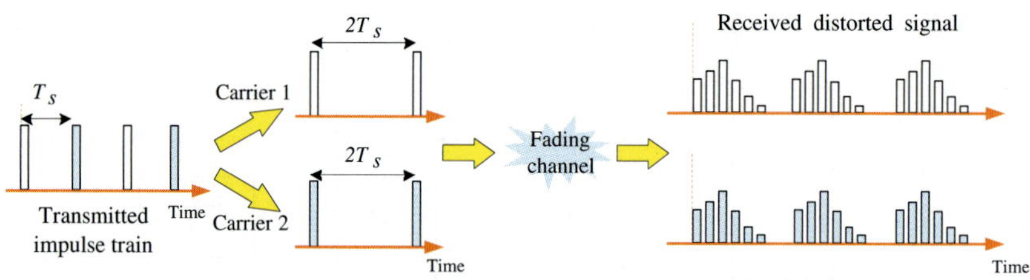

Fig. 2.16   Time-domain illustration of the benefits arising from lengthening the symbol duration.

The same conclusion can be drawn by examining the signal spectrum at the IDFT output. As shown in Fig. 2.17, the whole bandwidth is divided into two subchannels. If the latter are narrow enough compared to the channel coherence bandwidth, the channel frequency response turns out to be approximately flat over each subchannel. Hence, we may say that OFDM converts a frequency-selective channel into several adjacent flat fading subchannels.

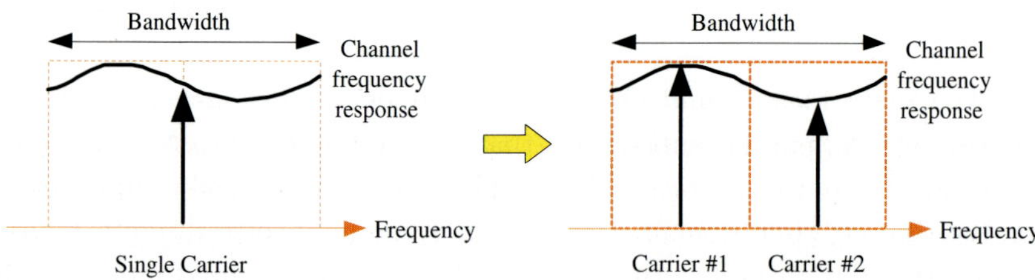

Fig. 2.17   Frequency-domain illustration of the benefits arising from lengthening the symbol duration.

From the ongoing discussion it appears that data transmission in OFDM systems is accomplished in a block-wise fashion, where each block conveys a number N_u of (possibly coded) data symbols. As a consequence of the time dispersion associated with the frequency-selective channel, contiguous blocks may partially overlap in the time-domain. This phenomenon results into interblock interference (IBI), with ensuing limitations of the system performance. The common approach to mitigate IBI is to introduce a guard interval of appropriate length among adjacent blocks. In practice, the guard interval is obtained by duplicating the last N_g samples of each IDFT output and, for this reason, is commonly referred to as *cyclic prefix* (CP). As illustrated in Fig. 2.18, the CP is appended in front of the corresponding IDFT output. This results into an extended block of $N_T = N + N_g$ samples which can totally remove the IBI as long as N_g is properly designed according to the channel delay spread.

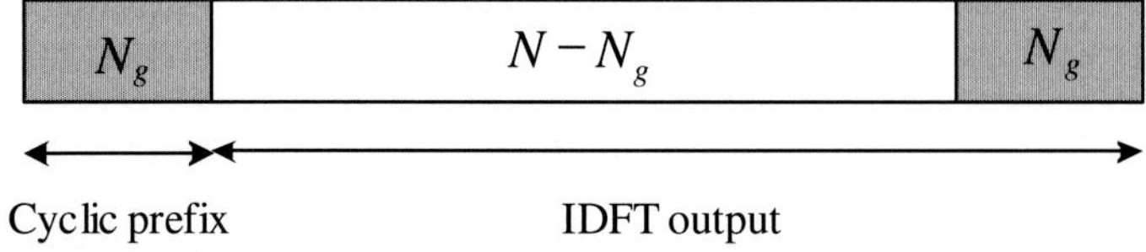

Fig. 2.18　Structure of an OFDM block with CP insertion.

Returning to Fig. 2.15 b), we see that the received samples are divided into adjacent segments of length N_T, each corresponding to a different block of transmitted data. Without loss of generality, in the ensuing discussion we concentrate on the ith segment. The first operation is the CP removal, which is simply accomplished by discarding the first N_g samples of the considered segment. The remaining N samples are fed to a discrete Fourier transform (DFT) unit and the corresponding output is subsequently passed to the channel equalizer. Assuming that synchronization has already been established and the CP is sufficiently long to eliminate the IBI, only a one-tap complex-valued multiplier is required to compensate for the channel distortion over each subcarrier. To better understand this fundamental property of OFDM, however, we need to introduce the mathematical model of the communication scheme depicted in Fig. 2.15.

2.3.2   *Discrete-time model of an OFDM system*

We denote $\boldsymbol{c}_i = [c_i(0), c_i(1), \ldots, c_i(N-1)]^T$ the ith block of data at the transmitter input, with $(\cdot)^T$ representing the transpose operator. Symbols $c_i(n)$ are taken from either a phase-shift keying (PSK) or quadrature amplitude modulation (QAM) constellation, while those corresponding to VCs are set to zero. After serial-to-parallel (S/P) conversion, vector $\boldsymbol{c}_i$ is fed to the IDFT unit. The corresponding output is given by

$$\boldsymbol{s}_i = \boldsymbol{F}^H \boldsymbol{c}_i, \tag{2.40}$$

where $\boldsymbol{F}$ is the N-point DFT matrix with entries

$$[\boldsymbol{F}]_{n,k} = \frac{1}{\sqrt{N}} \exp\left(\frac{-j2\pi nk}{N}\right), \quad \text{for} \quad 0 \le n, k \le N-1 \tag{2.41}$$

while the superscript $(\cdot)^H$ represents the Hermitian transposition.

Vector $\boldsymbol{s}_i$ is next parallel-to-serial (P/S) converted and its last N_g elements are copied in front of it as shown in Fig. 2.18. The resulting vector $\boldsymbol{s}_i^{(cp)}$ is modeled as

$$\boldsymbol{s}_i^{(cp)} = \boldsymbol{T}^{(cp)} \boldsymbol{s}_i, \tag{2.42}$$

where

$$\boldsymbol{T}^{(cp)} = \begin{bmatrix} \boldsymbol{P}_{N_g \times N} \\ \boldsymbol{I}_N \end{bmatrix}. \tag{2.43}$$

In the above equation, $\boldsymbol{I}_N$ represents the $N \times N$ identity matrix while $\boldsymbol{P}_{N_g \times N}$ is an $N_g \times N$ matrix collecting the last N_g rows of $\boldsymbol{I}_N$. The entries of $\boldsymbol{s}_i^{(cp)}$ are then fed to the D/A converter, which consists of an interpolation filter with signaling interval T_s. The latter produces a continuous-time waveform which is up-converted to a carrier frequency f_c and launched over the channel.

For presentational convenience, we consider a time-invariant frequency-selective channel with discrete-time impulse response $\boldsymbol{h} = [h(0), h(1), \ldots, h(L-1)]^T$, with L denoting the channel length expressed in signaling intervals. In practice, $\boldsymbol{h}$ represents the composite CIR encompassing the transmission medium as well as the transmit and receive filters. After down-conversion and low-pass filtering, the received waveform is sampled at rate $f_s = 1/T_s$. The resulting samples are mathematically expressed as the convolution between the transmitted blocks $\{\boldsymbol{s}_i^{(cp)}\}$ and $\boldsymbol{h}$. Assuming that the block duration is longer than the maximum delay spread and neglecting for simplicity the contribution of thermal noise, we can write the ith block of received samples as

$$r_i^{(cp)} = B^{(l)} s_i^{(cp)} + B^{(u)} s_{i-1}^{(cp)}, \qquad (2.44)$$

where $B^{(l)}$ and $B^{(u)}$ are $N_T \times N_T$ Toeplitz matrices given by

$$B^{(l)} = \begin{bmatrix} h(0) & 0 & 0 & \cdots & 0 \\ h(1) & h(0) & 0 & \cdots & 0 \\ h(2) & h(1) & h(0) & \cdots & 0 \\ \vdots & \vdots & \vdots & \vdots & \vdots \\ h(L-1) & h(L-2) & h(L-3) & \cdots & 0 \\ 0 & h(L-1) & h(L-2) & \cdots & 0 \\ \vdots & \vdots & \vdots & \vdots & \vdots \\ 0 & 0 & & \cdots & 0 \; h(0) \end{bmatrix} \qquad (2.45)$$

and

$$B^{(u)} = \begin{bmatrix} 0 \cdots & 0 & h(1) & h(2) & \cdots & h(L-1) \\ 0 \cdots & 0 & 0 & h(1) & \cdots & h(L-2) \\ \vdots & \vdots & \vdots & \ddots & \ddots & \vdots \\ 0 \cdots & \cdots & \cdots & \cdots & 0 & h(1) \\ 0 \cdots & \cdots & \cdots & \cdots & & 0 \\ \vdots & \vdots & \vdots & \ddots & \ddots & \vdots \\ 0 \cdots & \cdots & \cdots & \cdots & \cdots & 0 \end{bmatrix}. \qquad (2.46)$$

The second term in the right-hand-side of Eq. (2.44) is the IBI contribution, which is eliminated after discarding the CP. Defining the CP removal matrix as $R^{(cp)} = [\mathbf{0}_{N \times N_g} \; I_N]$ and using the identity $R^{(cp)} B^{(u)} = \mathbf{0}_{N \times N_T}$, we have

$$r_i = R^{(cp)} r_i^{(cp)} = B_c F^H c_i \qquad (2.47)$$

where $B_c = R^{(cp)} B^{(l)} T^{(cp)}$ is an $N \times N$ circulant matrix whose first column is $\left[h^T \; \mathbf{0}_{N-L}^T \right]^T$.

Vector r_i is serial-to-parallel converted and fed to the receive DFT unit. This produces

$$R_i =_c F B_c F^H c_i. \qquad (2.48)$$

Recalling the well-known diagonalization property of circulant matrices [92], we have

$$F B_c F^H = D_H, \qquad (2.49)$$

where D_H is a diagonal matrix with $H = \sqrt{N}Fh$ on its main diagonal. Hence, we may rewrite the DFT output as

$$R_i = D_H c_i, \qquad (2.50)$$

or, in scalar form,

$$R_i(n) = H(n)c_i(n), \qquad 0 \leq n \leq N - 1 \qquad (2.51)$$

where $R_i(n)$ and $c_i(n)$ are the nth entries of R_i and c_i, respectively, while $H(n)$ is the channel frequency response over the nth subcarrier, which reads

$$H(n) = \sum_{\ell=0}^{L-1} h(\ell)e^{-j2\pi n\ell/N}. \qquad (2.52)$$

Inspection of Eq. (2.51) indicates that OFDM can be viewed as a set of N non-interfering (orthogonal) parallel transmissions with different complex-valued attenuation factors $H(n)$. The transmitted symbols are recovered after pre-multiplying R_i by the inverse of D_H, *i.e.*,

$$\widehat{c}_i = D_H^{-1} R_i. \qquad (2.53)$$

Recalling that D_H is a diagonal matrix, the above equation can be rewritten in scalar form as

$$\widehat{c}_i(n) = \frac{R_i(n)}{H(n)}, \qquad 0 \leq n \leq N - 1 \qquad (2.54)$$

from which it is seen that channel equalization in OFDM is simply accomplished through a bank of one-tap complex-valued multipliers $1/H(n)$. In practice, due to the unavoidable presence of thermal noise and/or interference, the equalizer only provides soft estimates of the transmitted data symbols. The latter are eventually retrieved by passing the equalizer output to a data detection/decoding unit.

In the OFDM literature, the sequences at the IDFT input and DFT output are usually referred to as *frequency-domain* samples while those at the IDFT output and DFT input are called *time-domain* samples.

Example 2.5 For illustration purposes, we consider an OFDM system with only $N = 4$ subcarriers. The CP has length $N_g = 2$ and no VC is present. Transmission takes place over a multipath channel of length $L = 3$ and impulse response as in Example 2.4. The following two blocks of binary data symbols are fed to the IDFT unit

$$c_0 = \begin{bmatrix} 1 \\ -1 \\ -1 \\ 1 \end{bmatrix}, \qquad c_1 = \begin{bmatrix} -1 \\ -1 \\ -1 \\ 1 \end{bmatrix}. \qquad (2.55)$$

Then, the CP is appended in front of each IDFT output, thereby producing the vectors

$$\boldsymbol{s}_0^{(cp)} = \begin{bmatrix} 0 \\ \overline{1+j} \\ 0 \\ 1-j \\ 0 \\ 1+j \end{bmatrix}, \qquad \boldsymbol{s}_1^{(cp)} = \begin{bmatrix} -1 \\ j \\ -1 \\ -j \\ -1 \\ j \end{bmatrix}. \tag{2.56}$$

The received signal is distorted by frequency-selective fading. The time-domain samples corresponding to the second received OFDM block are expressed by

$$\boldsymbol{r}_1^{(cp)} = \begin{bmatrix} 0.815 & 0 & 0 & 0 & 0 & 0 \\ -0.495 & 0.815 & 0 & 0 & 0 & 0 \\ -0.3 & -0.495 & 0.815 & 0 & 0 & 0 \\ 0 & -0.3 & -0.495 & 0.815 & 0 & 0 \\ 0 & 0 & -0.3 & -0.495 & 0.815 & 0 \\ 0 & 0 & 0 & -0.3 & -0.495 & 0.815 \end{bmatrix} \begin{bmatrix} -1 \\ j \\ -1 \\ -j \\ -1 \\ j \end{bmatrix}$$

$$+ \begin{bmatrix} 0 & 0 & 0 & 0 & -0.495 & -0.3 \\ 0 & 0 & 0 & 0 & 0 & -0.495 \\ 0 & 0 & 0 & 0 & 0 & 0 \\ 0 & 0 & 0 & 0 & 0 & 0 \\ 0 & 0 & 0 & 0 & 0 & 0 \\ 0 & 0 & 0 & 0 & 0 & 0 \end{bmatrix} \begin{bmatrix} 0 \\ 1+j \\ 0 \\ 1-j \\ 0 \\ 1+j \end{bmatrix} = \begin{bmatrix} -1.31 - 0.495j \\ 0.195 + 0.515j \\ -0.515 - 0.495j \\ 0.495 - 1.115j \\ -0.515 + 0.495j \\ 0.495 + 1.115j \end{bmatrix}. \tag{2.57}$$

After CP removal, the received samples are fed to the DFT unit. From Eq. (2.52) we know that

$$\boldsymbol{H} = \sqrt{N}\boldsymbol{F}\boldsymbol{h} = \begin{bmatrix} 0.02 \\ 1.115 + 0.495j \\ 1.01 \\ 1.115 - 0.495j \end{bmatrix}, \tag{2.58}$$

and the data block $\boldsymbol{c}_1$ is thus retrieved as indicated in Eq. (2.53), *i.e.*,

$$\widehat{\boldsymbol{c}}_1 = \begin{bmatrix} 0.02 & 0 & 0 & 0 \\ 0 & 1.115 + 0.495j & 0 & 0 \\ 0 & 0 & 1.01 & 0 \\ 0 & 0 & 0 & 1.115 - 0.495j \end{bmatrix}^{-1} \begin{bmatrix} -0.02 \\ -1.115 - 0.495j \\ -1.01 \\ 1.115 - 0.495j \end{bmatrix}$$

$$= \begin{bmatrix} -1 \\ -1 \\ -1 \\ 1 \end{bmatrix}. \tag{2.59}$$

The above equation reveals that the transmitted symbols can ideally be recovered from the DFT output as long as the receiver has perfect knowledge of the channel response and the noise is vanishingly small. Also, we observe that channel distortion is easily compensated through a bank of four complex-valued multipliers while a time-domain equalizer with tens of taps is required in a conventional single-carrier system as that considered in Example 2.4.

2.4   Spectral efficiency

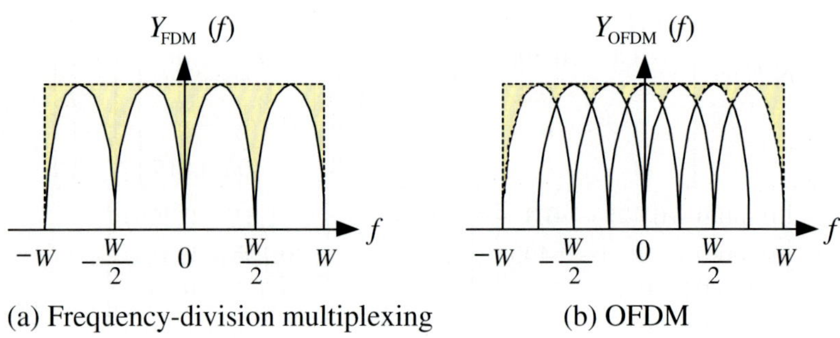

Fig. 2.19   Comparison between the spectral efficiencies of FDM and OFDM systems.

In addition to being robust against frequency-selective fading, another advantage of OFDM is the relatively high spectral efficiency as compared to conventional frequency-division multiplexing (FDM) systems. In these applications, the whole available bandwidth is divided into several subchannels and one data stream is transmitted over each subchannel. Figure 2.19 (a) depicts the spectrum of a typical FDM system employing four parallel subchannels. Here, the rectangular box spanning the frequency interval $[-W, W]$ represents the ideal signal spectrum that fully exploits the assigned bandwidth. It appears that FDM scheme suffers from some spectral inefficiency, as indicated by the large shaped area within the rectangular box.

As shown in Fig. 2.19 (b), in OFDM systems adjacent subchannels partially overlap in the frequency domain. As a result, OFDM has much higher spectral efficiency than conventional FDM schemes. To cope with the interference caused by spectra overlapping, carriers of different subchannels are mutually orthogonal. As we have seen, this goal is efficiently achieved by means of FFT/IFFT operations. It is evident from Fig. 2.19 (b) that the spectral efficiency improves as the number of subcarriers increases. On the other hand, employing more subcarriers on a fixed bandwidth results into narrower subchannels and longer OFDM blocks. This may greatly complicate the synchronization and channel equalization tasks since blocks of long duration are exposed to time-selective fading.

2.5 Strengths and drawbacks of OFDM

The main advantages of OFDM can be summarized as follows:

(1) Increased robustness against multipath fading, which is obtained by dividing the overall signal spectrum into narrowband flat-fading subchannels. As a result, channel equalization is accomplished through a simple bank of complex-valued multipliers, thereby avoiding the need for computationally demanding time-domain equalizers.
(2) High spectral efficiency due to partially overlapping subchannels in the frequency-domain.
(3) Interference suppression capability through the use of the cyclic prefix.
(4) Simple digital implementation by means of DFT/IDFT operations.
(5) Increased protection against narrowband interference which, if present, is expected to affect only a small percentage of the overall subcarriers.
(6) Opportunity of selecting the most appropriate coding and modulation scheme on each individual subcarrier according to the measured channel quality (adaptive modulation). In practice, higher order constellations are normally used on less attenuated subcarriers in order to increase the data throughput, while robust low-order modulations are employed over subcarriers characterized by low SNR values.

On the other hand, OFDM suffers from the following drawbacks as compared to conventional single-carrier (SC) transmissions:

(1) It is very sensitive to phase noise and frequency synchronization errors, which translates into more stringent specifications for local oscillators.
(2) It needs power amplifiers that behave linearly over a large dynamic range because of the relatively high peak-to-average power ratio (PAPR) characterizing the transmitted waveform.

(3) There is an inherent loss in spectral efficiency related to the use of the cyclic prefix.

2.6   OFDM-based multiple-access schemes

Conventional multiple-access techniques can be combined with OFDM to provide high-speed services to a number of simultaneously active users. Three prominent OFDM-based multiple-access schemes are available in the technical literature. They include OFDM with time-division multiple-access (OFDM-TDMA) [133], OFDM with code-division multiple-access (MC-CDMA) [53] and orthogonal frequency-division multiple-access (OFDMA) [141]. The main ideas behind these techniques are illustrated in Fig. 2.20 and are now briefly reviewed in order to highlight their main features.

OFDM-TDMA

In OFDM-TDMA, data transmission occurs into several consecutive time-slots, each comprehending one or more OFDM blocks. Since each slot is exclusively assigned to a specific user, no multiple-access interference (MAI) is present in the received data stream as long as a sufficiently long CP is appended in front of the transmitted blocks. A possible drawback of OFDM-TDMA is the need for very high power amplifiers at the transmit side due to the following reasons. First, because of its inherent TDMA structure, an OFDM-TDMA transmitter demands much higher instantaneous power than a frequency-division multiple-access (FDMA) system. Second, the transmit amplifier must exhibit a linear characteristic over a wide dynamic range due to the relatively high PAPR of the OFDM waveform [8]. Clearly, the need for highly linear power amplifiers increases the implementation cost of OFDM-TDMA transmitters.

MC-CDMA

MC-CDMA exploits the additional diversity gain provided by spread-spectrum techniques while inheriting the advantages of OFDM. In MC-CDMA systems, users spread their data symbol over M chips, which are then mapped onto a set of M distinct subcarriers out of a total of N. Each set of subcarriers is typically shared by a group of users which are separated by means of their specific spreading codes [42]. In order to achieve

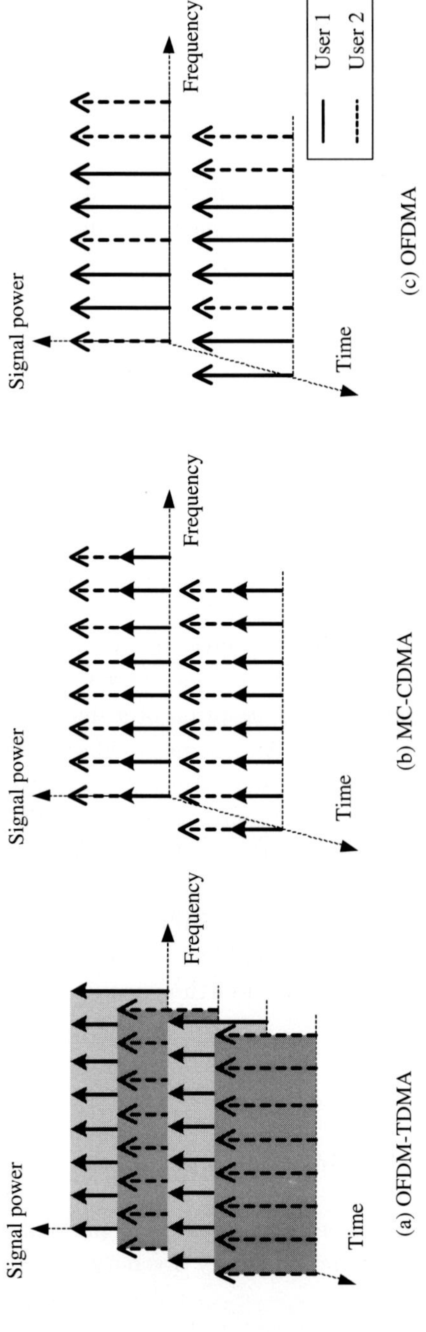

Fig. 2.20   Illustration of OFDM-based multiple-access schemes.

some form of frequency diversity, the M subcarriers can be interleaved over the whole signal spectrum so as to maximize their separation distance in the frequency domain. Similarly to CDMA, MC-CDMA signals are normally plagued by MAI when transmitted over a frequency-selective fading channel. Since subcarriers are subject to different channel attenuations, orthogonality among users will be destroyed even though an orthogonal code set is employed at the transmit side for spreading purposes. To alleviate the MAI problem, sophisticated channel estimation and interference cancellation techniques are needed in MC-CDMA systems [35].

OFDMA

The OFDMA concept is based on the inherent orthogonality of the OFDM subcarriers. The latter are divided into several disjoint clusters which are normally referred to as *subchannels*, and each user is exclusively assigned one or more subchannels depending on its requested data rate. Since all carriers are perfectly orthogonal, in case of ideal synchronization no MAI is present at the output of the receiver DFT unit. This property greatly simplifies the design of an OFDMA receiver by avoiding the need for computationally demanding detection techniques based on multiuser interference cancellation. In addition, the adoption of a dynamic subchannel assignment strategy offers to OFDMA systems an effective means to exploit the user-dependent frequency diversity. Actually, a specific carrier which appears in a deep fade to one user may exhibit a relatively small attenuation for another user. As a result, OFDMA can exploit channel state information to provide users with the "best" subcarriers that are currently available, thereby leading to remarkable gains in terms of achievable data throughput [172]. Thanks to its favorable features, OFDMA is widely recognized as a promising technique for fourth generation broadband wireless networks [149].

2.7   Channel coding and interleaving

Channel coding and interleaving are fundamental parts of any OFDM system as they allow to exploit the frequency diversity offered by the wireless channel.

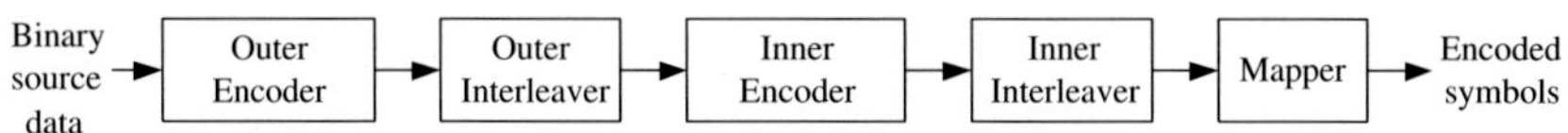

Fig. 2.21   Channel coding and interleaving in an OFDM transmitter.

Encoding

Figure 2.21 illustrates the generation process of the encoded symbols at the input of an OFDM system. The sequence of binary source data is divided into segments of k bits and fed to the outer encoder, where $n - k$ redundant bits are added to each segment to protect information against channel impairments and thermal noise. The encoder output is then passed to the outer interleaver, which is followed by the inner encoder. The output of the inner encoder is further interleaved before the encoded bits are mapped onto modulation symbols taken from a designated constellation. The most commonly used inner and outer coding architectures employ Reed–Solomon (RS) codes and convolutional codes, respectively [123]. The concatenated coding scheme of Fig. 2.21 is attractive due to its improved error correction capability and low decoding complexity.

Decoding

At the receiver, channel decoding and de-interleaving are accomplished as depicted in Fig. 2.22.

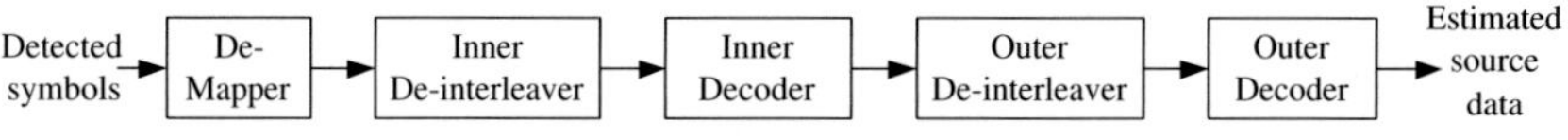

Fig. 2.22   Channel decoding and de-interleaving in an OFDM receiver.

The de-mapper converts the detected symbols into a sequence of bits. Since convolutional codes are very sensitive to burst errors, it is important that the inner de-interleaver can scatter the erroneous bits over the whole interleaving range before applying inner decoding. The convolutional inner decoder is efficiently implemented by means of the Viterbi algorithm [123]. After inner decoding, most bit errors in the received stream will be corrected. The output of the inner decoder is then de-interleaved before being passed to the outer decoder.

We recall that an RS code can correct up to $\lceil \frac{n-k}{2} \rceil$ erroneous bits in one encoded block of size n, where $\lceil x \rceil$ denotes the highest integer not larger than x. Therefore, if the outer de-interleaver scatters the remaining bit errors over multiple blocks and no more than $\lceil \frac{n-k}{2} \rceil$ bit errors are left in each block, all source data are correctly retrieved.

The above discussion indicates that bit interleaving and de-interleaving are essential in OFDM systems to fully exploit the correction capability of the employed code structures. However, these operations may result into large storage requirements, which are clearly undesirable in terms of implementation cost.

Chapter 3

Time and Frequency Synchronization

Synchronization plays a major role in the design of a digital communication system. Essentially, this function aims at retrieving some reference parameters from the received signal that are necessary for reliable data detection. In a multicarrier network, the following synchronization tasks can be identified.

(1) *sampling clock synchronization*: in practical systems the sampling clock frequency at the receiver is slightly different from the corresponding frequency at the transmitter. This produces interchannel interference (ICI) at the output of the receive DFT with a corresponding degradation of the system performance. The purpose of symbol clock synchronization is to limit this impairment to a tolerable level.

(2) *timing synchronization*: the goal of this operation is to identify the beginning of each received OFDMA block so as to find the correct position of the DFT window. In burst-mode transmissions timing synchronization is also used to locate the start of the frame (frame synchronization).

(3) *frequency synchronization*: a frequency error between the received carrier and the local oscillator used for signal demodulation results in a loss of orthogonality among subcarriers with ensuing limitations of the system performance. Frequency synchronization aims at restoring orthogonality by compensating for any frequency offset caused by oscillator inaccuracies or Doppler shifts.

We limit our discussion to timing and frequency synchronization without addressing the problem of sampling clock recovery in this chapter. The reason is that nowadays the accuracy of modern oscillators is in the order of some parts per million (ppm) and sample clock variations below 50 ppm have only marginal effects on the performance of practical multicarrier

51

systems [118].

In the ensuing discussion the synchronization task is separately addressed for the downlink and uplink case. As we shall see, while synchronization in the downlink can be achieved with the same methods employed in conventional OFDM transmissions, the situation is much more complicated in the uplink due to the possibly large number of parameters that the base station (BS) has to estimate and the inherent difficulty in correcting the time and frequency errors of each active user.

This chapter is organized as follows. The sensitivity of a multicarrier system to timing and frequency errors is discussed in Sec. 3.1. In Sec. 3.2 we illustrate several synchronization algorithms explicitly designed for downlink transmissions. The uplink case is treated in Sec. 3.3 and Sec. 3.4. In particular, timing and frequency estimation is studied in Sec. 3.3 while some schemes for compensating the synchronization errors at the BS are illustrated in Sec. 3.4.

3.1   Sensitivity to timing and frequency errors

Timing and frequency errors in multicarrier systems destroy orthogonality among subcarriers and may result in large performance degradations. To simplify the analysis, in the following we concentrate on a downlink transmission but we point out that the final results essentially apply also to the uplink case.

The time-domain samples of the ith OFDM block are given by

$$s_i^{(cp)}(k) = \frac{1}{\sqrt{N}} \sum_{n \in \mathcal{I}} c_i(n)\, e^{j2\pi nk/N}, \quad -N_g \leq k \leq N-1 \qquad (3.1)$$

where N is the size of the transmit IDFT unit, $\mathcal{I}$ denotes the set of modulated subcarriers, N_g is the length of the cyclic prefix (CP) in sampling periods and $c_i(n)$ is the symbol transmitted over the nth subcarrier. For notational simplicity, the superscript $(\cdot)^{(cp)}$ is neglected throughout this chapter.

The baseband-equivalent discrete-time signal transmitted by the BS is thus represented by

$$s_T(k) = \sum_i s_i(k - iN_T), \qquad (3.2)$$

where i counts the OFDM blocks and $N_T = N + N_g$ is the block length (included the CP).

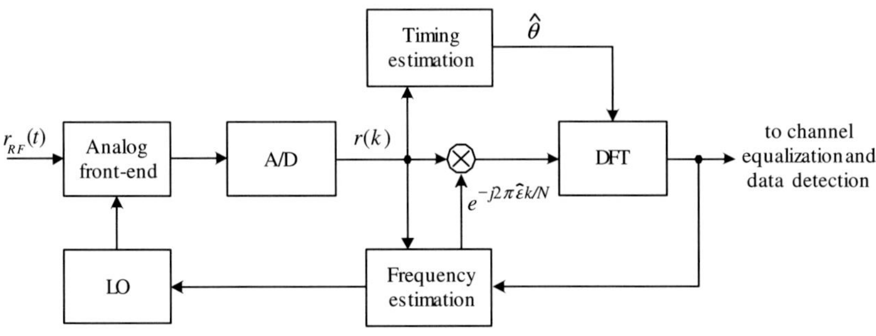

Fig. 3.1   Block diagram of an OFDM receiver.

The block diagram of the receiver is depicted in Fig. 3.1. In the analog front-end, the incoming waveform $r_{RF}(t)$ is filtered and down-converted to baseband using two quadrature sinusoids generated by a local oscillator (LO). The baseband signal is then passed to the A/D converter, where it is sampled with frequency $f_s = 1/T_s$.

Due to Doppler shifts and/or oscillator instabilities, the frequency f_{LO} of the LO is not exactly equal to the received carrier frequency f_c. The difference $f_d = f_c - f_{LO}$ is referred to as *carrier frequency offset* (CFO). In addition, since the time scales at the transmit and receive sides are not perfectly aligned, at the start-up the receiver does not know where the OFDM blocks start and, accordingly, the DFT window will be placed in a wrong position. As shown later, since small (fractional) timing errors do not produce any degradation of the system performance, it suffices to estimate the beginning of each received OFDM block within one sampling period.

In the following we denote θ the number of samples by which the receive time scale is shifted from its ideal setting. The samples from the A/D unit are thus expressed by

$$r(k) = e^{j2\pi\varepsilon k/N} \sum_{i} \sum_{\ell=0}^{L-1} h(\ell)s_i(k - \theta - \ell - iN_T) + w(k), \qquad (3.3)$$

where $\varepsilon = N f_d T_s$ is the frequency offset normalized to the subcarrier spacing $f_{cs} = 1/(NT_s)$, $\boldsymbol{h} = [h(0),\ h(1), \ldots, h(L-1)]^T$ is the discrete-time channel impulse response (CIR) encompassing the physical channel as well as the transmit/receive filters and, finally, $w(k)$ is complex-valued AWGN with variance σ_w^2. Since a carrier phase shift can be encapsulated into the CIR, it is normally compensated for during the channel equalization process.

The frequency and timing synchronization units shown in Fig. 3.1 employ the received samples $r(k)$ to compute estimates of ε and θ, say $\widehat{\varepsilon}$ and $\widehat{\theta}$. The former is used to adjust the frequency of the LO in a closed loop fashion or, alternatively, to counter-rotate $r(k)$ at an angular speed $2\pi\widehat{\varepsilon}/N$ (frequency correction), while the timing estimate is exploited to achieve the correct positioning of the receive DFT window (timing correction). Specifically, the samples $r(k)$ with indices $iN_T + \widehat{\theta} \le k \le iN_T + \widehat{\theta} + N - 1$ are fed to the DFT device and the corresponding output is used to detect the data symbols conveyed by the ith OFDM block. The DFT output can also be exploited to track and compensate for small short-term variations of the frequency error (fine-frequency estimation).

In the rest of this Section we assess the impact of uncompensated timing and frequency errors on the system performance.

3.1.1   *Effect of timing offset*

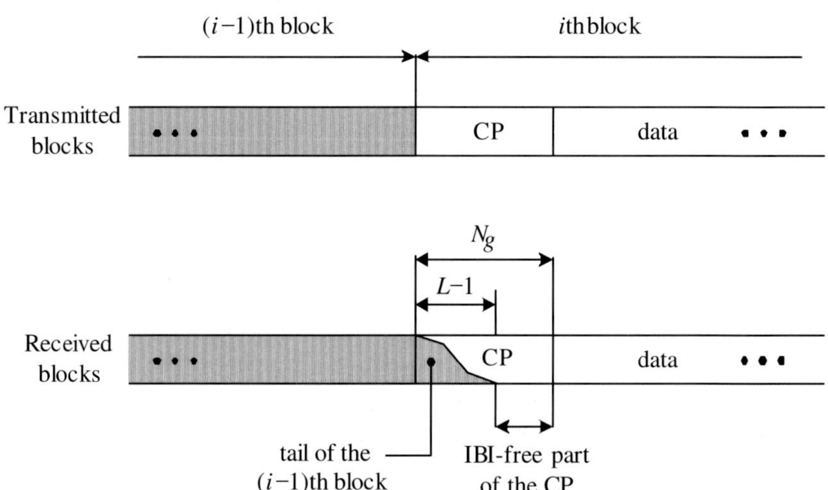

Fig. 3.2   Partial overlapping of received blocks due to multipath dispersion.

We assume perfect frequency synchronization (*i.e.*, $\varepsilon = 0$) and consider only the effect of a timing error $\Delta\theta = \widehat{\theta} - \theta$. As shown in Fig. 3.2, the tail of each received block extends over the first $L - 1$ samples of the subsequent block as a consequence of multipath dispersion. Since in a well designed system we must ensure that $N_g \geq L$, at the receiver a certain range of the guard interval is not affected by the previous block. As long as the DFT window starts anywhere in this range, no interblock interference (IBI) will be present at the DFT output.

To better explain this point, we see from Eqs. (3.1) and (3.3) that the mth received block (apart from thermal noise) is expressed by

$$s_{m,R}(k) = \sum_{\ell=0}^{L-1} h(\ell) s_m(k - \theta - \ell - mN_T), \qquad (3.4)$$

and is non-zero for $k'_m \leq k \leq k''_m$, where $k'_m = \theta + mN_T - N_g$ and $k''_m = \theta + (m+1)N_T - N_g + L - 2$.

This means that the last sample of the $(i-1)$th received block has index $k''_{i-1} = \theta + iN_T - N_g + L - 2$ while the first sample of the $(i+1)$th block occurs at $k'_{i+1} = \theta + iN_T + N$. Accordingly, samples $r(k)$ with index k in the set $[\theta + iN_T - N_g + L - 1; \; \theta + iN_T + N - 1]$ are only contributed by the ith OFDM block and, in consequence, do not suffer from IBI. Recalling that the DFT window for the detection of the ith block spans the interval $iN_T + \widehat{\theta} \leq k \leq iN_T + \widehat{\theta} + N - 1$, it follows that IBI is not present as long as $-N_g + L - 1 \leq \Delta\theta \leq 0$. In this case the DFT output over the nth subcarrier can be represented as

$$R_i(n) = e^{j2\pi n\Delta\theta/N} H(n) c_i(n) + W_i(n), \qquad (3.5)$$

where $W_i(n)$ is the noise contribution with power σ_w^2 and

$$H(n) = \sum_{\ell=0}^{L-1} h(\ell) \, e^{-j2\pi \ell n/N} \qquad (3.6)$$

is the channel frequency response over the considered subcarrier.

Inspection of Eq. (3.5) reveals that the timing offset appears as a linear phase across the DFT outputs and is compensated for by the channel equalizer, which cannot distinguish between phase shifts introduced by the channel and those deriving from the timing offset. This means that no single correct timing synchronization point exists in OFDM systems, since there are $N_g - L + 2$ values of $\widehat{\theta}$ for which interference is not present.

On the other hand, if the timing error is outside the interval $-N_g + L - 1 \leq \Delta\theta \leq 0$, the DFT output will be contributed not only by the ith

OFDM block, but also by the $(i-1)$th or $(i+1)$th block, depending on whether $\Delta\theta < -N_g + L - 1$ or $\Delta\theta > 0$. In addition to IBI, this results into a loss of orthogonality among subcarriers which, in turn, generates ICI. In this case the nth DFT output is affected by interference caused by data symbols transmitted over adjacent subcarriers and/or belonging to neighboring blocks, and reads

$$R_i(n) = e^{j2\pi n\Delta\theta/N}\alpha(\Delta\theta)H(n)c_i(n) + I_i(n,\Delta\theta) + W_i(n), \qquad (3.7)$$

where $I_i(n,\Delta\theta)$ accounts for IBI and ICI while $\alpha(\Delta\theta)$ is an attenuation factor which is well approximated by [148]

$$\alpha(\Delta\theta) = \sum_{\ell=0}^{L-1} |h(\ell|^2 \frac{N - \Delta\theta_\ell}{N}, \qquad (3.8)$$

with

$$\Delta\theta_\ell = \begin{cases} \Delta\theta - \ell, & \text{if} \quad \Delta\theta > \ell \\ \ell - N_g - \Delta\theta, & \text{if} \quad \Delta\theta < \ell - N_g \\ 0, & \text{otherwise.} \end{cases} \qquad (3.9)$$

The term $I_i(n,\Delta\theta)$ can reasonably be modeled as a zero-mean random variable whose power $\sigma_I^2(\Delta\theta)$ depends on the channel delay profile and timing error according to the following relation

$$\sigma_I^2(\Delta\theta) = C_2 \sum_{\ell=0}^{L-1} |h(\ell|^2 \left[2\frac{\Delta\theta_\ell}{N} + \left(\frac{\Delta\theta_\ell}{N}\right)^2 \right], \qquad (3.10)$$

where $C_2 = \mathrm{E}\{|c_i(n)|^2\}$ is the average power of the transmitted data symbols.

A useful indicator to evaluate the effect of timing errors on the system performance is the loss in signal-to-noise ratio (SNR). This quantity is defined as

$$\gamma(\Delta\theta) = \frac{SNR^{(ideal)}}{SNR^{(real)}}, \qquad (3.11)$$

where $SNR^{(ideal)}$ is the SNR across subcarriers in a perfectly synchronized system, while $SNR^{(real)}$ is the SNR in the presence of a timing offset. In the ideal case, the DFT output is given by

$$R_i^{(ideal)}(n) = H(n)c_i(n) + W_i(n), \qquad (3.12)$$

so that, for a channel with unit average power (*i.e.*, $\mathrm{E}\{|H(n)|^2\} = 1$), we have

$$SNR^{(ideal)} = C_2/\sigma_w^2. \qquad (3.13)$$

On the other hand, recalling that the three terms in the right-hand-side of Eq. (3.7) are statistically uncorrelated, it follows that

$$SNR^{(real)} = C_2\alpha^2(\Delta\theta)/\left[\sigma_w^2 + \sigma_I^2(\Delta\theta)\right].$$
(3.14)

Substituting the above results into Eq. (3.11) yields

$$\gamma(\Delta\theta) = \frac{1}{\alpha^2(\Delta\theta)}\left[1 + \frac{\sigma_I^2(\Delta\theta)}{\sigma_w^2}\right].$$
(3.15)

It is useful to express the SNR loss in terms of E_s/N_0, where E_s is the average received energy over each subcarrier while $N_0/2$ is the two-sided power spectral density of the ambient noise. For this purpose we collect Eqs. (3.10) and (3.15) and observe that $C_2/\sigma_w^2 = E_s/N_0$. This produces

$$\gamma(\Delta\theta) = \frac{1}{\alpha^2(\Delta\theta)}\left\{1 + \frac{E_s}{N_0}\sum_{\ell=0}^{L-1}|h(\ell|^2\left[2\frac{\Delta\theta_\ell}{N} + \left(\frac{\Delta\theta_\ell}{N}\right)^2\right]\right\}.$$
(3.16)

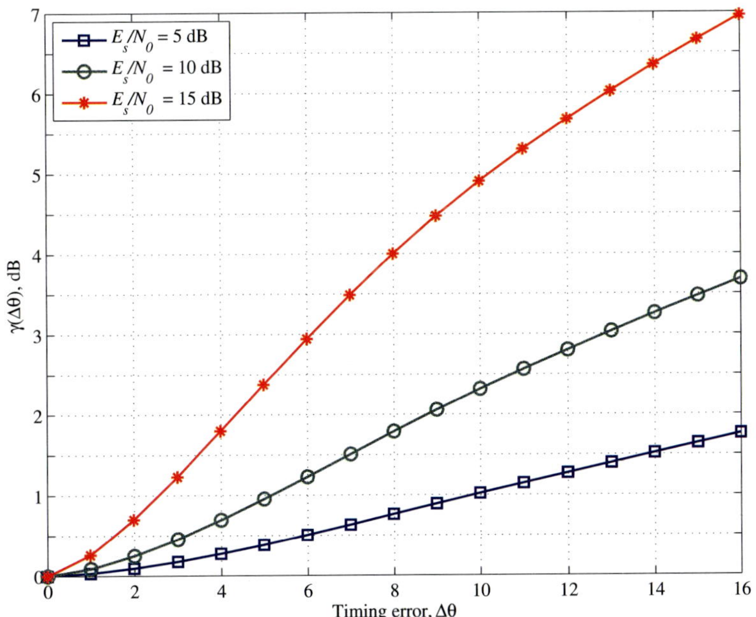

Fig. 3.3   SNR loss due to timing errors.

Figure 3.3 illustrates $\gamma(\Delta\theta)$ (in dB) versus the timing error $\Delta\theta$ for $N = 256$ and some values of E_s/N_0. The CIR has length $L = 8$ and the channel taps are modeled as circularly symmetric independent Gaussian random variables with zero-mean (Rayleigh fading) and power $E\{|h(\ell)|^2\} = \beta e^{-\ell/8}$, where β is a suitable factor that normalizes the average energy of the CIR to unity. At each simulation run, a new channel snapshot is generated and the results are obtained by numerically averaging the right-hand-side of Eq. (3.16) with respect to the channel statistics.

For a given timing error, we see that $\gamma(\Delta\theta)$ increases with E_s/N_0. This can be explained by observing that at low SNRs the system performance is mainly limited by thermal noise so that the impact of synchronization errors becomes less and less evident. The results in Fig. 3.3 indicate that in order to keep the SNR degradation to a tolerable level of less than 1.0 dB, the error $\Delta\theta$ after timing correction should be smaller than a few percents of the block length. As discussed earlier, the presence of the CP provides intrinsic protection against timing errors since no performance degradation occurs as long as $-N_g + L - 1 \le \Delta\theta \le 0$. The requirement of the timing synchronizer is thus determined by the number of samples by which the CP exceeds the CIR duration. This provides the designer with a trade-off tool. Using a longer CP results into a relaxation of the timing synchronization requirements, but inevitably increases the system overhead.

3.1.2 *Effect of frequency offset*

We now assess the impact of a frequency error on the system performance. For simplicity, we assume ideal timing synchronization and let $\widehat{\theta} = \theta = 0$. At the receiver, the DFT output for the ith OFDM block is computed as

$$R_i(n) = \frac{1}{\sqrt{N}} \sum_{k=0}^{N-1} r(k + iN_T)\, e^{-j2\pi nk/N}, \quad 0 \le n \le N - 1 \qquad (3.17)$$

and is not affected by IBI as long as $N_g \ge L - 1$. Substituting Eq. (3.3) into Eq. (3.17) and performing standard manipulations yields

$$R_i(n) = e^{j\varphi_i} \sum_{m \in \mathcal{I}} H(m)c_i(m)\, e^{j\pi(N-1)(\varepsilon+m-n)/N} f_N(\varepsilon + m - n) + W_i(n),$$

$$(3.18)$$

where $W_i(n)$ is thermal noise, $\varphi_i = 2\pi i\varepsilon N_T/N$ and

$$f_N(x) = \frac{\sin(\pi x)}{N\sin(\pi x/N)}. \qquad (3.19)$$

We begin by considering the situation in which the frequency offset is a multiple of the subcarrier spacing f_{cs}. In this case ε is integer-valued and

Eq. (3.18) reduces to

$$R_i(n) = e^{j\varphi_i} H\left(|n - \varepsilon|_N\right) c_i\left(|n - \varepsilon|_N\right) + W_i(n), \qquad (3.20)$$

where $|n - \varepsilon|_N$ is the value of $n - \varepsilon$ reduced to the interval $[0, N-1]$. This equation indicates that an integer frequency offset does not destroy orthogonality among subcarriers and only results into a shift of the subcarrier indices by a quantity ε. In this case the nth DFT output is an attenuated and phase-rotated version of $c_i\left(|n - \varepsilon|_N\right)$ rather than of $c_i(n)$. Vice versa, when ε is not integer-valued the subcarriers are no longer orthogonal and ICI does occur. In this case it is convenient to isolate the contribution of $c_i(n)$ in the right-hand-side of Eq. (3.18) to obtain

$$R_i(n) = e^{j[\varphi_i + \pi\varepsilon(N-1)/N]} H(n)c_i(n)\, f_N(\varepsilon) + I_i(n,\varepsilon) + W_i(n), \qquad (3.21)$$

where $I_i(n,\varepsilon)$ accounts for ICI and reads

$$I_i(n,\varepsilon) = e^{j\varphi_i} \sum_{m \neq n} H(m)c_i(m)\, e^{j\pi(N-1)(\varepsilon+m-n)/N} f_N(\varepsilon + m - n). \qquad (3.22)$$

Letting $\mathrm{E}\{|H(n)|^2\} = 1$ and assuming independent and identically distributed data symbols with zero-mean and power C_2, from Eq. (3.22) we see that $I_i(n,\varepsilon)$ has zero-mean and power

$$\sigma_I^2(\varepsilon) = C_2 \sum_{m \neq n} f_N^2(\varepsilon + m - n). \qquad (3.23)$$

A more concise expression of $\sigma_I^2(\varepsilon)$ is found when all N available subcarriers are used for data transmission, *i.e.*, $\mathcal{I} = \{0, 1, \ldots, N-1\}$. In this case the above equation becomes

$$\sigma_I^2(\varepsilon) = C_2\left[1 - f_N^2(\varepsilon)\right], \qquad (3.24)$$

where we have used the identity

$$\sum_{m=0}^{N-1} f_N^2(\varepsilon + m - n) = 1, \qquad (3.25)$$

which holds true independently of ε.

The impact of the frequency error on the system performance is still assessed in terms of the SNR loss, which is defined as

$$\gamma(\varepsilon) = \frac{SNR^{(ideal)}}{SNR^{(real)}}, \qquad (3.26)$$

where $SNR^{(ideal)}$ is the SNR of a perfectly synchronized system as given in Eq. (3.13), while

$$SNR^{(real)} = C_2 f_N^2(\varepsilon) / \left[\sigma_w^2 + \sigma_I^2(\varepsilon)\right] \qquad (3.27)$$

is the SNR in the presence of a frequency offset ε. Substituting Eqs. (3.13) and (3.27) into Eq. (3.26) and recalling that $C_2/\sigma_w^2 = E_s/N_0$, we have

$$\gamma(\varepsilon) = \frac{1}{f_N^2(\varepsilon)} \left[1 + \frac{E_s}{N_0}(1 - f_N^2(\varepsilon)) \right], \qquad (3.28)$$

where we have also borne in mind Eq. (3.24). For small values of ε, the above equation can be simplified using the Taylor series expansion of $f_N^2(\varepsilon)$ around $\varepsilon = 0$. This produces

$$\gamma(\varepsilon) \approx 1 + \frac{1}{3}\frac{E_s}{N_0}(\pi\varepsilon)^2, \qquad (3.29)$$

from which it follows that the loss in SNR is approximately related to the square of the normalized frequency offset.

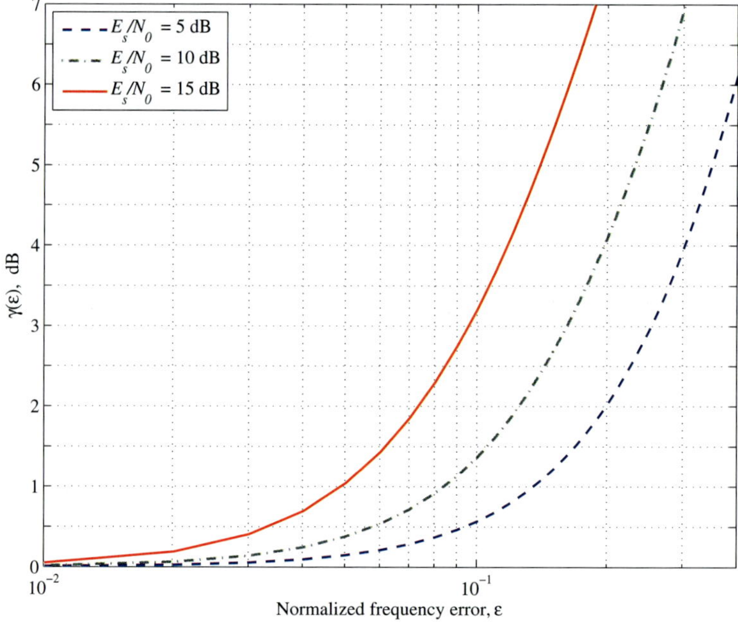

Fig. 3.4   SNR loss due to frequency errors.

Equation (3.28) is plotted in Fig. 3.4 as a function of ε for some values of E_s/N_0 and $N = 256$. This diagram indicates that the frequency offset should be kept as low as 4-5% of the subcarrier distance to avoid a severe degradation of the system performance. For example, in the IEEE

802.16 standard for wireless MANs, the subcarrier spacing is 11.16 kHz and the maximum tolerable frequency error is thus in the order of 500 Hz. Assuming a carrier frequency of 5 GHz, this corresponds to an oscillator instability of 0.1 ppm. Since the accuracy of low-cost oscillators for mobile terminals usually does not meet the above requirement, an estimate $\widehat{\varepsilon}$ of the frequency offset must be computed at each terminal and used to counter-rotate the samples at the input of the DFT device so as to reduce the residual frequency error $\Delta\varepsilon = \varepsilon - \widehat{\varepsilon}$ within a tolerable range.

3.2 Synchronization for downlink transmissions

Synchronization for OFDMA downlink transmissions is a relatively simple task that can be accomplished with the same methods employed in conventional single-user OFDM systems. Here, each terminal exploits the broadcast signal transmitted by the BS to get timing and frequency estimates, which are then exploited to control the position of the DFT window and to adjust the frequency of the local oscillator.

The synchronization process is typically split into an acquisition step followed by a tracking phase. During acquisition, pilot blocks with a particular repetitive structure are normally exploited to get initial estimates of the synchronization parameters [76, 95, 96, 99, 142, 146, 178]. Since in this phase the time- and frequency-scales of the receiving terminal are still to be aligned to the incoming signal, synchronization algorithms must be found that can cope with large synchronization errors. The tracking phase is devoted to the refinement of the initial timing and frequency estimates as well as to counteract short-term variations that may occur due to oscillator drifts and/or time-varying Doppler shifts. For this purpose, several techniques exploiting either the redundancy of the CP or pilot tones multiplexed in the frequency-domain are available in the literature [24, 29, 163]. Alternatively, blind methods operating over the DFT output can be used [30, 98].

In this Section we investigate timing and frequency estimation in a downlink scenario. Both the acquisition and tracking phases are considered and separately discussed. As standardized in many commercial systems including DAB [39], DVB-T [40] and HIPERLAN/II [41], the transmission is organized in frames, each containing some known reference blocks to assist the synchronization process.

A possible example of frame structure is depicted in Fig. 3.5. Here, a null block where nothing is transmitted (no signal power) is placed at

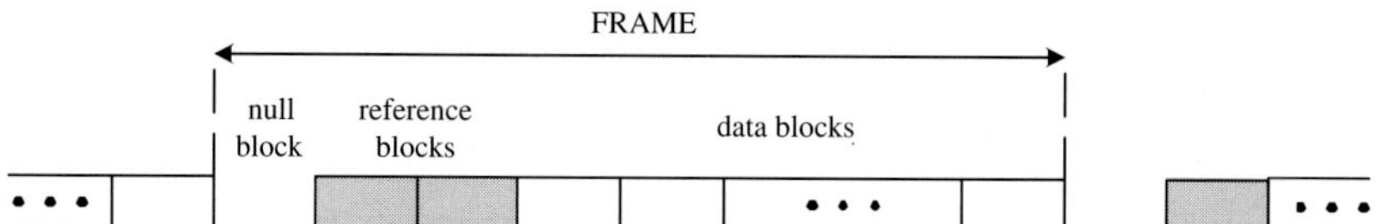

Fig. 3.5   Example of frame structure in the downlink.

the beginning of the frame, followed by a given number of reference and
data blocks. In addition, some pilot tones carrying known symbols are
normally placed within data blocks at some specified subcarriers in order
to track possible variations of the synchronization parameters. The null
block is exploited for interference and noise power estimation. Furthermore,
it provides a simple means to achieve coarse frame synchronization. In
this case, the drop of power corresponding to the null block is revealed
by a power detector and used as a rough estimate of the start of a new
frame [107]. Fine frame synchronization is next achieved using information
provided by the timing synchronization unit.

3.2.1   *Timing acquisition*

In most multicarrier applications, timing acquisition represents the first
step of the downlink synchronization process. This operation has two main
objectives. First, it detects the presence of a new frame in the received
data stream. Second, once the frame has been detected, it provides a
coarse estimate of the timing error so as to find the correct position of the
receive DFT window. Since the CFO is usually unknown in this phase, it is
desirable that the timing recovery scheme be robust against possibly large
frequency offsets.

One of the first timing acquisition algorithms for OFDM transmissions
was proposed by Nogami and Nagashima [107], and was based on the idea
of searching for a null reference block in the received frame. Unfortunately,
this method provides highly inaccurate timing estimates. Also, it is not
suited for burst-mode applications since the null block cannot be distinguished by the idle period between neighboring bursts. A popular approach
to overcome these difficulties makes use of some reference blocks exhibiting
a repetitive structure in the time domain. In this case, a robust timing estimator can be designed by searching for the peak of the correlation among

the repetitive parts. This approach was originally proposed by Schmidl and Cox (S&C) in [142], where a reference block with two identical halves of length $N/2$ is transmitted at the beginning of each frame and exploited for timing and frequency acquisition.

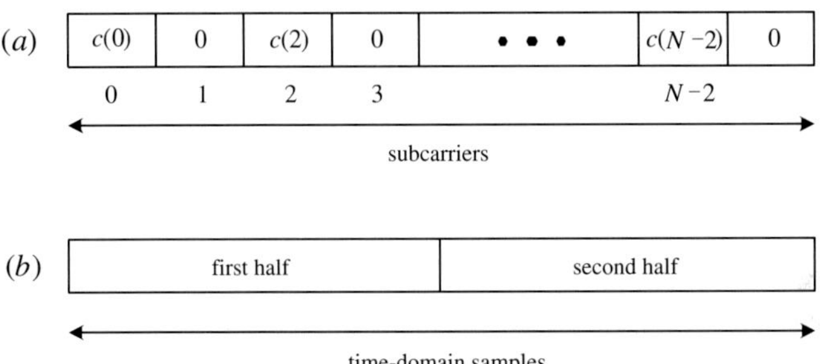

Fig. 3.6   S&C reference block in the frequency-domain (a) and in the time-domain (b).

As shown in Fig. 3.6, the reference block can easily be generated in the frequency domain by modulating the subcarriers with even indices by a pseudonoise (PN) sequence $\boldsymbol{c} = [c(0), c(2), \ldots, c(N-2)]^T$ while setting to zero the remaining subcarriers with odd indices. As long as the CP is not shorter than the CIR duration, the two halves of the reference block will remain identical after passing through the transmission channel except for a phase difference caused by the CFO. Hence, if the received samples corresponding to the first half are given by

$$r(k) = s_R(k)e^{j2\pi\varepsilon k/N} + w(k), \qquad \theta \le k \le \theta + N/2 - 1 \tag{3.30}$$

with $s_R(k)$ being the useful signal and $w(k)$ denoting the thermal noise, then the samples in the second half take the form

$$r(k+N/2) = s_R(k)e^{j2\pi\varepsilon k/N}e^{j\pi\varepsilon} + w(k+N/2), \quad \theta \le k \le \theta + N/2 - 1. \tag{3.31}$$

In this case, the magnitude of a sliding window correlation of lag $N/2$ provides useful information about the timing error since a peak is expected when the sliding window is perfectly aligned with the reference block. This approach leads to the timing estimate [142]

$$\widehat{\theta} = \arg\max_{\tilde{\theta}} \left\{ \left| \Gamma(\tilde{\theta}) \right| \right\}, \tag{3.32}$$

where $\Gamma(\widetilde{\theta})$ is the following normalized $N/2$-lag autocorrelation of the received samples

$$\Gamma(\widetilde{\theta}) = \frac{\displaystyle\sum_{q=0}^{N/2-1} r(q + N/2 + \widetilde{\theta})r^*(q + \widetilde{\theta})}{\displaystyle\sum_{q=0}^{N/2-1} \left| r(q + N/2 + \widetilde{\theta}) \right|^2}. \tag{3.33}$$

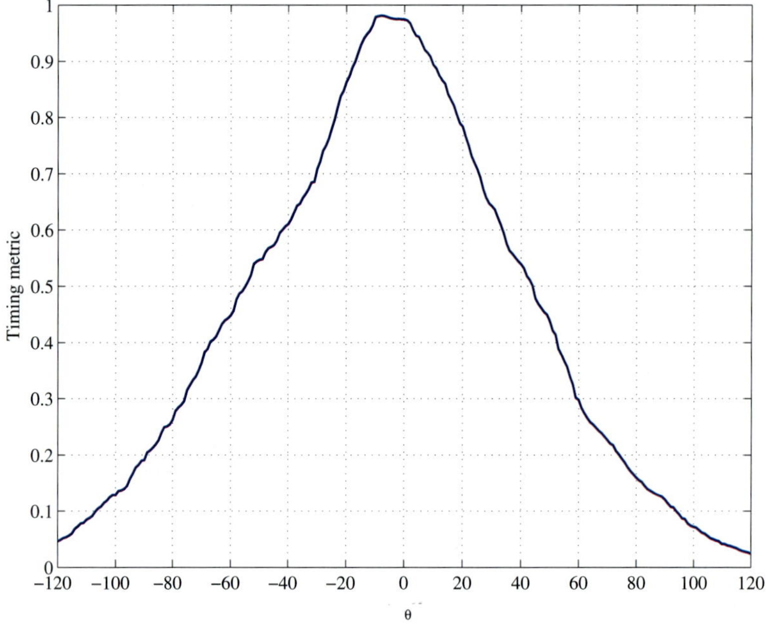

Fig. 3.7   Example of timing metric for the S&C algorithm.

Figure 3.7 shows an example of timing metric $\left|\Gamma(\widetilde{\theta})\right|$ as a function of the difference $\delta_\theta = \widetilde{\theta} - \theta$. The results are obtained numerically over a Rayleigh multipath channel with $L = 8$ taps. The number of subcarriers is $N = 256$ and the CP has length $N_g = 16$. The signal-to-noise ratio over the received samples is defined as $SNR = \sigma_s^2/\sigma_w^2$ with $\sigma_s^2 = \mathrm{E}\{|s_R(k)|^2\}$, and is set to 20 dB.

As mentioned before, the first step of the timing acquisition process is represented by the detection of a new frame in the received data stream.

For this purpose, $\left|\Gamma(\widetilde{\theta})\right|$ is continuously monitored and the start of a frame is declared whenever it overcomes a given threshold λ. The latter must properly be designed by taking into account the statistics of the timing metric so as to achieve a reasonably trade-off between false alarm and mis-detection probabilities. Once the presence of a new frame has been detected, a timing estimate $\widehat{\theta}$ is computed by searching for the maximum of $\left|\Gamma(\widetilde{\theta})\right|$ as indicated in Eq. (3.32).

Unfortunately, we see from Fig. 3.7 that the timing metric of the S&C algorithm exhibits a large "plateau" that may greatly reduce the estimation accuracy. Solutions to this problem are proposed in some recent works, where reference blocks with suitably designed patterns are exploited to obtain sharper timing metric trajectories [95, 146]. For instance, Shi and Serpedin (S&S) use a training block composed of four repetitive parts $[+B +B -B +B]$ with a sign inversion in the third segment [146]. As depicted in Fig. 3.8, a sliding window of length N spans the received time-domain samples with indices $\widetilde{\theta} \leq k \leq \widetilde{\theta}+N-1$, and collects them into four vectors $\boldsymbol{r}_j(\widetilde{\theta}) = \{r(k + jN/4 + \widetilde{\theta}) \; ; 0 \leq k \leq N/4 - 1\}$ with $j = 0, 1, 2, 3$.

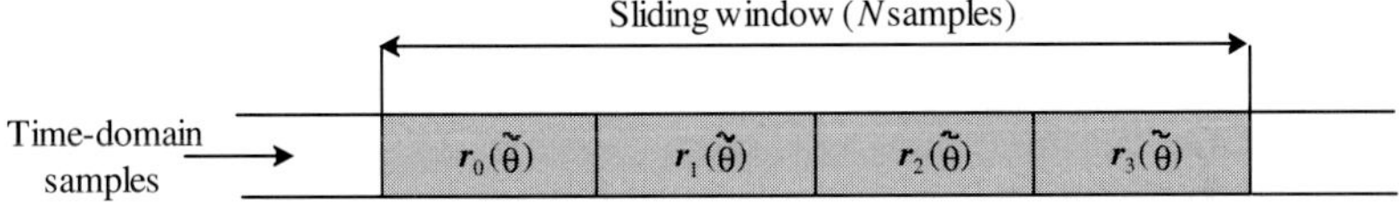

Fig. 3.8   Sliding window used in the S&S timing acquisition scheme.

The timing metric is then computed as

$$\Gamma_{SS}(\widetilde{\theta}) = \frac{\left|\Lambda_1(\widetilde{\theta})\right| + \left|\Lambda_2(\widetilde{\theta})\right| + \left|\Lambda_3(\widetilde{\theta})\right|}{\frac{3}{2}\sum_{j=0}^{3}\left\|\boldsymbol{r}_j(\widetilde{\theta})\right\|^2}, \tag{3.34}$$

where

$$\Lambda_1(\widetilde{\theta}) = \boldsymbol{r}_0^H(\widetilde{\theta})\boldsymbol{r}_1(\widetilde{\theta}) - \boldsymbol{r}_1^H(\widetilde{\theta})\boldsymbol{r}_2(\widetilde{\theta}) - \boldsymbol{r}_2^H(\widetilde{\theta})\boldsymbol{r}_3(\widetilde{\theta}), \tag{3.35}$$

$$\Lambda_2(\widetilde{\theta}) = \boldsymbol{r}_1^H(\widetilde{\theta})\boldsymbol{r}_3(\widetilde{\theta}) - \boldsymbol{r}_0^H(\widetilde{\theta})\boldsymbol{r}_2(\widetilde{\theta}), \tag{3.36}$$

$$\Lambda_3(\widetilde{\theta}) = \boldsymbol{r}_0^H(\widetilde{\theta})\boldsymbol{r}_3(\widetilde{\theta}). \tag{3.37}$$

Figure 3.9 illustrates $\Gamma_{SS}(\widetilde{\theta})$ as obtained in the same operating conditions of Fig. 3.7. Since the plateau region associated with the S&C metric

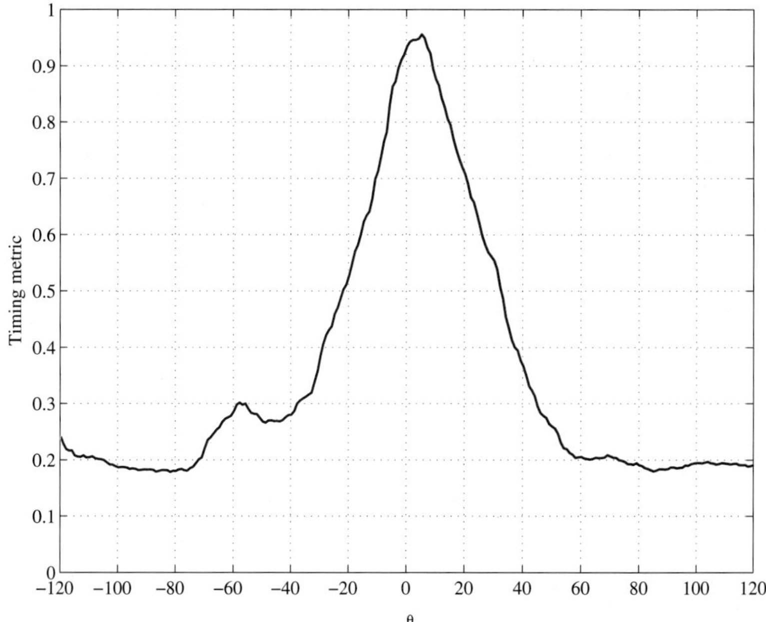

Fig. 3.9   Example of timing metric for the S&S algorithm.

is significantly reduced, more accurate timing estimates are expected. As indicated in [95], reference blocks with more than four repetitive segments can be designed to further increase the sharpness of the timing trajectory.

Simulation results obtained with both S&C and S&S algorithms indicate that the residual timing error $\Delta\theta$ takes on positive values with non-negligible probability. In this case the system performance may severely be degraded by IBI since the DFT window includes samples of the current OFDM block as well as of the next block. Appending a short *cyclic postfix* at the end of each transmitted block is a viable solution to mitigate the effect of small positive timing errors. Alternatively, we can pre-advance the estimate $\widehat{\theta}$ by some samples θ_c to obtain a final timing estimate in the form [95]

$$\widehat{\theta}^{(f)} = \widehat{\theta} - \theta_c, \tag{3.38}$$

where $\widehat{\theta}$ is still given in Eq. (3.32) while θ_c is designed so as to maximize the probability that $\Delta\theta^{(f)} = \widehat{\theta}^{(f)} - \theta$ lies in the interval $N_g + L - 1 \leq \Delta\theta^{(f)} \leq 0$ in order to mitigate IBI.

3.2.2 *Fine timing tracking*

If the transmit and receive clock oscillators are adequately stable, the timing estimate computed at the beginning of the downlink frame on the basis of the reference block can be used for data detection over the entire frame. In certain applications, however, the presence of non-negligible errors in the sampling clock frequency results in a short-term variation of the timing error $\Delta\theta$ which must be tracked in some way.

One straightforward solution is found by considering $\Delta\theta$ as introduced by the *physical channel* rather than by the oscillator drift. This amounts to absorbing $\Delta\theta$ into the CIR vector or, equivalently, to replacing $\boldsymbol{h} = [h(0), h(1), \ldots, h(L-1)]^T$ by its time-shifted version $\boldsymbol{h}'(\Delta\theta) = [h(\Delta\theta), h(1+\Delta\theta), \ldots, h(L-1+\Delta\theta)]^T$. Therefore, in the presence of small sampling frequency offsets, channel estimates computed over different OFDMA blocks are differently delayed as a consequence of the long-term fluctuations of $\Delta\theta$. A possible method to track these fluctuations is to look for the delay of the first significant tap in the estimated CIR vector. This approach is adopted in [178], where the integer part of the timing estimate is used by the DFT controller to adjust the DFT window position, while the fractional part appears as a linear phase across subcarriers and is compensated for by the channel equalization unit.

Alternative schemes to track residual timing errors make use of suitable correlations computed either in the time- or frequency-domain. For instance, the method proposed in [168] exploits known pilot tones multiplexed into the transmitted data stream, which are correlated at the output of the receive DFT with the transmitted pilot pattern. A time-domain approach is discussed in [163] and [76], where the autocorrelation properties induced by the use of the CP on the received time-domain samples is exploited for fine timing tracking. In this case, the following N-lag autocorrelation function is used as a timing metric

$$\gamma(k) = \sum_{q=0}^{N_g-1} r(k-q)r^*(k-q-N), \qquad (3.39)$$

where k is the time index of the currently received sample.

Since the CP is just a duplication of the last N_g samples of the OFDM block, we expect that $\gamma(k)$ may periodically exhibit peaks whenever the samples $r(k-q-N)$ with $0 \le q \le N_g-1$ belong to the CP. This intuition is confirmed by the experimental results of Fig. 3.10, where $\gamma(k)$ is shown versus the time index k for a Rayleigh multipath channel with CIR duration

$L = 8$ and $SNR = 20$ dB. The number of subcarriers is $N = 256$ while $N_g = 16$.

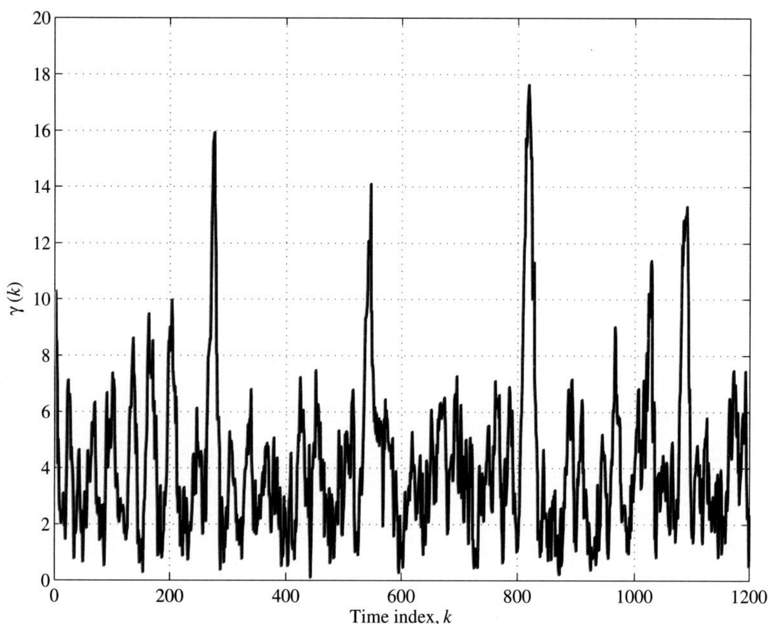

Fig. 3.10   Timing metric based on the CP correlation properties.

Figure 3.10 indicates the presence of peaks at a regular distance of N_T samples, which can be used to continuously track the residual timing offset. It should be observed that accurate timing estimation may be difficult in the presence of strong interference and/or noise due to the relatively short integration window employed in Eq. (3.39). A possible remedy to this drawback is suggested in [163], where the timing metric is smoothed by means of a first-order infinite impulse response (IIR) filter. This yields the following modified metric

$$\overline{\gamma}(k) = \alpha\overline{\gamma}(k - N_T) + (1 - \alpha)\gamma(k), \qquad (3.40)$$

in which $\gamma(k)$ is still given in Eq. (3.39) and $0 < \alpha < 1$ is a forgetting factor which is designed so as to achieve a reasonable trade-off between estimation accuracy and tracking capabilities. The location of the peaks in $\overline{\gamma}(k)$ indicate the start of the received blocks and are used to control the position of the DFT window.

3.2.3 *Frequency acquisition*

After frame detection and timing acquisition, each terminal must compute a coarse frequency estimate to align its local oscillator to the received carrier frequency. This operation is referred to as frequency acquisition and is normally accomplished at each new received frame by exploiting the same reference blocks used for timing acquisition, in addition to possibly other dedicated blocks. As mentioned previously, the reference blocks are normally composed by some repetitive parts which remain identical after passing through the channel except for a phase shift caused by the frequency error. The latter is thus estimated by measuring the induced phase shift. This approach has been employed by Moose in [96], where the phase shift between two successive identical blocks is measured in the frequency-domain at the DFT output. More precisely, assume that timing acquisition has already been achieved and let $R_1(n)$ and $R_2(n)$ be the nth DFT output corresponding to the two reference blocks. Then, we may write

$$R_1(n) = S_R(n) + W_1(n), \qquad (3.41)$$

and

$$R_2(n) = S_R(n)e^{j2\pi\varepsilon N_T/N} + W_2(n), \qquad (3.42)$$

where $S_R(n)$ is the signal component (the same over the two blocks as long as the channel is static) while $W_1(n)$ and $W_2(n)$ are noise terms. The above equations indicate that an estimate of ε can be derived as

$$\widehat{\varepsilon} = \frac{1}{2\pi(N_T/N)} \arg\left\{ \sum_{n=0}^{N-1} R_2(n)R_1^*(n) \right\}. \qquad (3.43)$$

One major drawback of this scheme is the relatively short acquisition range. Actually, since the $\arg\{\cdot\}$ function returns values in the range $[-\pi, \pi)$, we see from Eq. (3.43) that $|\widehat{\varepsilon}| \leq N/(2N_T)$, which is less than one half of the subcarrier spacing.

A viable method to enlarge the frequency acquisition range is proposed by Schmidl and Cox (S&C) in [142]. Similarly to Moose, they perform frequency acquisition by exploiting two reference blocks which are suitably designed so as to guarantee an acquisition range of several subcarrier spacings.

As depicted in Fig. 3.11, the first block is the same used for timing acquisition and is composed of two identical halves in the time-domain (each of length $N/2$). The second block contains a differentially encoded pseudo-noise sequence PN1 on the even subcarriers and another pseudo-noise sequence PN2 on the odd subcarriers. In describing the S&C method, we

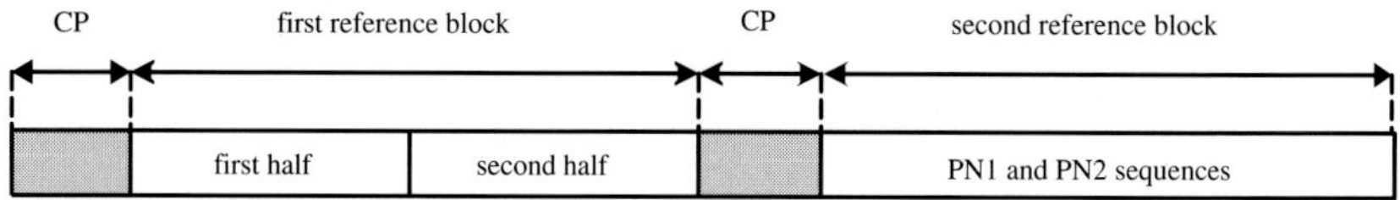

Fig. 3.11   Reference blocks employed by the S&C frequency acquisition scheme.

assume for simplicity that the timing acquisition phase has been success-fully completed and the receiver has perfect knowledge of the timing offset θ. Also, we decompose the frequency error into a *fractional part*, less than $1/T$ in magnitude, plus an integer part which is multiple of $2/T$, where $T = NT_s$ is the length of the OFDM block (excluded the CP). Hence, we may write the normalized frequency error as

$$\varepsilon = \nu + 2\eta, \tag{3.44}$$

where $\nu \in (-1, 1]$ and η is an integer.

The S&C algorithm exploits the first reference block to get an estimate of ν. For this purpose, the following $N/2$-lag autocorrelation is computed

$$\Psi = \sum_{k=\theta}^{\theta+N/2-1} r(k + N/2)r^*(k), \tag{3.45}$$

where $r(k)$ and $r(k + N/2)$ are time-domain samples in the two halves of the first reference block as expressed in Eqs. (3.30) and (3.31), respectively. Apart from thermal noise, $r(k)$ and $r(k + N/2)$ are identical except for a phase shift of $\pi\nu$. Hence, an estimate of ν is obtained as

$$\widehat{\nu} = \frac{1}{\pi} \arg \left\{ \sum_{k=\theta}^{\theta+N/2-1} r(k + N/2)r^*(k) \right\}. \tag{3.46}$$

This equation indicates that timing information is necessary to compute $\widehat{\nu}$. In practice, the quantity θ in Eq. (3.46) is replaced by its corresponding estimate $\widehat{\theta}$ as given in Eq. (3.32).

In order to compensate for the fractional part of the CFO, the time-domain samples are counter-rotated at an angular speed $2\pi\widehat{\nu}/N$ and fed to the DFT unit. We denote $R_1(n)$ and $R_2(n)$ $(n = 0, 1, \ldots, N - 1)$ the DFT outputs corresponding to the first and second reference blocks, respectively. Although no ICI will be present on $R_1(n)$ and $R_2(n)$ as long as $\widehat{\nu} \approx \nu$, the DFT outputs will be shifted from their correct position if $\eta \neq 0$ due to the uncompensated integer frequency error. Bearing in mind Eq. (3.20), we may write

$$R_1(n) = e^{j\varphi_1} H\left(\left|n - 2\eta\right|_N\right) c_1\left(\left|n - 2\eta\right|_N\right) + W_1(n), \tag{3.47}$$

and

$$R_2(n) = e^{j(\varphi_1 + 4\pi\eta N_T/N)} H\left(\left|n - 2\eta\right|_N\right) c_2\left(\left|n - 2\eta\right|_N\right) + W_2(n), \tag{3.48}$$

where $\left|n - 2\eta\right|_N$ is the value $n - 2\eta$ reduced to the interval $[0, N - 1]$, $H(n)$ is the channel response and $c_i(n)$ the symbol transmitted over the nth subcarrier and belonging to the ith block. Neglecting for simplicity the noise terms and calling $d(n) = c_2(n)/c_1(n)$ the differentially-modulated PN sequence on the even subcarriers of the second block, from Eqs. (3.47) and (3.48) we see that $R_2(n) \approx e^{j4\pi\eta N_T/N} d\left(\left|n - 2\eta\right|_N\right) R_1(n)$ if n is even. An estimate of η is thus calculated by looking for the integer $\widehat{\eta}$ that maximizes the following metric

$$B(\widetilde{\eta}) = \frac{\left|\sum_{n \in J} R_2(n) R_1^*(n) d^*\left(\left|n - 2\widetilde{\eta}\right|_N\right)\right|}{\sum_{n \in J} \left|R_2(n)\right|^2}, \tag{3.49}$$

where J is the set of indices for the even subcarriers and $\widetilde{\eta}$ varies over the range of possible frequency offsets. Bearing in mind Eq. (3.44), the estimated CFO is finally given by

$$\widehat{\varepsilon} = \widehat{\nu} + 2\widehat{\eta}, \tag{3.50}$$

and its mean-square error (MSE) can reasonably be approximated as [142]

$$\text{MSE}\left\{\widehat{\varepsilon}\right\} = \frac{2(SNR)^{-1}}{\pi^2 N}, \tag{3.51}$$

where $SNR = \sigma_s^2/\sigma_w^2$ is the signal-to-noise ratio over the received time-domain samples.

Appealing features of the S&C method are its simplicity and robustness, which make it well suited for burst-mode transmissions where accurate estimates of the synchronization parameters must be obtained as fast as possible. An extension of the S&C algorithm has been proposed by Morelli and Mengali (M&M) in [99] by considering a reference block composed by $Q \geq 2$ repetitive parts, each comprising $P = N/Q$ time-domain samples. In the M&M algorithm the estimated CFO is computed as

$$\widehat{\varepsilon} = \frac{Q}{2\pi} \sum_{q=1}^{Q/2} \chi(q) \arg\left\{\Psi(q)\Psi^*(q-1)\right\}, \tag{3.52}$$

where $\chi(q)$ are suitable weighting coefficients given by

$$\chi(q) = \frac{12(Q-q)(Q-q+1) - Q^2}{2Q(Q^2-1)}, \tag{3.53}$$

while $\Psi(q)$ is the following qP-lag autocorrelation

$$\Psi(q) = \sum_{k=\theta}^{\theta+N-qP-1} r(k+qP)r^*(k) \quad q = 1, 2, \ldots, Q/2. \tag{3.54}$$

The M&M scheme gives unbiased estimates as long as $|\varepsilon| \leq Q/2$ and the SNR is adequately high. Hence, if Q is designed such that the possible frequency offsets lie in the interval $[-Q/2, Q/2]$, the CFO can be estimated without the need for a second reference block as required by the S&C method, thereby allowing a substantial reduction of the system overhead.

The MSE of the estimate Eq. (3.52) is given by [99]

$$\mathrm{MSE}\{\widehat{\varepsilon}\} = \frac{3(SNR)^{-1}}{2\pi^2 N(1 - 1/Q^2)}, \tag{3.55}$$

and for $Q > 2$ is lower than the corresponding result Eq. (3.51) obtained with the S&C method.

Figure 3.12 compares the S&C and M&M algorithms in terms of MSE versus SNR. The number of available subcarriers is $N = 256$ and the channel has $L = 8$ taps. The latter are Gaussian distributed with zero-mean and an exponentially decaying power delay profile. Parameter Q with the M&M scheme has been fixed to 8. The dashed lines represent theoretical analysis as given by Eqs. (3.51) and (3.55) while marks indicate simulation results. We see that the theoretical MSEs are validated only at large SNR values. The best results are obtained with the M&M algorithm, which achieves a gain of approximately 0.8 dB over the S&C.

3.2.4 *Frequency tracking*

The CFO estimate obtained during the acquisition phase is used to adjust the frequency of the LO or, alternatively, to counter-rotate the baseband received samples $r(k)$ at an angular speed $2\pi\widehat{\varepsilon}/N$ so as to produce the new sequence $r'(k) = r(k)e^{-j2\pi k\widehat{\varepsilon}/N}$. Due to estimation inaccuracies and/or time-varying Doppler shifts, $r'(k)$ may still be affected by a residual frequency error $\Delta\varepsilon = \varepsilon - \widehat{\varepsilon}$. The latter induces a phase shift that varies linearly in time with a slope proportional to $\Delta\varepsilon$. As long as $\Delta\varepsilon$ is adequately small, the phase shift can be absorbed into the channel frequency response and compensated for during the channel equalization process. However, if $\Delta\varepsilon$

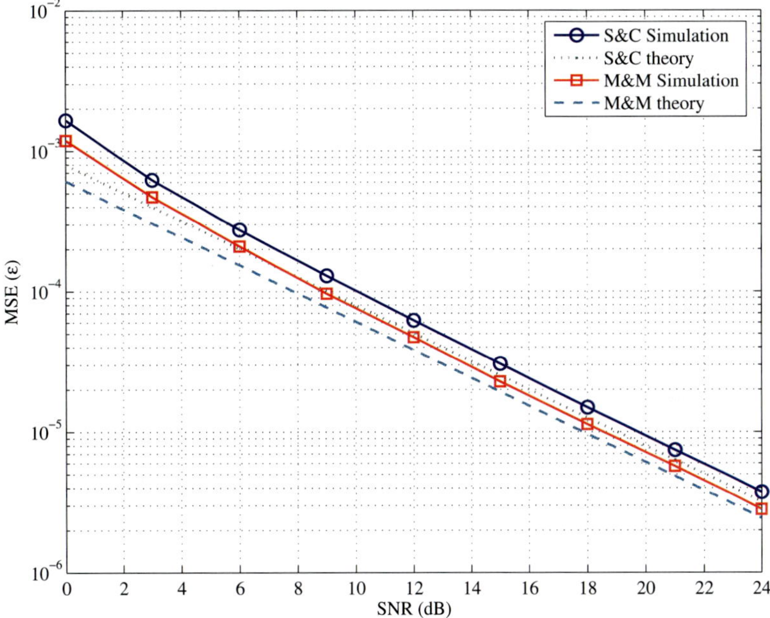

Fig. 3.12   Accuracy of the frequency estimates vs. SNR with S&C and M&M algorithms.

exceeds a few percent of the subcarrier spacing, the DFT output will be affected by non-negligible ICI. In such a case frequency tracking becomes mandatory to avoid severe degradation of the system performance. This operation is typically accomplished on a block-by-block basis using a closed-loop architecture as that depicted in Fig. 3.13.

Here, the sequence $r'_i(m)$ $(-N_g \leq m \leq N-1)$ collects the samples $r'(k)$ belonging to the ith received OFDM block (included the CP) while e_i is an error signal which is proportional to the residual frequency offset. This signal is computed at each new received block and fed to the loop filter, which updates the frequency estimate according to the following recursive equation

$$\Delta\widehat{\varepsilon}_{i+1} = \Delta\widehat{\varepsilon}_i + \alpha e_i, \tag{3.56}$$

where $\Delta\widehat{\varepsilon}_i$ is the estimate of $\Delta\varepsilon$ over the ith block and α is a design parameter (step-size) that controls the convergence speed of the tracking loop. Increasing α improves the tracking capabilities but inevitably degrades the estimation accuracy in the steady-state. Thus, convergence

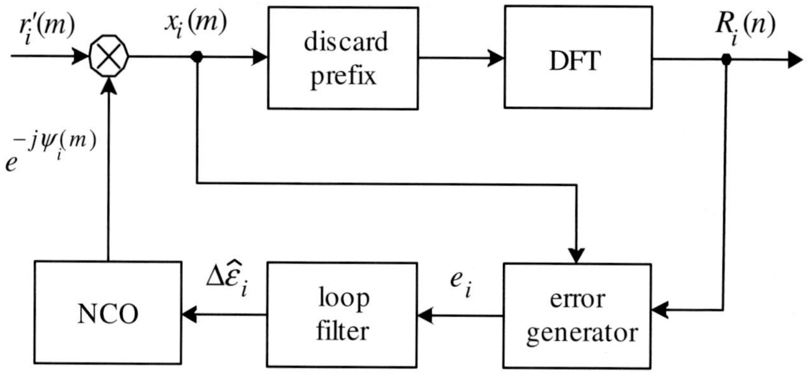

Fig. 3.13   Closed-loop architecture for tracking a residual CFO.

speed and tracking accuracy are contrasting goals which must be traded-off by a proper selection of the step-size.

Returning to Fig. 3.13, we see that $\Delta\widehat{\varepsilon}_i$ is fed to a numerically controlled oscillator (NCO) that generates the exponential term $e^{-j\psi_i(m)}$. The phase $\psi_i(m)$ is recursively computed as

$$\psi_i(m) = \psi_i(m-1) + 2\pi\Delta\widehat{\varepsilon}_i/N, \quad -N_g \le m \le N-1 \qquad (3.57)$$

where $\psi_i(-N_g - 1)$ is set equal to $\psi_{i-1}(N-1)$ in order to avoid any phase jump between the last sample of the $(i-1)$th block and the first sample of the ith block. Inspection of Eq. (3.57) indicates that $\psi_i(m)$ varies linearly in time with a slope proportional to the current frequency estimate $\Delta\widehat{\varepsilon}_i$. The exponential term is then used to obtain the frequency-corrected samples $x_i(m) = r'_i(m)e^{-j\psi_i(m)}$ for $-N_g \le m \le N-1$. After discarding the CP, the latter are finally fed to the DFT device which generates the frequency-domain samples $R_i(n)$ $(0 \le n \le N-1)$.

Several frequency tracking schemes available in the literature rely on the closed-loop structure of Fig. 3.13 and only differ in the adopted error signal e_i [29, 30, 98, 163]. In particular, we distinguish between frequency-domain and time-domain algorithms, depending on whether e_i is computed using the DFT output $R_i(n)$ or the samples $x_i(m)$ at the input of the DFT device. For example, the schemes proposed in [29] and [163] operate in the time-domain and exploit the redundancy offered by the CP to obtain an error signal of the type

$$e_i = \frac{1}{N_g}\Im m \left\{ \sum_{m=-N_g}^{-1} x_i(m+N)x_i^*(m) \right\}, \qquad (3.58)$$

where $x_i(m)$ $(-N_g \leq m \leq -1)$ are samples taken from the CP of the ith received block.

To explain the rationale behind Eq. (3.58), we temporarily neglect the effect of thermal noise as well as any interference on $x_i(m)$ caused by channel echoes. Then, in the presence of a residual frequency offset $\Delta\varepsilon - \Delta\widehat{\varepsilon}_i$, the samples $x_i(m)$ and $x_i(m+N)$ only differ for a phase shift and we can reasonably write $x_i(m+N) \approx x_i(m)e^{j2\pi(\Delta\varepsilon-\Delta\widehat{\varepsilon}_i)}$ for $-N_g \leq m \leq -1$. Substituting this relation into Eq. (3.58) indicates that e_i is proportional to $\sin\left[2\pi(\Delta\varepsilon - \Delta\widehat{\varepsilon}_i)\right]$ and can be used in Eq. (3.56) to improve the accuracy of the frequency estimate as it is now explained. To fix the ideas, assume that $\Delta\widehat{\varepsilon}_i$ is (slightly) smaller than the true offset $\Delta\varepsilon$. Since in this case e_i is a positive quantity, from Eq. (3.56) it follows that $\Delta\widehat{\varepsilon}_{i+1} > \Delta\widehat{\varepsilon}_i$, which results into a reduction of the estimation error. The situation $\Delta\widehat{\varepsilon}_i > \Delta\varepsilon$ can be dealt with similar arguments and leads to the same final conclusion. The equilibrium point is achieved in a perfectly synchronized system where $\Delta\widehat{\varepsilon}_i = \Delta\varepsilon$. Indeed, in this case $e_i = 0$ and from Eq. (3.56) we have $\Delta\widehat{\varepsilon}_{i+1} = \Delta\widehat{\varepsilon}_i$, meaning that the frequency estimate is kept fixed at its current value. In practice, the estimate will fluctuate around the equilibrium point due to the unavoidable presence of thermal noise and interference.

As mentioned previously, the error signal can also be computed in the frequency-domain by exploiting the quantities $R_i(n)$ at the output of the DFT unit (see Fig. 3.13). An example in this sense is given in [30], where e_i is derived using a maximum likelihood (ML) approach and reads

$$e_i = \Re e \left\{ \sum_{n\in\mathcal{I}} R_i^*(n)\left[R_i(n+1) - R_i(n-1)\right] \right\}. \qquad (3.59)$$

A similar method with improved performance is proposed in [98] and employs the following error signal

$$e_i = \Re e \left\{ \sum_{n\in\mathcal{I}} \frac{R_i^*(n)\left[R_i(n+1) - R_i(n-1)\right]}{1 + \beta\left|R_i(n)\right|^2} \right\}, \qquad (3.60)$$

where β is a suitable parameter that depends on the operating SNR.

It is worth noting that all the considered schemes for computing e_i are blind in that they do not exploit any pilot symbols embedded into the transmitted data stream.

3.3   Synchronization for uplink transmissions

In a typical multiuser system, each terminal computes timing and frequency estimates by exploiting the downlink signal broadcasted by the BS. This operation reduces the synchronization errors to a tolerable level and, in case of multicarrier transmissions, can easily be accomplished using the techniques described in the previous section. The estimated parameters are then employed by each user not only to detect the downlink data stream, but also as synchronization references for the uplink transmission. Due to Doppler shifts and propagation delays, however, the uplink signals arriving at the BS may still be affected by residual frequency and timing errors. To see how this comes about, we denote $T_B = N_T T_s$ the length of each OFDM block (including the CP) and assume that the BS starts to transmit the ℓth downlink block at $t = \ell T_B$ ($\ell = 0, 1, 2, \ldots$) on the carrier frequency f_c. The block is received by the mth user at $t = \ell T_B + \tau_m$ on the frequency $f_c + \Delta f_m$, where τ_m and Δf_m are the line-of-sight (LOS) propagation delay and Doppler shift of the considered user, respectively. The latter are expressed by

$$\tau_m = \frac{d_m}{c}, \tag{3.61}$$

and

$$\Delta f_m = \frac{f_c v_m}{c}, \tag{3.62}$$

where c is the speed of light, v_m represents the speed of the mth mobile terminal and d_m is the separation distance between the considered terminal and the BS.

During the uplink phase, each user transmits according to the timing and frequency references established on the basis of the downlink broadcast channel. Assuming that the synchronization parameters have been perfectly estimated, the OFDM uplink blocks are transmitted by the mth user at instants $t = iT_B + \tau_m$ ($i = 0, 1, 2, \ldots$) on the frequency $f_c + \Delta f_m + F$, where F is the nominal separation between the uplink and downlink carrier frequencies (clearly, $F = 0$ in time-division-duplex systems). Because of the propagation delay and Doppler shift, the BS receives the blocks from the mth user at instants $iT_B + 2\tau_m$ on the frequency $f_c + 2\Delta f_m + F$, which results into timing and frequency errors of $2\tau_m$ and $2\Delta f_m$, respectively. The foregoing discussion indicates that synchronization performed at each terminal during the downlink phase may be sufficient to avoid any further synchronization in the uplink as long as the Doppler shift is adequately smaller than the subcarrier spacing and the duration of the CP is so large to accommodate both the CIR duration and the two-way propagation delay $2\tau_m$. If the above conditions are not simultaneously met, however, the uplink signals loose their orthogonality and multiple-access interference (MAI) arises in addition to ICI and IBI. In such a case synchronization at the BS becomes mandatory to avoid severe degradations of the system performance.

Intuitively speaking, synchronization in a multiuser uplink scenario is much more challenging than in the downlink. The reason is that while in the downlink each terminal must estimate and compensate only for its own synchronization parameters, the uplink waveform arriving at the BS is a mixture of signals transmitted by different users, each characterized by different timing and frequency offsets. The latter cannot be estimated with the same methods employed in the downlink because each user must be separated from the others before the synchronization process can be started. The separation method is closely related to the particular carrier assignment scheme (CAS) adopted in the system, *i.e.*, the strategy according to which subcarriers are distributed among the active users.

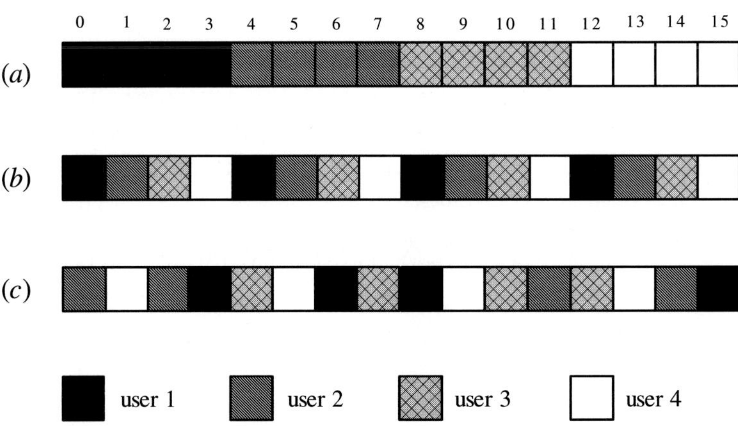

Fig. 3.14 Examples of subcarrier allocation schemes: subband CAS (a), interleaved CAS (b) and generalized CAS (c).

Commonly adopted carrier assignment schemes are the subband and interleaved CAS as depicted in Fig. 3.14 (a) and (b), where a total of

$N = 16$ subcarriers is assumed for illustration purposes. As is seen, in the subband CAS users are provided with groups of adjacent subcarriers while in the interleaved CAS the subcarriers of each user are interleaved over the signal bandwidth in order to fully exploit the frequency diversity of the multipath channel. However, the current trend in OFDMA favors a more flexible allocation scheme called generalized CAS (see Fig. 3.14 (c)), in which users can select the best subcarriers (namely, those exhibiting the highest channel gains) that are currently available.

In the rest of this section, the problem of timing and frequency estimation in the OFDMA uplink is addressed separately for systems employing subband, interleaved or generalized CAS. How to use the estimated synchronization parameters for MAI mitigation is the subject of Sec. 3.4.

3.3.1 *Uplink signal model with synchronization errors*

Without loss of generality, we adopt a baseband-equivalent discrete-time signal model with sampling period T_s. The time-domain samples of the mth user during the ith OFDM block are expressed by

$$s_{m,i}(k) = \frac{1}{\sqrt{N}} \sum_{n \in \mathcal{I}_m} c_{m,i}(n) \, e^{j 2\pi n k/N}, \quad -N_g \leq k \leq N - 1 \qquad (3.63)$$

where $\mathcal{I}_m$ is the set of subcarriers assigned to the considered user while $c_{m,i}(n)$ is the symbol transmitted over the nth subcarrier. To avoid that a given subcarrier can be shared by different users, we must ensure that $\mathcal{I}_m \cap \mathcal{I}_j = \emptyset$ if $m \neq j$. Clearly, the signal transmitted by the mth terminal consists of several adjacent blocks and is given by

$$s_m(k) = \sum_i s_{m,i}(k - iN_T). \qquad (3.64)$$

We assume that M users are simultaneously active in the system and transmit their data streams to the BS receiver. Each stream $s_m(k)$ $(m = 1, 2, \ldots, M)$ propagates through a multipath channel with impulse response $\boldsymbol{h}_m = [h_m(0), h_m(1), \ldots, h_m(L_m - 1)]^T$ and arrives at the BS with a timing offset θ_m and a frequency error ε_m (normalized to the subcarrier spacing). After baseband conversion and sampling, the received samples are modeled as

$$r(k) = \sum_{m=1}^{M} r_m(k) + w(k), \qquad (3.65)$$

where $w(k)$ represents complex-valued AWGN with variance σ_w^2 while $r_m(k)$ is the signal from the mth user and reads

$$r_m(k) = e^{j2\pi\varepsilon_m k/N} \sum_{\ell=0}^{L_m-1} h_m(\ell)s_m(k - \theta_m - \ell). \qquad (3.66)$$

As mentioned previously, timing and frequency errors cause the loss of orthogonality among subcarriers of different users and give rise to multiple-access interference. Since the latter significantly degrades the system performance, the BS must compute estimates of θ_m and ε_m for each active user. The estimates are then used to restore orthogonality among the uplink signals. As is intuitively clear, this multiple-parameter estimation problem can be solved only after the users' signals are properly separated at the BS.

A simple way to counteract the effects of users' timing errors is to select the length of the CP so as to accommodate both the channel delay spread and timing offsets. This results into a *quasi-synchronous* scenario [6] where the two-way propagation delays are viewed as part of the channel impulse responses and the received samples can thus be rewritten as

$$r_m(k) = e^{j2\pi\varepsilon_m k/N} \sum_{\ell=0}^{L-1} h'_m(\ell)s_m(k - \ell), \qquad (3.67)$$

where $\boldsymbol{h}'_m = [h'_m(0), h'_m(1), \ldots, h'_m(L-1)]^T$ is the mth *extended* channel vector, with entries

$$h'_m(\ell) = h_m(\ell - \theta_m), \qquad 0 \le \ell \le L-1 \qquad (3.68)$$

and length $L = \max_m \{L_m + \theta_m\}$. In practice, a quasi-synchronous system is equivalent to a perfectly time-synchronized network in which the duration of the mth CIR (expressed in sampling periods) is artificially extended from L_m to L.

The situation is depicted in Fig. 3.15, where OFDMA blocks of different users arrive at the receiver with different delays depending on the distances between the user terminals and the BS. As is seen, each CP is decomposed into two segments. The first one (colored in black) has length $L_m - 1$ and is affected by interference from the previous block due to channel dispersion. The second segment (colored in gray) accommodates the last $N_g - L_m + 1$ samples of the CP and is free from IBI. The vertical line on the left represents the starting point of the ith OFDMA block in the BS timescale, while the ith receive DFT window starts at $t = iN_T$. If the length N_g of the CP is not shorter than $L-1$, the samples $r_{m,i}(k) = r_m(k + iN_T)$

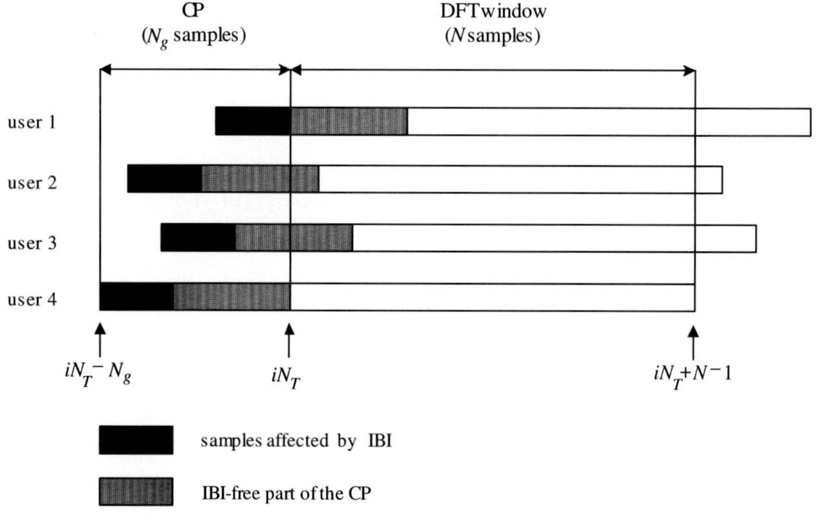

Fig. 3.15   Uplink received signals and DFT window in a quasi-synchronous scenario.

$(0 \leq k \leq N - 1)$ falling within the ith DFT window are immune to IBI and, accordingly, are expressed by

$$r_{m,i}(k) = e^{j2\pi\varepsilon_m(k+iN_T)/N} \sum_{\ell=0}^{L-1} h'_m(\ell)s_{m,i}(k - \ell), \quad 0 \leq k \leq N - 1 \quad (3.69)$$

with $s_{m,i}(k)$ as given in Eq. (3.63). Substituting Eq. (3.63) into Eq. (3.69) yields

$$r_{m,i}(k) = \frac{1}{\sqrt{N}} e^{j2\pi\varepsilon_m k/N} \sum_{n\in\mathcal{I}_m} \widetilde{H}_{m,i}(n)c_{m,i}(n) \, e^{j2\pi nk/N}, \quad (3.70)$$

for $0 \leq k \leq N - 1$, where $\widetilde{H}_{m,i}(n) = H'_m(n) \, e^{j2\pi\varepsilon_m iN_T/N}$ and

$$H'_m(n) = \sum_{\ell=0}^{L-1} h'_m(\ell) \, e^{-j2\pi n\ell/N}, \quad 0 \leq n \leq N - 1 \quad (3.71)$$

is the N-point DFT of $\{h'_m(\ell)\}$. Finally, from Eq. (3.65) we see that the samples $r_i(k) = r(k + iN_T)$ $(0 \leq k \leq N - 1)$ of the superimposed uplink signals within the ith receive DFT window are given by

$$r_i(k) = \sum_{m=1}^{M} r_{m,i}(k) + w_i(k), \quad (3.72)$$

with $w_i(k) = w(k + iN_T)$.

The fact that propagation delays are absorbed by the extended channel vectors makes quasi-synchronous systems extremely appealing since timing errors simply appear as phase shifts at the DFT output and are compensated for by the channel equalization process. Timing estimation is thus unnecessary and the BS has only to estimate the frequency offsets ε_m, thereby reducing the number of synchronization parameters by a factor of two. The price for this simplification is a certain loss of efficiency due to the extended CP. To keep the loss to a tolerable level, the length of the CP must be maintained within a small fraction of the block duration. This poses an upper limit to the maximum admissible value of θ_m, say $\theta_{\max}$, which must be adequately smaller than N. Since each θ_m is proportional to the two-way propagation delay, the distances between the users' terminals and the BS receiver cannot exceed a certain value $d_{\max}$. In particular, recalling that $\theta_m \approx 2\tau_m/T_s$ and bearing in mind Eq. (3.61), we obtain $d_{\max} = cT_s\theta_{\max}/2$.

3.3.2 *Timing and frequency estimation for systems with subband CAS*

In OFDMA systems with subband CAS, the available spectrum is divided into several groups of adjacent subcarriers (subbands) and each user is exclusively assigned to one ore more groups. In the presence of frequency errors, subbands of different users are shifted in frequency from their nominal positions so that subcarriers located at the edges of a given group may experience significant ICI. To mitigate this problem, it is expedient to separate subbands pertaining to different users by means of suitable guard intervals comprising a specified number of unmodulated subcarriers.

Assigning groups of adjacent subcarriers to each user facilitates the task of separating the uplink signals at the BS. As shown in Fig. 3.16, it suffices to pass the received samples through a bank of digital band-pass filters, each selecting one group of subcarriers. If the users' frequency offsets are adequately smaller than the guard intervals among adjacent subbands, the filtering operation roughly separates the uplink signals and allows the BS to perform timing and frequency estimation independently for each user. Clearly, perfect users' separation is not possible since this would require ideal brickwall filters and/or very large guard intervals. Hence, the output from the filter tuned on the mth subband takes the form

$$x_m(k) = r_m(k) + I_m(k) + w_m(k), \tag{3.73}$$

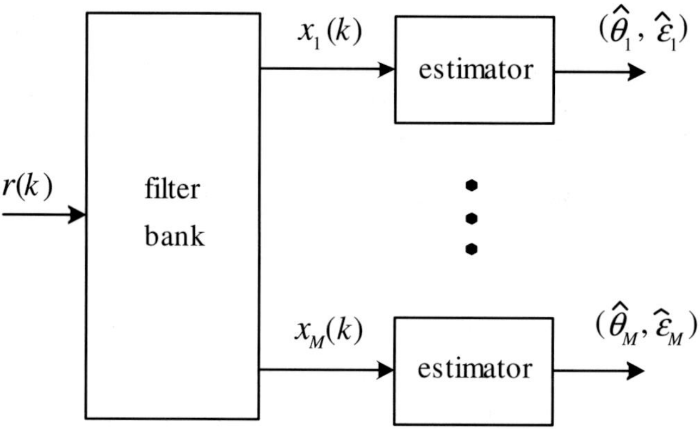

Fig. 3.16   Timing and frequency estimation for an OFDMA uplink receiver with sub-band CAS.

where $r_m(k)$ is the mth uplink signal as given in Eq. (3.66), $w_m(k)$ is the contribution of thermal noise and, finally, $I_m(k)$ is an interference term that accounts for imperfect users' separation. As is intuitively clear, estimates of θ_m and ε_m can be obtained from $x_m(k)$ applying any timing and frequency estimation schemes suitable for single-user OFDM systems. One possibility is to adopt the method discussed in [163], which exploits the correlation induced on $x_m(k)$ by the use of the CP. In this case timing and frequency estimates are obtained in the form

$$\widehat{\theta}_m = \arg\max_{\tilde{\theta}}\{\gamma_m(\tilde{\theta})\}, \tag{3.74}$$

and

$$\widehat{\varepsilon}_m = \frac{1}{2\pi}\arg\{\gamma_m(\widehat{\theta}_m)\}, \tag{3.75}$$

where

$$\gamma_m(\tilde{\theta}) = \sum_{k=\tilde{\theta}-N_g}^{\tilde{\theta}-1} x_m(k+N)x_m^*(k) \tag{3.76}$$

is the N-lag autocorrelation of the sequence $x_m(k)$.

A slightly modified version of this algorithm is used in [162], where it is shown that the estimator's performance is heavily affected by the number of subcarriers in one subband and deteriorates as this number becomes smaller and smaller due to the increased correlation among the received time-domain samples. A second factor that may limit the estimation accuracy is the amount of residual MAI and ICI arising from imperfect separation of the users' signals. A simple way to improve the system performance consists of averaging $\gamma_m(\widetilde{\theta})$ over Q successive OFDM blocks. This yields a new metric

$$\overline{\gamma}_m(\widetilde{\theta}) = \sum_{q=0}^{Q-1} \gamma_m(\widetilde{\theta} + qN_T), \tag{3.77}$$

which can be used in Eqs. (3.74) and (3.75) in place of $\gamma_m(\widetilde{\theta})$. In spite of its effectiveness, this solution may provide the receiver with outdated estimates of the synchronization parameters due to the enlarged estimation window. In practice, it can be adopted on condition that timing and frequency offsets do not change significantly over a time interval comprising Q OFDM blocks.

An alternative scheme to obtain estimates of θ_m and ε_m from the sequence $\{x_m(k)\}$ is discussed in [6]. This method exploits unmodulated (virtual) subcarriers inserted in each user subband and updates the timing and frequency estimates until the average energy of the DFT outputs corresponding to the virtual carriers achieves a minimum. Mathematically, we have

$$\left(\widehat{\theta}_m, \widehat{\varepsilon}_m\right) = \arg \min_{\widetilde{\theta}_m, \widetilde{\varepsilon}_m} \left\{ J(\widetilde{\theta}_m, \widetilde{\varepsilon}_m) \right\}, \tag{3.78}$$

where $\widetilde{\theta}_m$ and $\widetilde{\varepsilon}_m$ represent trial values of θ_m and ε_m, respectively, while the cost function $J(\widetilde{\theta}_m, \widetilde{\varepsilon}_m)$ is proportional to the average energy of the time- and frequency-corrected samples $x_m(k + \widetilde{\theta}_m)e^{j2\pi\widetilde{\varepsilon}_m k/N}$ falling across the virtual carriers. As is seen, computing $\widehat{\theta}_m$ and $\widehat{\varepsilon}_m$ directly from Eq. (3.78) requires a complicated bidimensional (2D) grid search over the set spanned by $\widetilde{\theta}_m$ and $\widetilde{\varepsilon}_m$. A certain reduction of complexity is possible if the minimum of $J(\widetilde{\theta}_m, \widetilde{\varepsilon}_m)$ is approached through a 2D steepest-descent algorithm.

As mentioned previously, the main advantage of the subband CAS is the possibility of separating signals from different users through a simple filter bank even in a completely asynchronous scenario with arbitrarily large timing errors. On the other hand, grouping the subcarriers together prevents the possibility of optimally exploiting the channel diversity since a deep fade might hit a substantial number of subcarriers of a given user if they are close together. Interleaving the subcarriers over the available spectrum is a viable method to provide the users with some form of frequency diversity. As it is now shown, however, this approach greatly complicates the synchronization task.

3.3.3 *Timing and frequency estimation for systems with interleaved CAS*

In OFDMA systems with interleaved CAS, the N available subcarriers are divided into R subchannels, where R is the maximum number of users that the system can simultaneously support. Each subchannel has $P = N/R$ subcarriers that are uniformly spaced in the frequency domain at a distance R from each other. In particular, the subchannel assigned to the mth user is composed of subcarriers with indices in the set $\mathcal{I}_m = \{i_m + pR \; ; \; 0 \le p \le P - 1\}$, where i_m may be any integer in the interval $[0, R - 1]$.

Compared to the subband CAS, the interleaved CAS is clearly more robust against frequency-selective fading by exploiting the frequency diversity. However, separating the uplink signals in an interleaved OFDMA system is much more difficult than in subband transmissions. The reason is that in the presence of frequency errors the users' signals overlap in the frequency-domain and cannot simply be separated through a filter bank. As it is now shown, however, the interleaved CAS provides the uplink signals with an inherent periodic structure that can be exploited for synchronization purposes.

For simplicity, in the following the timing and frequency estimation tasks are separately addressed. The reason is that in an interleaved OFDMA system the joint estimation of all synchronization parameters appears as a formidable problem for which no feasible solution is available in the open literature. Accordingly, for the time being we consider a quasi-synchronous scenario and limit our attention to the frequency estimation problem. A method for estimating the timing offsets of the active users is illustrated later.

We concentrate on the ith received OFDMA block and consider the samples $r_{m,i}(k)$ $(0 \le k \le N - 1)$ of the mth uplink signal falling within the ith receive DFT window. Since $c_{m,i}(n)$ is non-zero only for $n = i_m + pR$ $(0 \le p \le P - 1)$, we may rewrite Eq. (3.70) in the equivalent form

$$r_{m,i}(k) = \frac{1}{\sqrt{N}} e^{j2\pi\xi_m k/P} \sum_{p=0}^{P-1} S_{m,i}(p) \; e^{j2\pi pk/P}, \qquad (3.79)$$

where $S_{m,i}(p) = \widetilde{H}_{m,i}(i_m + pR)c_{m,i}(i_m + pR)$, while ξ_m is defined as

$$\xi_m = \frac{i_m + \varepsilon_m}{R}. \qquad (3.80)$$

Inspection of Eq. (3.79) reveals that

$$r_{m,i}(k) = e^{j2\pi\ell\xi_m} r_{m,i}(k + \ell P), \qquad (3.81)$$

from which it follows that each OFDMA block has a periodic structure that repeats every P samples. This inner structure can be exploited for frequency estimation. A solution in this sense is proposed in [11] by resorting to subspace-based methods. The resulting procedure is called the Cao-Tureli-Yao Estimator (CTYE) and operates in the following way:

The Cao-Tureli-Yao Estimator (CTYE)

(1) arrange the received samples $r_i(k)$ $(k = 0, 1, \ldots, N - 1)$ into the following $R \times P$ matrix

$$\boldsymbol{M}_i = \begin{bmatrix} r_i(0) & \cdots & r_i(P-1) \\ r_i(P) & \cdots & r_i(2P-1) \\ \vdots & \ddots & \vdots \\ r_i(N-P) & \cdots & r_i(N-1) \end{bmatrix}; \qquad (3.82)$$

(2) Compute the $R \times R$ sample-correlation matrix

$$\boldsymbol{Z}_i = \frac{1}{P} \boldsymbol{M}_i \boldsymbol{M}_i^H; \qquad (3.83)$$

(3) Determine the noise subspace by finding the $R - M$ smallest eigenvalues of $\boldsymbol{Z}_i$ and arrange the corresponding eigenvectors into an $R \times (R - M)$ matrix $\boldsymbol{U}_i$;

(4) Compute estimates $\left\{ \widehat{\xi}_m \right\}_{m=1}^{M}$ of the quantities ξ_m by locating the M largest peaks of the following metric

$$\Gamma(\widetilde{\xi}) = \frac{1}{\left\| \boldsymbol{U}_i^H \boldsymbol{a}(\widetilde{\xi}) \right\|^2}, \qquad (3.84)$$

where $\boldsymbol{a}(\widetilde{\xi}) = \begin{bmatrix} 1, & e^{j2\pi\widetilde{\xi}}, & e^{j4\pi\widetilde{\xi}}, \ldots, e^{j2\pi(R-1)\widetilde{\xi}} \end{bmatrix}^T$;

(5) Use Eq. (3.80) and the quantities $\left\{ \widehat{\xi}_m \right\}_{m=1}^{M}$ to compute frequency estimates in the form

$$\widehat{\varepsilon}_m = R\widehat{\xi}_m - i_m, \quad 0 \le m \le M - 1. \qquad (3.85)$$

This structure-based algorithm is reminiscent of the multiple signal classification (MUSIC) technique [143], and provides estimates of the users' CFOs without requiring neither training blocks nor channel knowledge. The only requirement is that the CFOs cannot exceed one half of the subcarrier spacing since otherwise the uncertainty intervals of the quantities

ξ_m are partially overlapping and in such a case there is no way of matching each $\widehat{\xi}_m$ with the corresponding user. Luckily, the above requirement does not represent a serious problem since the uplink CFOs are mainly due to Doppler shifts and in a well-designed system they are typically confined within 20 or 30% of the subcarrier spacing.

The main drawback of the CTYE is that in its original form it cannot be applied to a fully-loaded system in which the number M of active users is equal to the number R of subchannels. The reason is that the rank of the $R \times (R-M)$ matrix $\boldsymbol{U}_i$ must be at least one, which means that $M \leq R-1$. This limitation may be overcome by extending the length of the CP from N_g to $N_g + hP$, where h is a suitable integer. The first N_g samples are used as a guard interval among blocks to avoid IBI. The last hP samples are free from IBI and are exploited by CTYE together with the remaining N samples to estimate the frequency offsets. This results into a matrix $\boldsymbol{U}_i$ of dimensions $(R+h) \times (R+h-M)$ and the algorithm can thus work even with $M = R$.

It is shown in [11] that the performance of CTYE degrades as the number of active users becomes large. A simple way to improve the estimation accuracy is to enlarge the observation window so as to comprehend a specified number I of adjacent OFDMA blocks. In this case the CTYE proceeds as indicated earlier, except that the sample correlation matrix $\boldsymbol{Z}_i$ is now computed as

$$\boldsymbol{Z}_i = \frac{1}{PI} \sum_{k=i}^{i+I-1} \boldsymbol{M}_k \boldsymbol{M}_k^H . \tag{3.86}$$

A major assumption for the application of the CTYE is that the OFDMA uplink signals are quasi-synchronous. As discussed previously, this poses an upper limit to the maximum distance between the BS and the mobile terminals, which may prevent the use of CTYE in a number of applications, including cellular networks with relatively large cell radii (on the order of some kilometers). A possible solution to this problem relies on the transmission of some training blocks at the beginning of each uplink frame. These blocks are exploited for synchronization purposes and can be equipped with long CPs comprising both the channel delay spread and the propagation delay. In this way the uplink signals are quasi-synchronous during the training period, thereby allowing the use of CTYE for frequency estimation. To reduce unnecessary overhead, however, it is desirable that data blocks have a shorter prefix (on the order of the channel response duration). Thus, accurate estimation of the timing offsets is necessary to align all users in time and avoid IBI over the data section of the frame. A simple method for obtaining timing estimates is based on knowledge of the users' channel responses and is now explained by reconsidering the mth extended channel vector $\boldsymbol{h}'_m$ defined in Eq. (3.68).

We begin by observing that

$$\boldsymbol{h}'_m = \begin{bmatrix} \boldsymbol{0}^T_{\theta_m} & \boldsymbol{h}^T_m & \boldsymbol{0}^T_{L-\theta_m-L_m} \end{bmatrix}^T, \tag{3.87}$$

where $\boldsymbol{h}_m = [h_m(0), h_m(1), \ldots, h_m(L_m - 1)]^T$ while $\boldsymbol{0}_K$ is a K-dimensional column vector with all zero entries. Next, we assume that an estimate of $\boldsymbol{h}'_m$ is available at the BS receiver in the form

$$\widehat{\boldsymbol{h}}'_m = \boldsymbol{h}'_m + \boldsymbol{\eta}_m, \tag{3.88}$$

where $\boldsymbol{\eta}_m$ accounts for the estimation error. In practice, $\widehat{\boldsymbol{h}}'_m$ can be computed by exploiting the training blocks transmitted at the beginning of the uplink frame using one of the methods described in the next chapter. Combining Eqs. (3.87) and (3.88) produces

$$\widehat{\boldsymbol{h}}'_m = \boldsymbol{A}_m(\theta_m)\boldsymbol{h}_m + \boldsymbol{\eta}_m, \tag{3.89}$$

where $\boldsymbol{A}_m(\theta_m)$ is an $L \times L_m$ matrix with entries

$$[\boldsymbol{A}_m(\theta_m)]_{\ell,k} = \begin{cases} 1 & \text{if } \ell - k = \theta_m \\ 0 & \text{otherwise .} \end{cases} \tag{3.90}$$

Vector $\widehat{\boldsymbol{h}}'_m$ is now exploited to compute estimates of θ_m and $\boldsymbol{h}_m$ by looking for the minimum of the following least-squares (LS) cost function

$$\Lambda(\widetilde{\theta}, \widetilde{\boldsymbol{h}}) = \left\| \widehat{\boldsymbol{h}}'_m - \boldsymbol{A}_m(\widetilde{\theta})\widetilde{\boldsymbol{h}} \right\|^2 . \tag{3.91}$$

Minimizing with respect to $\widetilde{\boldsymbol{h}}$ and observing that $\boldsymbol{A}^T_m(\widetilde{\theta})\boldsymbol{A}_m(\widetilde{\theta})$ is the identity matrix yields $\widehat{\boldsymbol{h}}_m(\widetilde{\theta}) = \boldsymbol{A}^T_m(\widetilde{\theta})\widehat{\boldsymbol{h}}'_m$. Inserting this result back into Eq. (3.91) and minimizing with respect to $\widetilde{\theta}$ gives an estimate of θ_m in the form

$$\widehat{\theta}_m = \arg\max_{\widetilde{\theta}} \left\{ \left\| \boldsymbol{A}^T_m(\widetilde{\theta})\widehat{\boldsymbol{h}}'_m \right\|^2 \right\}, \tag{3.92}$$

or equivalently,

$$\widehat{\theta}_m = \arg\max_{\widetilde{\theta}} \left\{ \sum_{\ell=\widetilde{\theta}}^{L_m+\widetilde{\theta}-1} \left| \widehat{h}'_m(\ell) \right|^2 \right\}. \tag{3.93}$$

The above equation indicates that the timing estimator looks for the maximum of the energy of $\widehat{\boldsymbol{h}}'_m$ over a sliding window of length L_m equal to the duration of the mth CIR $\boldsymbol{h}_m$.

3.3.4   *Frequency estimation for systems with generalized CAS*

The generalized CAS is a dynamic resource allocation scheme in which subchannels are assigned to users according to their actual channel quality and requested data rates. The fact that each user can select the best subcarriers that are currently available makes this allocation strategy more flexible than subband or interleaved schemes. In particular, the generalized CAS provides the system with some form of *multiuser diversity* [87] since a subcarrier that appears in a deep fade to one user may exhibit a relatively large gain for another user. On the other hand, the absence of any rigid structure in the allocation policy makes the synchronization task even more challenging than with interleaved CAS.

A method for estimating the timing and frequency errors of a new user entering an OFDMA network with generalized CAS has been proposed in [97]. This scheme has potentially good performance but relies on the fact that all other active users have already been synchronized, an assumption that may be too stringent in practical applications. Alternative solutions described in [125] and [126] are based on the ML principle and provide estimates of the synchronization parameters by exploiting a training block transmitted by each user at the beginning of the uplink frame. These methods are now revisited assuming a quasi-synchronous scenario wherein the CP of the training block is made sufficiently long to comprise both the channel delay spread and propagation delays incurred by users' signals. In the ensuing discussion we limit our attention to the joint ML estimation of the channel responses and frequency errors. If needed, timing estimates can be obtained from the channel responses as indicated in the previous section.

Without loss of generality, we assume that the training block has index $i = 0$ and denote $p_m(n)$ $(n \in \mathcal{I}_m)$ the pilot symbols transmitted by the mth user over its assigned subcarriers. The corresponding time-domain samples can thus be written as

$$b_m(k) = \frac{1}{\sqrt{N}} \sum_{n \in \mathcal{I}_m} p_m(n)\, e^{j2\pi nk/N}, \quad -N_g \le k \le N - 1. \qquad (3.94)$$

At the BS receiver, the CP is removed and the remaining samples are expressed by

$$r(k) = \sum_{m=1}^{M} e^{j2\pi \varepsilon_m k/N} \sum_{\ell=0}^{L-1} h'_m(\ell) b_m(k-\ell) + w(k), \quad 0 \le k \le N - 1 \qquad (3.95)$$

where $w(k)$ represents thermal noise, $h'_m(\ell)$ is defined in Eq. (3.68) and M is the number of simultaneously active users.

Collecting the received samples into an N-dimensional vector $\boldsymbol{r} = [r(0), r(1), \ldots, r(N-1)]^T$, we may rewrite Eq. (3.95) into the equivalent form

$$\boldsymbol{r} = \sum_{m=1}^{M} \boldsymbol{r}_m + \boldsymbol{w}, \tag{3.96}$$

where $\boldsymbol{w} = [w(0), w(1), \ldots, w(N-1)]^T$ is a Gaussian vector with zero-mean and covariance matrix $\sigma_w^2 \boldsymbol{I}_N$, while

$$\boldsymbol{r}_m = \boldsymbol{\Gamma}(\varepsilon_m) \boldsymbol{B}_m \boldsymbol{h}'_m, \tag{3.97}$$

where

$$\boldsymbol{h}'_m = [h'_m(0), h'_m(1), \ldots, h'_m(L-1)]^T \tag{3.98}$$

is the mth extended channel vector given in Eq. (3.87) and $\boldsymbol{\Gamma}(\varepsilon_m)$ is a diagonal matrix

$$\boldsymbol{\Gamma}(\varepsilon_m) = \mathrm{diag}\left\{ 1, \ e^{j2\pi\varepsilon_m/N}, \ldots, e^{j2\pi(N-1)\varepsilon_m/N} \right\}, \tag{3.99}$$

and $\boldsymbol{B}_m$ is an $N \times L$ matrix with known entries $[\boldsymbol{B}_m]_{k,\ell} = b_m(k-\ell)$ for $0 \le k \le N-1$ and $0 \le \ell \le L-1$.

The received vector $\boldsymbol{r}$ is now exploited to jointly estimate the frequency offsets $\boldsymbol{\varepsilon} = [\varepsilon_1, \varepsilon_2, \ldots, \varepsilon_M]^T$ and channel responses $\boldsymbol{h}' = [\boldsymbol{h}_1'^T, \boldsymbol{h}_2'^T, \ldots, \boldsymbol{h}_M'^T]^T$ of all active users. In doing so we adopt an ML approach and rewrite Eqs. (3.96) and (3.97) in a more concise form as

$$\boldsymbol{r} = \boldsymbol{Q}(\boldsymbol{\varepsilon}) \boldsymbol{h}' + \boldsymbol{w}, \tag{3.100}$$

with

$$\boldsymbol{Q}(\boldsymbol{\varepsilon}) = [\boldsymbol{\Gamma}(\varepsilon_1)\boldsymbol{B}_1 \ \ \boldsymbol{\Gamma}(\varepsilon_2)\boldsymbol{B}_2 \ \cdots \ \boldsymbol{\Gamma}(\varepsilon_M)\boldsymbol{B}_M]. \tag{3.101}$$

Then, the log-likelihood function for the unknown set of parameters is given by

$$\Lambda(\widetilde{\boldsymbol{\varepsilon}}, \widetilde{\boldsymbol{h}}') = -N \ln(\pi\sigma_w^2) - \frac{1}{\sigma_w^2} \left\| \boldsymbol{r} - \boldsymbol{Q}(\widetilde{\boldsymbol{\varepsilon}})\widetilde{\boldsymbol{h}}' \right\|^2, \tag{3.102}$$

where $\widetilde{\boldsymbol{\varepsilon}}$ and $\widetilde{\boldsymbol{h}}'$ are trial values of $\boldsymbol{\varepsilon}$ and $\boldsymbol{h}'$, respectively.

The joint ML estimates of $\boldsymbol{\varepsilon}$ and $\boldsymbol{h}'$ are obtained by searching for the global maximum of $\Lambda(\widetilde{\boldsymbol{\varepsilon}}, \widetilde{\boldsymbol{h}}')$. This yields

$$\widehat{\boldsymbol{\varepsilon}} = \arg\max_{\widetilde{\boldsymbol{\varepsilon}}} \left\{ \| \boldsymbol{P}(\widetilde{\boldsymbol{\varepsilon}})\boldsymbol{r} \|^2 \right\}, \tag{3.103}$$

and

$$\widehat{\boldsymbol{h}}' = \left[\boldsymbol{Q}^H(\widehat{\boldsymbol{\varepsilon}})\boldsymbol{Q}(\widehat{\boldsymbol{\varepsilon}})\right]^{-1}\boldsymbol{Q}^H(\widehat{\boldsymbol{\varepsilon}})\boldsymbol{r}, \tag{3.104}$$

with $\boldsymbol{P}(\widetilde{\boldsymbol{\varepsilon}})$ being defined as

$$\boldsymbol{P}(\widetilde{\boldsymbol{\varepsilon}}) = \boldsymbol{Q}(\widetilde{\boldsymbol{\varepsilon}})\left[\boldsymbol{Q}^H(\widetilde{\boldsymbol{\varepsilon}})\boldsymbol{Q}(\widetilde{\boldsymbol{\varepsilon}})\right]^{-1}\boldsymbol{Q}^H(\widetilde{\boldsymbol{\varepsilon}}). \tag{3.105}$$

From the above equations it appears that the estimates of $\boldsymbol{\varepsilon}$ and $\boldsymbol{h}'$ are decoupled, meaning that the former is computed first and is then exploited to get the latter. Unfortunately, the maximization in Eq. (3.103) requires a grid-search over the multidimensional domain spanned by $\boldsymbol{\varepsilon}$, which would be too intense even in the presence of few active users. A viable solution to this problem is proposed in [125] and [126] by resorting to the space-alternating projection expectation-maximization algorithm (SAGE) [45] .

Similarly to the well known EM algorithm [34] , this technique operates in an iterative fashion where the original measurements are replaced with some *complete data set* from which the original measurements can be obtained through a many-to-one mapping. The SAGE algorithm alternates between an E-step, calculating the log-likelihood function of the complete data, and an M-step, maximizing that expectation with respect to the unknown parameters. At any iteration the parameter estimates are updated and the process continues until no significant changes in the updates are observed. Compared to the classical EM algorithm, the SAGE has the advantage of a faster convergence rate. The reason is that the maximizations in the EM are simultaneously performed with respect to all unknown parameters, which results into a slow process that requires searches over spaces with many dimensions. Vice versa, the maximizations required in the SAGE are performed varying small groups of parameters at a time. In the following the SAGE algorithm is applied to our problem without further explanations. The interested reader is referred to [45] for details.

Returning to the estimation of $\boldsymbol{\varepsilon}$ and $\boldsymbol{h}'$, we apply the SAGE so as to reduce the M-dimensional maximization problem in Eq. (3.103) to a series of simpler maximizations. The resulting procedure consists of *iterations* and *cycles*. An iteration is made of M cycles and each cycle updates the parameters of a single user while keeping those of the others at their most updated values. Specifically, we call $\widehat{\varepsilon}_m^{(j)}$ and $\widehat{\boldsymbol{h}}_m'^{(j)}$ the estimates of ε_m and $\boldsymbol{h}_m'$ after the jth iteration, respectively. Given initial estimates $\widehat{\varepsilon}_m^{(0)}$ and $\widehat{\boldsymbol{h}}_m'^{(0)}$, the BS computes the following M vectors, one for each user

$$\widehat{\boldsymbol{r}}_m^{(0)} = \boldsymbol{\Gamma}(\widehat{\varepsilon}_m^{(0)})\boldsymbol{B}_m\widehat{\boldsymbol{h}}_m'^{(0)}, \qquad 1 \le m \le M. \tag{3.106}$$

Then, during the mth cycle of the jth iteration the SAGE algorithm proceeds as follows:

SAGE-based frequency estimator

- **E-step** Compute

$$\boldsymbol{y}_m^{(j)} = \boldsymbol{r} - \sum_{k=1}^{m-1} \widehat{\boldsymbol{r}}_k^{(j)} - \sum_{k=m+1}^{M} \widehat{\boldsymbol{r}}_k^{(j-1)}, \qquad (3.107)$$

where a notation of the type $\sum_{\ell}^{u}$ is zero whenever $u < \ell$.

- **M-step** Compute estimates of ε_m and $\boldsymbol{h}_m'$ by locating the minimum of the following cost function

$$\Lambda^{(j)}(\widetilde{\varepsilon}_m, \widetilde{\boldsymbol{h}}_m') = \left\| \boldsymbol{y}_m^{(j)} - \boldsymbol{\Gamma}(\widetilde{\varepsilon}_m)\boldsymbol{B}_m \widetilde{\boldsymbol{h}}_m' \right\|^2 \qquad (3.108)$$

with respect to $\widetilde{\varepsilon}_m$ and $\widetilde{\boldsymbol{h}}_m'$. This yields

$$\widehat{\varepsilon}_m^{(j)} = \arg\max_{\widetilde{\varepsilon}_m} \left\{ \left\| \boldsymbol{P}_m \boldsymbol{\Gamma}^H(\widetilde{\varepsilon}_m) \boldsymbol{y}_m^{(j)} \right\|^2 \right\}, \qquad (3.109)$$

and

$$\widehat{\boldsymbol{h}}_m'^{(j)} = \left(\boldsymbol{B}_m^H \boldsymbol{B}_m \right)^{-1} \boldsymbol{B}_m^H \boldsymbol{\Gamma}^H(\widehat{\varepsilon}_m^{(j)}) \boldsymbol{y}_m^{(j)}, \qquad (3.110)$$

where $\boldsymbol{P}_m = \boldsymbol{B}_m \left(\boldsymbol{B}_m^H \boldsymbol{B}_m \right)^{-1} \boldsymbol{B}_m^H$ is a matrix that can be pre-computed and stored in the receiver as it only depends on the pilot symbols transmitted by the mth user . The estimated parameters are used to obtain the following vector

$$\widehat{\boldsymbol{r}}_m^{(j)} = \boldsymbol{\Gamma}(\widehat{\varepsilon}_m^{(j)}) \boldsymbol{B}_m \widehat{\boldsymbol{h}}_m'^{(j)}, \qquad (3.111)$$

which is then exploited in the E-step of the next cycle or iteration.

In the ensuing discussion, the estimator based on Eq. (3.109) is referred to as the Alternating-Projection Frequency Estimator (APFE). A physical interpretation of this algorithm is of interest. From Eqs. (Eq. (3.96)) and (3.97) we see that the signal component in $\boldsymbol{r}$ results from the contributions $\boldsymbol{r}_k$ of several users $(1 \leq k \leq M)$, each depending on a set of parameters $(\varepsilon_k, \boldsymbol{h}_k')$. If all the sets were known except for $(\varepsilon_m, \boldsymbol{h}_m')$, the contributions of the users with indices $k \neq m$ could be subtracted from $\boldsymbol{r}$, yielding a MAI-free vector

$$\boldsymbol{y}_m = \boldsymbol{r} - \sum_{k \neq m} \boldsymbol{r}_k \qquad (3.112)$$

or, bearing in mind Eqs. (3.96) and (3.97),

$$\boldsymbol{y}_m = \boldsymbol{\Gamma}(\varepsilon_m)\boldsymbol{B}_m \boldsymbol{h}'_m + \boldsymbol{w}. \qquad (3.113)$$

Then, the issue would arise of estimating $(\varepsilon_m, \boldsymbol{h}'_m)$ based on the observation of $\boldsymbol{y}_m$. Unfortunately, $\boldsymbol{y}_m$ is not available at the BS since from Eq. (3.112) we see that its computation would entail perfect knowledge of the interfering signals $\boldsymbol{r}_k$. However, a comparison between Eqs. (3.107) and (3.112) reveals that $\boldsymbol{y}_m^{(j)}$ can be considered as a reasonable approximation of $\boldsymbol{y}_m$. In this respect, we may write

$$\boldsymbol{y}_m^{(j)} = \boldsymbol{\Gamma}(\varepsilon_m)\boldsymbol{B}_m \boldsymbol{h}'_m + \boldsymbol{d}_m + \boldsymbol{w}, \qquad (3.114)$$

where $\boldsymbol{d}_m$ is a disturbance term that accounts for imperfect cancellation of the interfering signals. Vector $\boldsymbol{y}_m^{(j)}$ is thus used in place of the true $\boldsymbol{y}_m$ to compute LS estimates of $(\varepsilon_m, \boldsymbol{h}'_m)$ as indicated in Eqs. (3.109) and (3.110). In light of the above arguments, the algorithm based on Eqs. (3.109) and (3.110) is recognized as a recursive approximation to the ML estimator in which previous estimates of the synchronization parameters are exploited to cancel out the MAI. Compared to the true ML estimator, the APFE is much simpler to implement as it splits the multidimensional maximization problem Eq. (3.103) into a series of mono-dimensional grid searches.

A possible shortcoming of EM-type algorithms comes from the fact that the log-likelihood function $\Lambda(\widetilde{\boldsymbol{\varepsilon}},\widetilde{\boldsymbol{h}}')$ is not guaranteed to have a unique absolute maximum. Indeed, it might exhibit several local peaks that can attract the APFE toward spurious locks. False locks occur since the algorithm tends to settle on the local peak immediately uphill from the initial estimates $\widehat{\boldsymbol{\varepsilon}}^{(0)} = [\widehat{\varepsilon}_1^{(0)}, \widehat{\varepsilon}_2^{(0)}, \ldots, \widehat{\varepsilon}_M^{(0)}]^T$. This indicates that the APFE has a higher chance to converge to the global maximum of $\Lambda(\widetilde{\boldsymbol{\varepsilon}},\widetilde{\boldsymbol{h}}')$ if an accurate estimate $\widehat{\boldsymbol{\varepsilon}}^{(0)}$ is used for the initialization task. Two methods can be used to obtain $\widehat{\boldsymbol{\varepsilon}}^{(0)}$. One possibility is to simply initialize the frequency estimates to zero. Alternatively, one can compute the N-point DFT of $\boldsymbol{r}$ and select the DFT outputs corresponding to the set $\mathcal{I}_m$ of subcarriers assigned to the mth user while putting to zero all the others. After returning in the time-domain through an IDFT operation, the resulting samples are exploited to get an estimate $\widehat{\varepsilon}_m^{(0)}$ by resorting to the frequency estimator proposed in [100] and suitable for single-user transmissions. As is intuitively clear, computing the DFT of $\boldsymbol{r}$ and forcing to zero the subcarriers allocated to interfering users is a viable method to partially mitigate the MAI. Albeit more computationally demanding, this approach is expected to provide better initialization values and faster convergence rate than simply putting $\widehat{\varepsilon}_m^{(0)} = 0$.

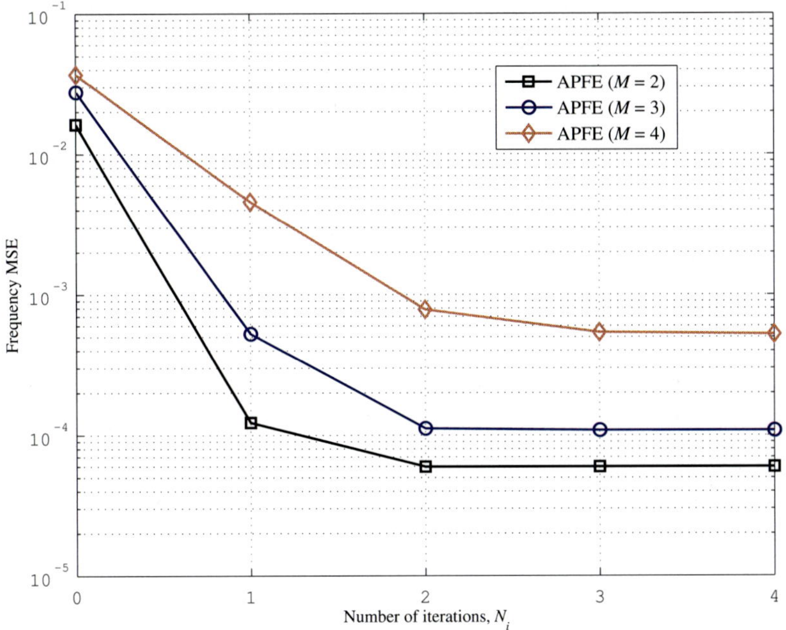

Fig. 3.17   Convergence rate of APFE.

The performance of APFE has been assessed for an OFDMA system with $N = 128$ subcarriers operating in the 5 GHz frequency band. The channel response of each user has length $L_m = 8$, and the channel coefficients are modeled as independent and complex-valued Gaussian random variables with zero-mean (Rayleigh fading) and an exponential power delay profile. The normalized CFOs are uniformly distributed over the interval $[-0.3, 0.3]$ and vary at each new simulation run. We assume a quasi-synchronous system where the CP of the training block is sufficiently long to accommodate both the channel response and the maximum propagation delay. Each user transmits data over 32 distinct subcarriers, which are randomly assigned in order to demonstrate the applicability of APFE in conjunction with a generalized CAS. Without loss of generality, only results for the first user are illustrated.

Figure 3.17 shows the MSE of the frequency estimates $\mathrm{E}\{[\hat{\varepsilon}_1 - \varepsilon_1]^2\}$ as a function of the number N_i of iterations in case of $M =2$, 3 or 4 active users. The latter have equal power with $E_s/N_0 = 20$ dB and the frequency estimates are initialized to zero to reduce the system complexity. We see that APFE achieves convergence in only two iterations and no further gains are observed with $N_i > 2$.

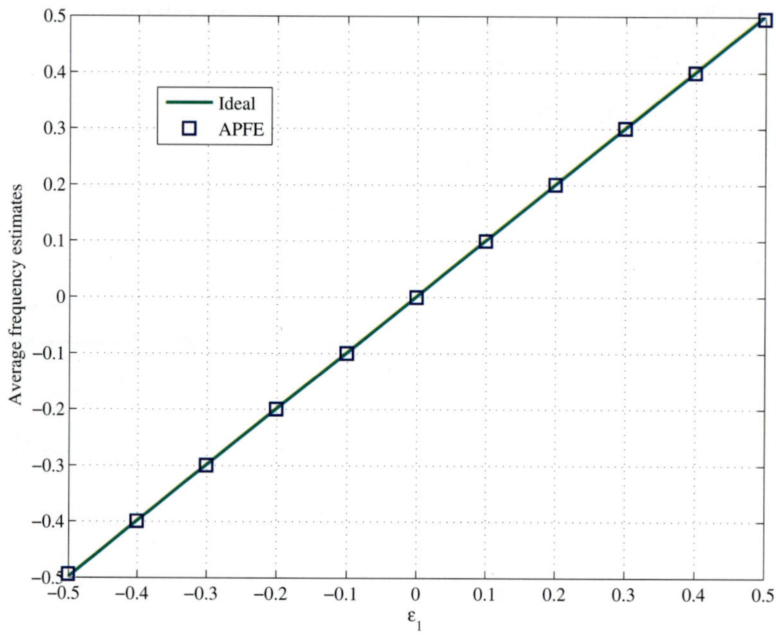

Fig. 3.18   Average frequency estimates of APFE vs. ε_1.

The average frequency estimates are shown in Fig. 3.18 as a function of ε_1 assuming that three users are active in the system. Here, ε_1 is kept fixed at each new simulation run while the frequency offsets of the other users vary independently over the range $[-0.3, 0.3]$. The ideal line $\mathrm{E}\{\widehat{\varepsilon}_1\} = \varepsilon_1$ is also drawn for comparison. These results indicate that APFE provides unbiased estimate over the interval $|\varepsilon_1| < 0.5$.

Figure 3.19 illustrates the frequency MSE as a function of E_s/N_0 in case of two active users. The tick solid line represents the Cramer–Rao lower bound (CRLB) for frequency estimation in quasi-synchronous OFDMA uplink transmissions [125] and is shown as a benchmark. The simulation set up is the same as in Fig. 3.17, except that now an interleaved CAS is adopted in order to make comparisons with the CTYE discussed in the previous subsection. We see that APFE achieves the CRLB for $E_s/N_0 > 15$ dB.

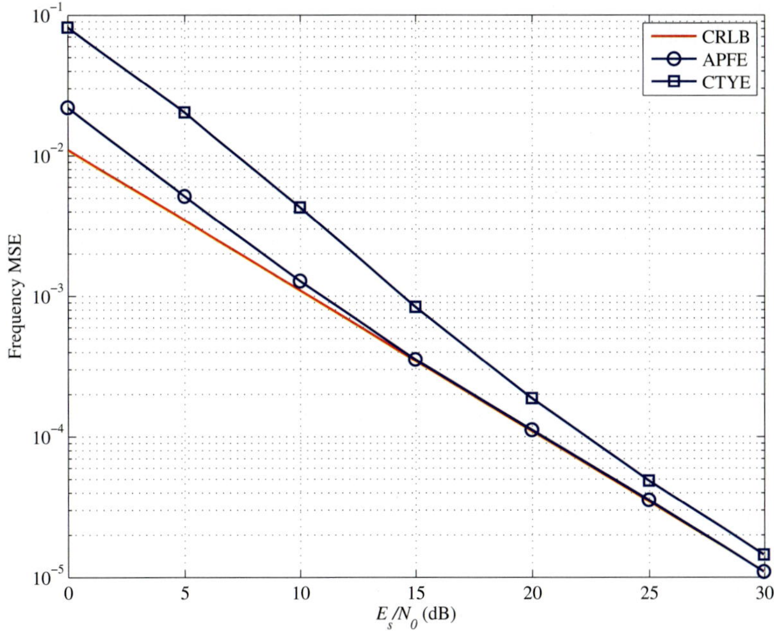

Fig. 3.19   Accuracy of APFE and CTYE vs. E_s/N_0.

The CTYE exhibits good performance at high SNR values, but a certain degradation is observed with respect to APFE for $E_s/N_0 < 15$ dB.

3.4   Timing and frequency offset compensation in uplink transmissions

Once the uplink timing and frequency offsets have been estimated, they must be employed by the BS receiver to restore orthogonality among sub-carriers. This operation is known as timing and frequency correction and represents the final stage of the synchronization process. In downlink transmissions, frequency correction is typically achieved by counter-rotating the time-domain samples at an angular speed $2\pi\widehat{\varepsilon}_m/N$, while timing adjustment is accomplished by shifting the DFT window by a number $\widehat{\theta}_m$ of sampling intervals. Unfortunately, these methods cannot be used in an uplink scenario. The reason is that the uplink signals arriving at the BS are affected by different synchronization errors, so that the correction of one user's time and frequency offset would misalign other initially aligned users. A solution to this problem is presented in [162] and [97], where estimates of the users' offsets are returned to the active terminals via a downlink control channel and exploited by each user to properly adjust its transmitted signal. In a time-varying scenario, however, users should be periodically provided with updated estimates of their synchronization parameters, which may result into an excessive extra load for the downlink transmission and outdated adjustment due to the intrinsic feedback delay. An interesting alternative is to use advanced signal processing techniques to compensate for synchronization errors *directly* at the BS, *i.e.*, without the need of returning timing and frequency estimates back to the active terminals. Solutions derived along this line of reasoning are largely inherited from the multiuser detection area and are subject to the particular subcarrier allocation scheme adopted in the system.

In the rest of this section we first concentrate on the problem of timing and frequency correction for an OFDMA system with subband CAS. A more flexible generalized CAS is next considered to illustrate how linear multiuser detection and interference cancellation schemes can be employed to compensate for the users' CFOs.

3.4.1 *Timing and frequency compensation with subband CAS*

In OFDMA systems with subband CAS the uplink signals arriving at the BS can be separated by a bank of band-pass filters if suitable guard intervals are inserted between adjacent subbands. The receiver can thus estimate and correct the synchronization errors independently for each active user. A solution in this sense is depicted in Fig. 3.20. After users' separation, each uplink signal $x_m(k)$ $(1 \leq m \leq M)$ is exploited to get estimates $\widehat{\theta}_m$ and $\widehat{\varepsilon}_m$ of the timing and frequency offsets using one of the methods described in Sec. 3.3.2. The estimated parameters are then employed to compensate for the synchronization errors of each signal by resorting to conventional single-user techniques. In particular, the samples $x_m(k)$ are multiplied by the exponential term $e^{-j2\pi k\widehat{\varepsilon}_m/N}$ to cancel out any phase rotation induced by the CFO whereas the timing estimate $\widehat{\theta}_m$ is used to select the N samples that are next processed by the DFT unit. After channel equalization (not shown in the figure), the DFT outputs corresponding to the mth subchannel are finally passed to the data detection unit.

The receiver architecture shown in Fig. 3.20 relies on the fact that the

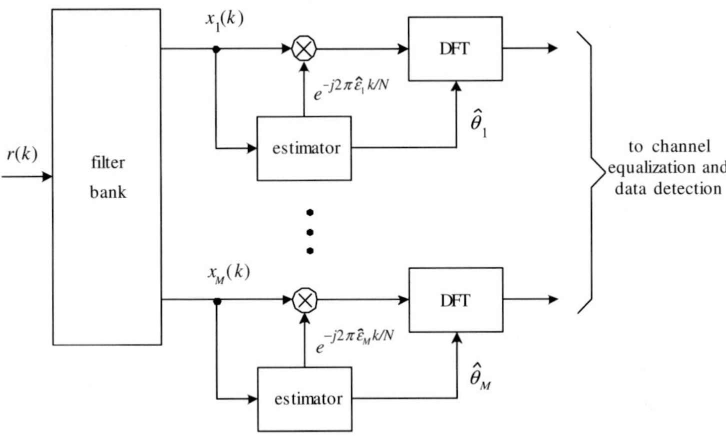

Fig. 3.20 Timing and frequency synchronization for an OFDMA uplink receiver with subband CAS.

uplink signals are perfectly separated at the output of the filter bank. In practice, however, perfect separation is not possible even in the presence of ideal brick-wall filters due to the frequency leakage among adjacent sub-channels caused by synchronization errors. This means that some residual MAI will be present at the DFT output, with ensuing limitations of the error-rate performance. In addition, compensating for the frequency errors in the time-domain as depicted in Fig. 3.20 requires an N-point DFT operation for each active user. Since the complexity involved with the DFT represents a major concern for system implementation, the receiver structure of Fig. 3.20 may be too computationally demanding in practical applications, especially when the number M of simultaneously active users and/or the number N of available subcarriers are relatively large.

An alternative scheme for uplink frequency correction in subband OFDMA systems is sketched in Fig. 3.21. This solution has been proposed in [18] and is referred to as the Choi–Lee–Jung–Lee (CLJL) method in the ensuing discussion. Its main advantage is that it avoids the need for multiple DFT operations, but can only be applied to a quasi-synchronous system where the uplink signals are time aligned within the length of the CP and timing correction is thus unnecessary.

To explain the rationale behind CLJL, we reconsider the N samples $r_i(k)$ $(0 \leq k \leq N-1)$ falling within the ith receive DFT window. Collecting

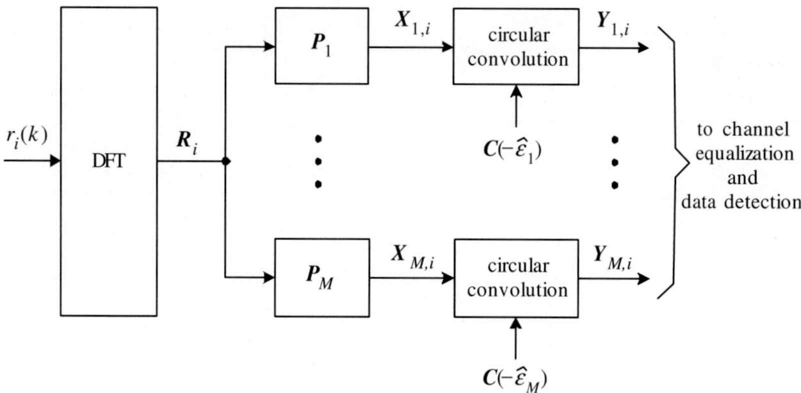

Fig. 3.21 Frequency correction by means of circular convolutions applied at the DFT output.

Eqs. (3.70) and (3.72) we may write

$$r_i(k) = \sum_{m=1}^{M} z_{m,i}(k)\, e^{j2\pi\varepsilon_m k/N} + w_i(k), \qquad 0 \le k \le N-1 \qquad (3.115)$$

with

$$z_{m,i}(k) = \frac{1}{\sqrt{N}} \sum_{n\in\mathcal{I}_m} \widetilde{H}_{m,i}(n) c_{m,i}(n)\, e^{j2\pi nk/N}. \qquad (3.116)$$

For convenience, the N-point DFT of the sequences $r_i(k)$, $z_{m,i}(k)$ and $w_i(k)$ are arranged into three N-dimensional vectors $\boldsymbol{R}_i$, $\boldsymbol{Z}_{m,i}$ and $\boldsymbol{W}_i$, respectively. Then, recalling that a multiplication in the time-domain corresponds to a circular convolution in the frequency-domain, from Eq. (3.115) we have

$$\boldsymbol{R}_i = \sum_{m=1}^{M} \boldsymbol{Z}_{m,i} \otimes \boldsymbol{C}(\varepsilon_m) + \boldsymbol{W}_i, \qquad (3.117)$$

where $\otimes$ denotes the N-point circular convolution, $\boldsymbol{Z}_{m,i}$ has entries

$$Z_{m,i}(n) = \begin{cases} \widetilde{H}_{m,i}(n) c_{m,i}(n) & \text{if } n \in \mathcal{I}_m \\ 0 & \text{otherwise} \end{cases} \qquad (3.118)$$

and, finally, $\boldsymbol{C}(\varepsilon_m)$ is the N-point DFT of $\left\{ e^{j2\pi\varepsilon_m k/N} \;;\; 0 \le k \le N-1 \right\}$ with entries

$$C(\varepsilon_m, n) = \frac{\sin\left[\pi\left(n - \varepsilon_m\right)\right]}{\sin\left[\pi\left(n - \varepsilon_m\right)/N\right]}\, e^{-j\pi(N-1)(n-\varepsilon_m)/N}, \quad 0 \le n \le N-1.$$

$$(3.119)$$

Returning to Fig. 3.21, we see that for each active user an N-dimensional vector $\boldsymbol{X}_{m,i}$ $(1 \leq m \leq M)$ is obtained from the DFT output by putting to zero all entries of $\boldsymbol{R}_i$ that do not correspond to the subcarriers of the considered user. This amounts to setting $\boldsymbol{X}_{m,i} = \boldsymbol{P}_m \boldsymbol{R}_i$, where $\boldsymbol{P}_m$ is a diagonal matrix with entries

$$[\boldsymbol{P}_m]_{n,n} = \begin{cases} 1 & \text{if } n \in \mathcal{I}_m \\ 0 & \text{otherwise.} \end{cases} \tag{3.120}$$

In practice, $\boldsymbol{P}_m$ acts as a band-pass filter that aims at isolating the contribution of the mth uplink signal at the DFT output. Bearing in mind Eq. (3.117) and assuming perfect signal separation, we may write

$$\boldsymbol{X}_{m,i} \approx \boldsymbol{Z}_{m,i} \otimes \boldsymbol{C}(\varepsilon_m) + \boldsymbol{W}_{m,i}, \tag{3.121}$$

where $\boldsymbol{W}_{m,i} = \boldsymbol{P}_m \boldsymbol{W}_i$ is the noise contribution. The above equation indicates that $\boldsymbol{X}_{m,i}$ can reasonably be assumed free from MAI. However, it is still affected by residual ICI due to the uncompensated frequency error ε_m.

Instead of performing frequency correction in the time-domain as illustrated in Fig. 3.20, we can equivalently compensate for ε_m in the frequency-domain using a suitable circular convolution followed by band-pass filtering [18]. This produces

$$\boldsymbol{Y}_{m,i} = \boldsymbol{P}_m \left[\boldsymbol{X}_{m,i} \otimes \boldsymbol{C}(-\widehat{\varepsilon}_m) \right], \tag{3.122}$$

where $\boldsymbol{C}(-\widehat{\varepsilon}_m)$ is a vector that collects the N-point DFT of the sequence $\left\{ e^{-j2\pi\widehat{\varepsilon}_m k/N} \; ; \; 0 \leq k \leq N-1 \right\}$ and whose entries are obtained from Eq. (3.119) after replacing ε_m by $-\widehat{\varepsilon}_m$.

Substituting Eq. (3.121) into Eq. (3.122) and assuming ideal frequency estimation (*i.e.*, $\widehat{\varepsilon}_m = \varepsilon_m$), yields

$$\boldsymbol{Y}_{m,i} = \boldsymbol{Z}_{m,i} + \boldsymbol{P}_m \left[\boldsymbol{W}_{m,i} \otimes \boldsymbol{C}(-\widehat{\varepsilon}_m) \right], \tag{3.123}$$

where we have used the identity $\boldsymbol{Z}_{m,i} \otimes \boldsymbol{C}(\varepsilon_m) \otimes \boldsymbol{C}(-\varepsilon_m) = \boldsymbol{Z}_{m,i}$. The above equation, together with Eq. (3.118), indicates that $\boldsymbol{Y}_{m,i}$ is free from interference except for channel distortion and thermal noise. In practice, however, non-ideal frequency compensation and imperfect users' separation will generate residual ICI and MAI on $\boldsymbol{Y}_{m,i}$, thereby resulting in some performance degradation with respect to the ideal setting described by Eq. (3.123).

As mentioned previously, a favorable feature of CLJL is that it only needs a single DFT operation. This result is achieved by operating over the frequency-domain samples $\boldsymbol{R}_i$ and leads to a significant reduction of complexity as compared to the receive architecture of Fig. 3.20, where a separate DFT operation is required for each user.

3.4.2 *Frequency compensation through interference cancellation*

The CLJL scheme discussed in the previous subsection is only suited for OFDMA systems with subband CAS. The reason is that the bank of matrices $\boldsymbol{P}_m$ $(1 \le m \le M)$ in Fig. 3.21 provides accurate users' separation as long as the subcarriers of a given user are grouped together and sufficiently large guard intervals are inserted among adjacent subchannels. When used in conjunction with an interleaved or a generalized CAS, however, the CLJL cannot significantly reduce the MAI induced by frequency errors. In this case, alternative approaches must be resorted to. One possibility is offered by the concept of multiuser detection [164]. The latter includes all advanced signal processing techniques for the joint demodulation of mutually interfering data streams.

Multiuser detection schemes are largely categorized into linear or interference cancellation (IC) architectures. In this subsection we limit our attention to the latter class. In particular, we show how the IC concept can be applied to CLJL in order to reduce the residual interference present on $\boldsymbol{Y}_{m,i}$. The resulting scheme has been derived by Huang and Letaief (HL) in [55] and operates in an iterative fashion.

Calling $\widehat{\boldsymbol{Y}}_{m,i}^{(j)}$ the mth restored signal after the jth iteration, the HL proceeds as follows:

The HL algorithm

- **Initialization**

  Compute the CLJL vectors defined in Eq. (3.122), *i.e.*,

  $$\boldsymbol{Y}_{m,i} = \boldsymbol{P}_m \left[(\boldsymbol{P}_m \boldsymbol{R}_i) \otimes \boldsymbol{C}(-\widehat{\varepsilon}_m) \right], \quad 1 \le m \le M \qquad (3.124)$$

  and set $\widehat{\boldsymbol{Y}}_{m,i}^{(0)} = \boldsymbol{Y}_{m,i}$ for $m = 1, 2, \ldots, M$.

- **jth iteration (j=1,2,...)**

  For each active user $(m = 1, 2, \ldots, M)$ perform interference cancellation in the form

  $$\widetilde{\boldsymbol{Y}}_{m,i}^{(j)} = \boldsymbol{R}_i - \sum_{k=1, k \ne m}^{M} \widehat{\boldsymbol{Y}}_{k,i}^{(j-1)} \otimes \boldsymbol{C}(\widehat{\varepsilon}_k), \quad 1 \le m \le M \qquad (3.125)$$

  and remove the effect of ε_m following a CLJL approach

  $$\widehat{\boldsymbol{Y}}_{m,i}^{(j)} = \boldsymbol{P}_m \left[\left(\boldsymbol{P}_m \widetilde{\boldsymbol{Y}}_{m,i}^{(j)} \right) \otimes \boldsymbol{C}(-\widehat{\varepsilon}_m) \right], \quad 1 \le m \le M. \qquad (3.126)$$

As indicated in Eq. (3.125), at each iteration circular convolutions are employed to regenerate interference, which is then subtracted from the original DFT output $\boldsymbol{R}_i$. The expurgated vectors $\widetilde{\boldsymbol{Y}}_{m,i}^{(j)}$ are next used to obtain the restored signals $\widehat{\boldsymbol{Y}}_{m,i}^{(j)}$ according to Eq. (3.126). In this respect, the HL can be regarded as a parallel interference cancellation (PIC) scheme. In contrast to the conventional PIC, however, HL does not suffer from error propagation since orthogonality among the received signals is tentatively restored without employing any data decision.

Simulation results reported in [55] indicate that HL performs much better than CLJL after just a few iterations. In particular, its increased robustness against ICI and MAI makes it suited for any CAS, whereas CLJL can only be used in conjunction with a subband CAS. It is worth noting that the windowing function $\boldsymbol{P}_m$ employed in Eqs. (3.124) and (3.126) aims at removing all the energy present on subcarriers allocated to other users. Albeit useful to reduce interference, this operation entails some performance loss in the presence of relatively large CFOs since in this case the undesignated subcarriers might contain a significant portion of the user's energy which is definitely discarded by HL.

3.4.3 *Frequency compensation through linear multiuser detection*

Linear multiuser detection can be used as an alternative to IC-based solutions for mitigating interference caused by uplink CFOs. An example in this sense is provided by the Cao-Tureli-Yao-Honan (CTYH) scheme discussed in [12]. This method is suited for any CAS, but can only operate in a quasi-synchronous scenario where no IBI is present. The CTYH is now derived following a two-step procedure. We begin by establishing a new convenient signal model for the DFT output $\boldsymbol{R}_i$. Orthogonality among subcarriers is subsequently restored by means of linear transformations applied to $\boldsymbol{R}_i$.

In deriving the new signal model we make the following assumptions without loss of generality:

(1) each user transmits its data over $P = N/R$ subcarriers, where R is the maximum number of simultaneously active users in the system under consideration;

(2) the indices of subcarriers assigned to the mth user belong to the set $\mathcal{I}_m = \{q_m(p); 0 \leq p \leq P - 1\}$.

Bearing in mind Eq. (3.70), we may rewrite the samples $r_{m,i}(k)$ of the

mth received uplink signal as

$$r_{m,i}(k) = \frac{1}{\sqrt{N}} e^{j2\pi\varepsilon_m k/N} \sum_{p=0}^{P-1} S_{m,i}(p)\, e^{j2\pi q_m(p)k/N}, \quad 0 \le k \le N-1$$

$$(3.127)$$

where

$$S_{m,i}(p) = \widetilde{H}_{m,i}(q_m(p))c_{m,i}(q_m(p)) \qquad (3.128)$$

is an attenuated and phase-rotated version of the symbol transmitted over the $q_m(p)$th subcarrier. For convenience, we define a vector $\boldsymbol{R}_{m,i} = [R_{m,i}(0), R_{m,i}(1), \ldots, R_{m,i}(N-1)]^T$ whose entries are the DFT of $r_{m,i}(k)$, i.e.,

$$R_{m,i}(n) = \frac{1}{\sqrt{N}} \sum_{k=0}^{N-1} r_{m,i}(k)\, e^{-j2\pi nk/N}, \quad 0 \le n \le N-1. \qquad (3.129)$$

Then, substituting Eq. (3.127) into Eq. (3.129) and letting $\boldsymbol{S}_{m,i} = [S_{m,i}(0), S_{m,i}(1), \ldots, S_{m,i}(P-1)]^T$, yields

$$\boldsymbol{R}_{m,i} = \boldsymbol{\Pi}_m(\varepsilon_m)\boldsymbol{S}_{m,i}, \qquad (3.130)$$

where $\boldsymbol{\Pi}_m(\varepsilon_m)$ is an $N \times P$ matrix with elements

$$[\boldsymbol{\Pi}_m(\varepsilon_m)]_{n,p} = f_N\left[q_m(p) + \varepsilon_m - n\right]\, e^{j\pi(N-1)(q_m(p)+\varepsilon_m-n)/N}, \qquad (3.131)$$

for $0 \le n \le N-1$ and $0 \le p \le P-1$, with $f_N(x)$ defined as in Eq. (3.19).

As shown in Eq. (3.72), the samples $r_i(k)$ of the ith received OFDMA block are the superposition of all uplink signals plus thermal noise. The output of the receive DFT unit is thus given by

$$\boldsymbol{R}_i = \sum_{m=1}^{M} \boldsymbol{R}_{m,i} + \boldsymbol{W}_i, \qquad (3.132)$$

where $\boldsymbol{W}_i$ is a complex-valued Gaussian vector with zero-mean and covariance matrix $\sigma_w^2 \boldsymbol{I}_N$. Finally, substituting Eq. (3.130) into Eq. (3.132) and letting $\boldsymbol{S}_i = \begin{bmatrix} \boldsymbol{S}_{1,i}^T & \boldsymbol{S}_{2,i}^T & \cdots & \boldsymbol{S}_{M,i}^T \end{bmatrix}^T$, we obtain the desired signal model for $\boldsymbol{R}_i$ in the form

$$\boldsymbol{R}_i = \boldsymbol{\Pi}(\boldsymbol{\varepsilon})\boldsymbol{S}_i + \boldsymbol{W}_i, \qquad (3.133)$$

where $\boldsymbol{\Pi}(\boldsymbol{\varepsilon}) = \begin{bmatrix} \boldsymbol{\Pi}_1(\varepsilon_1) & \boldsymbol{\Pi}_2(\varepsilon_2) & \cdots & \boldsymbol{\Pi}_M(\varepsilon_M) \end{bmatrix}^T$ is an $N \times MP$ matrix whose elements are related to the users' frequency offsets $\boldsymbol{\varepsilon} = [\varepsilon_1, \varepsilon_2, \ldots, \varepsilon_M]^T$.

Inspection of Eq. (3.128) reveals that the entries of $\boldsymbol{S}_i$ are the transmitted data symbols multiplied by the corresponding channel frequency response. Accordingly, $\boldsymbol{S}_i$ is the vector that would be present at the DFT output in the absence of any interference and thermal noise. The purpose of CTYH is to obtain an estimate of $\boldsymbol{S}_i$ starting from $\boldsymbol{R}_i$. As illustrated in Fig. 3.22 , this goal is achieved by means of a linear transformation applied to $\boldsymbol{R}_i$. The estimated vector $\widehat{\boldsymbol{S}}_i$ is then fed to the channel equalizer and data detection unit, which provides decisions on the transmitted data symbols.

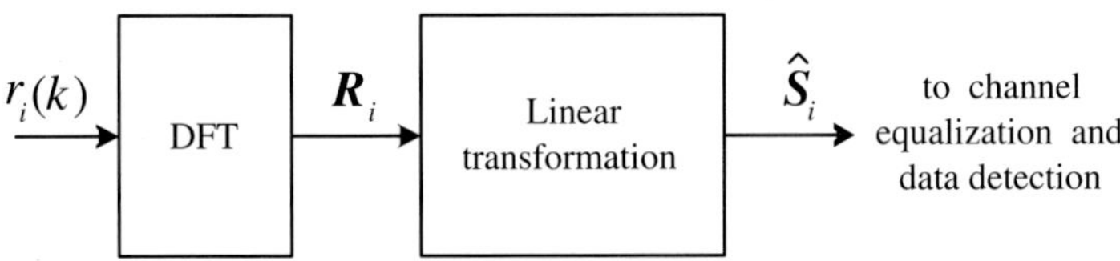

Fig. 3.22   Frequency correction by means of a linear transformation at the DFT output.

Two possible methods for computing $\widehat{\boldsymbol{S}}_i$ are illustrated in [12]. The first one is based on the LS approach and is equivalent to the well known *linear decorrelating detector* (LDD) [164]

$$\widehat{\boldsymbol{S}}_{i,LDD} = \boldsymbol{\Pi}^{\dagger}(\varepsilon)\boldsymbol{R}_i, \tag{3.134}$$

where $\boldsymbol{\Pi}^{\dagger}(\varepsilon) = \left[\boldsymbol{\Pi}^H(\varepsilon)\boldsymbol{\Pi}(\varepsilon)\right]^{-1}\boldsymbol{\Pi}^H(\varepsilon)$ denotes the Moore-Penrose generalized inverse of $\boldsymbol{\Pi}(\varepsilon)$.

Substituting Eq. (3.133) into Eq. (3.134) yields

$$\widehat{\boldsymbol{S}}_{i,LDD} = \boldsymbol{S}_i + \boldsymbol{\Pi}^{\dagger}(\varepsilon)\boldsymbol{W}_i, \tag{3.135}$$

meaning that the decorrelating detector can totally suppress any interference caused by frequency errors. As it is known, the price for this result is a certain enhancement of the output noise level.

The second solution is based on the MMSE approach and aims at minimizing the overall effect of interference plus ambient noise. The resulting scheme is known as the linear MMSE detector [164] and reads

$$\widehat{\boldsymbol{S}}_{i,MMSE} = \boldsymbol{Q}(\varepsilon,\sigma_w^2)\boldsymbol{R}_i, \tag{3.136}$$

with $\boldsymbol{Q}(\varepsilon,\sigma_w^2) = \left[\boldsymbol{\Pi}^H(\varepsilon)\boldsymbol{\Pi}(\varepsilon) + \sigma_w^2\boldsymbol{I}_{MP}\right]^{-1}\boldsymbol{\Pi}^H(\varepsilon)$. Although the output of the MMSE detector is still affected by some residual MAI, the noise enhancement phenomenon is greatly reduced as compared to the LDD.

The main drawback of CTYH is the relatively huge complexity required to evaluate $\boldsymbol{\Pi}^{\dagger}(\varepsilon)$ or $\boldsymbol{Q}(\varepsilon,\sigma_w^2)$. Note that these matrices cannot be precomputed and stored in the receiver as they do depend on the actual CFOs and noise power. Since the quantities ε and σ_w^2 are not perfectly known at the BS, in practice they are replaced by suitable estimates $\widehat{\varepsilon}$ and $\widehat{\sigma}_w^2$. It is observed in [12] that $\mathbf{\Pi}^\dagger(\varepsilon)$ and $\mathbf{Q}(\varepsilon,\sigma_w^2)$ are banded matrices with non-zero elements only in the vicinity of their main diagonal. This property can be exploited to reduce the complexity involved with their computation.

3.4.4   *Performance of frequency correction schemes*

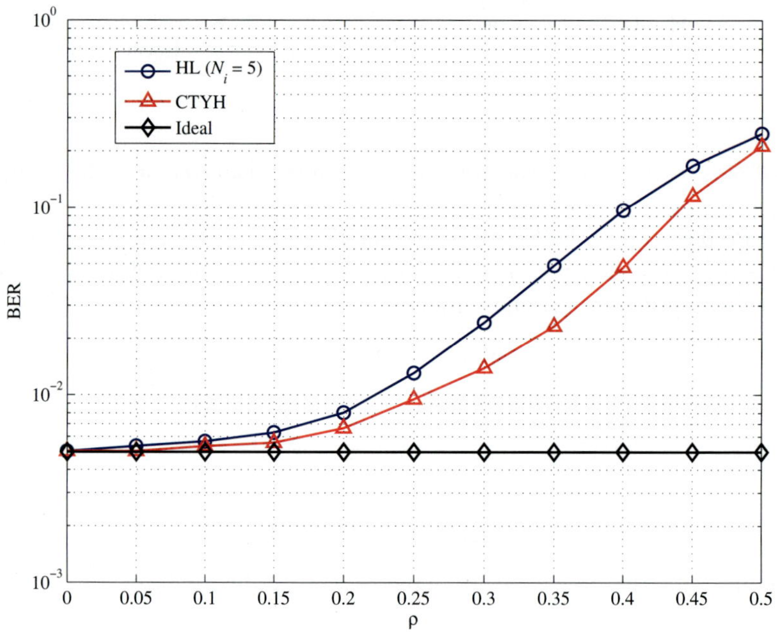

Fig. 3.23   BER performance of HL and CTYH vs. ρ for an uncoded QPSK transmission with $E_s/N_0 = 20$ dB.

It is interesting to compare the performance of HL and CTYH in terms of bit-error-rate (BER) in a quasi-synchronous uplink scenario. For this purpose, we consider an OFDMA system with $N = 128$ subcarriers and a generalized carrier assignment policy. Each subchannel is composed by 32 subcarriers, so that the maximum number of simultaneously active users is limited to $R = 4$. We assume a fully-loaded system in which $M = R = 4$ and let $\varepsilon = \rho\left[1, -1, 1, -1\right]^T$, where ρ is a deterministic parameter belonging to the interval $[0, 0.5]$ and known as *frequency attenuation factor*

[55]. A new channel snapshot is generated at each simulation run and kept fixed over an entire frame. Ideal frequency and channel estimates are assumed throughout simulations. Five iterations are performed by HL while CTYH employs the decorrelating matrix $\mathbf{\Pi}^\dagger(\boldsymbol{\varepsilon})$ as in Eq. (3.134).

Figure 3.23 illustrates the BER performance as a function of ρ for an uncoded QPSK transmission. Users have equal power with $E_s/N_0 = 20$ dB. The curve labeled "ideal" is obtained by assuming that all CFOs have been perfectly corrected at the mobile terminals, *i.e.*, $\varepsilon_m = 0$ for $m = 1, 2, 3, 4$. This provides a benchmark for the BER performance since in this case the users' signals are perfectly orthogonal and no interference is present at the DFT output. We see that the BER degrades with ρ due to the increased amount of ICI and MAI. As mentioned previously, the latter is mitigated by CTYH at the price of non-negligible noise enhancement, while the windowing functions used by HL leads to a significant loss of signal energy in the presence of relatively large CFOs.

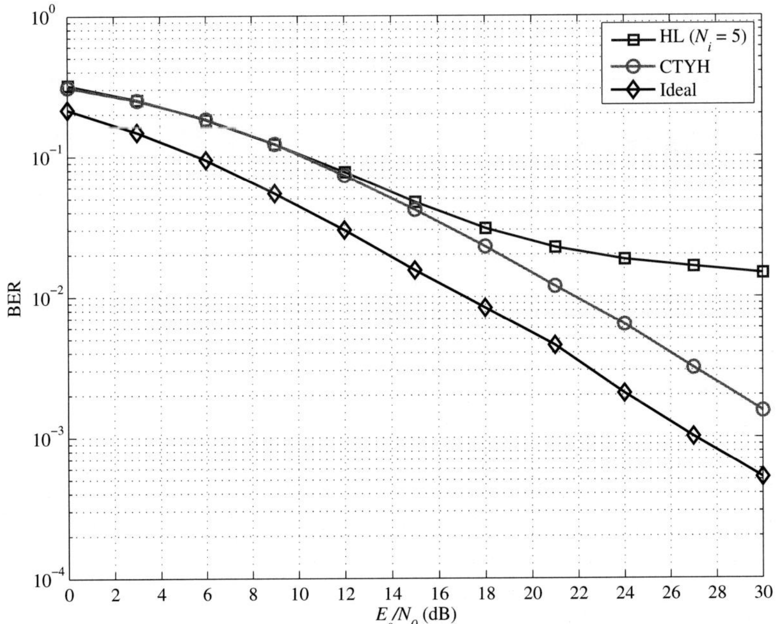

Fig. 3.24 BER performance of HL and CTYH vs. E_s/N_0 for an uncoded QPSK transmission with $\rho = 0.3$.

Figure 3.24 shows the BER of the considered schemes vs. E_s/N_0 for an uncoded QPSK transmission. Users have equal power and ρ is set to 0.3. Again, we see that CTYH provides the best performance. In particular, at an error rate of 10^{-2} the loss of CTYH with respect to the ideal system is approximately 4 dB. As for HL, it performs poorly and exhibits an error floor at high SNR values.

Chapter 4

Channel Estimation and Equalization

In OFDM transmissions, the effect of channel distortion on each subcarrier is represented by a single complex-valued coefficient that affects the amplitude and phase of the relevant information symbol. Coherent detection of the transmitted data can be performed only after this multiplicative distortion has been properly compensated for. This operation is known as *channel equalization*, and can easily be accomplished in the frequency-domain if an estimate of the channel response is available at the receiver. An alternative to coherent detection is offered by the use of differential encoding techniques. In this case information data are transmitted as phase variations between adjacent subcarriers and are recovered at the receiver through differential demodulation, thereby eliminating the need for channel knowledge. The price for this simplification is a certain loss of power efficiency as compared to coherent detection.

In this Chapter we present some popular schemes to recover channel state information (CSI) in OFDM systems. One common approach is based on the periodic insertion of pilot symbols within the transmitted signal. This idea has been adopted in many OFDM standards and has led to the development of so-called pilot-aided schemes. Although the use of pilot symbols may facilitate the channel estimation task to a great extent, it inevitably leads to some reduction of the data throughput because of the required extra overhead. This problem has motivated intense research activity on blind channel identification and equalization techniques, where the inherent redundancy present in the transmitted signal is exploited at the receiver to get CSI with the aid of only a few pilots or using no pilots at all.

The Chapter has the following outline. Section 4.1 illustrates the concept of frequency-domain channel equalization. Combining schemes are also presented for receivers equipped with multiple antenna elements. The idea of pilot-aided channel estimation is discussed in Sec. 4.2. After illustrating some popular pilot insertion patterns adopted in commercial systems, we show how the minimum allowable distance between pilots is related to the statistical parameters of the wireless channel. Several techniques for pilots' interpolation are also discussed. Section 4.3 illustrates recent advances in the area of blind and semi-blind channel estimation and equalization. Here, two different approaches are considered. The first one relies on the concept of subspace decomposition, while in the other the expectation-maximization (EM) algorithm is applied to couple the channel estimation/equalization task with the decision making process.

4.1 Channel equalization

Channel equalization is the process through which a coherent receiver tries to compensate for any distortion induced by frequency-selective fading. For the sake of simplicity, ideal timing and frequency synchronization is considered throughout this chapter. The channel is assumed static over each OFDM block, but can vary from block to block. Under these assumptions, the output of the receive DFT unit during the ith block is given by

$$R_i(n) = H_i(n)c_i(n) + W_i(n), \qquad 0 \le n \le N - 1 \qquad (4.1)$$

where $H_i(n)$ is the channel frequency response over the nth subcarrier, $c_i(n)$ is the relevant data symbol and, finally, $W_i(n)$ represents the frequency-domain noise contribution with zero-mean and variance σ_w^2.

One appealing feature of OFDM is that channel equalization can independently be performed over each subcarrier by means of a bank of one-tap multipliers. In practice, the nth DFT output $R_i(n)$ is weighted by a complex-valued quantity $p_i(n)$ in an attempt of compensating for the channel-induced attenuation and phase rotation. As shown in Fig. 4.1, the equalized sample $Y_i(n) = p_i(n)R_i(n)$ is subsequently passed to the detection unit, which delivers final decisions $\widehat{c}_i(n)$ on the transmitted data.

A popular approach for the design of the equalizer coefficients relies on the minimum mean-square error (MMSE) criterion . In this case $p_i(n)$ is chosen so as to minimize the following quantity

$$J_i(n) = \mathrm{E}\left\{ |p_i(n)R_i(n) - c_i(n)|^2 \right\}, \qquad (4.2)$$

which represents the mean-square error (MSE) between the equalizer output $Y_i(n)$ and the transmitted symbol $c_i(n)$.

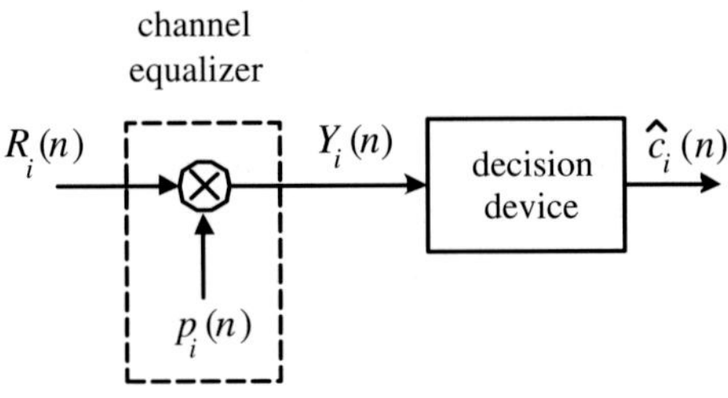

Fig. 4.1   Equalization and data detection over the nth subcarrier.

From the orthogonality principle [72], we know that the optimal weights $\{p_i(n)\}$ are such that the error $Y_i(n) - c_i(n)$ is orthogonal to the relevant DFT output, *i.e.*,

$$\mathrm{E}\left\{[p_i(n)R_i(n) - c_i(n)]\, R_i^*(n)\right\} = 0. \tag{4.3}$$

Substituting Eq. (4.1) into Eq. (4.3) and computing the expectation with respect to thermal noise and data symbols (the latter are assumed to be statistically independent with zero-mean and power C_2), yields

$$p_i(n) = \frac{H_i^*(n)}{|H_i(n)|^2 + \rho}, \tag{4.4}$$

where $\rho = \sigma_w^2/C_2$ is the inverse of the operating signal-to-noise ratio (SNR).

As indicated by Eq. (4.4), computing the MMSE equalization coefficients requires knowledge of $H_i(n)$ and σ_w^2. A suboptimum solution is obtained by designing parameter ρ for a *fixed* nominal noise power $\bar{\sigma}_w^2$, thereby allowing the equalizer to operate in a mismatched mode whenever $\sigma_w^2 \neq \bar{\sigma}_w^2$. The resulting scheme dispenses from knowledge of σ_w^2 and only needs channel state information. This simplified approach also includes the well-known Zero-Forcing (ZF) equalization criterion, which corresponds to setting $\bar{\sigma}_w^2 = 0$. In this case the equalizer performs a pure channel inversion and its coefficients are given by

$$p_i(n) = \frac{1}{H_i(n)}, \tag{4.5}$$

while the DFT output takes the form

$$Y_i(n) = c_i(n) + \frac{W_i(n)}{H_i(n)}, \qquad 0 \leq n \leq N - 1. \tag{4.6}$$

This equation indicates that ZF equalization is capable of totally compensating for any distortion induced by the wireless channel. However, the noise power at the equalizer output is given by $\sigma_w^2/|H_i(n)|^2$ and may be excessively large over deeply faded subcarriers characterized by low channel gains.

It is worth noting that the equalization coefficients in Eqs. (4.4) and (4.5) only differ for a positive multiplicative factor $1 + \rho/|H_i(n)|^2$, so that the phase of the equalized sample $Y_i(n)$ is the same in both cases. An interesting consequence of this fact is that ZF and MMSE equalizers are perfectly equivalent in the presence of a pure phase modulation (as occurs with PSK data symbols) since in this case the decision on $c_i(n)$ is solely based on the argument of $Y_i(n)$.

All the above results can easily be extended to OFDM receivers equipped with $Q > 1$ antenna elements for diversity reception. In such a situation, the contributions from all receive antennas may properly be combined to improve the reliability of data decisions. As is intuitively clear, the best performance is obtained when the combining strategy is integrated with the channel equalization process in a single functional unit. To see how this comes about, denote $H_i^{(q)}(n)$ the frequency response of the channel viewed by the qth receiving antenna and let

$$R_i^{(q)}(n) = H_i^{(q)}(n)c_i(n) + W_i^{(q)}(n), \qquad 0 \le n \le N - 1 \qquad (4.7)$$

be the DFT output over the corresponding diversity branch.

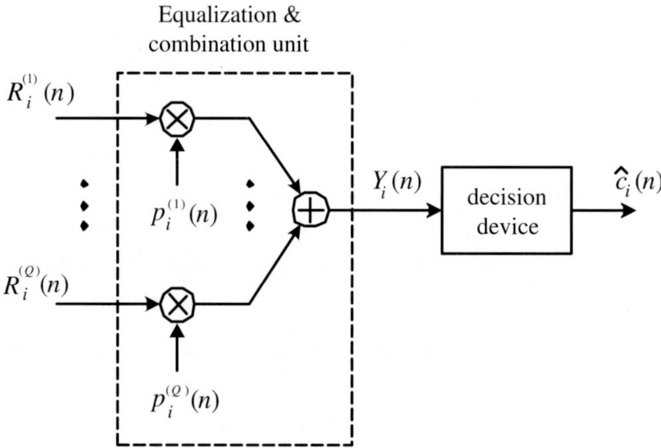

Fig. 4.2   Equalization and data detection over the nth subcarrier in the presence of multiple receiving antennas.

As illustrated in Fig. 4.2, the decision statistic for $c_i(n)$ is obtained by linearly combining the DFT outputs from the Q available antennas, *i.e.*,

$$Y_i(n) = \sum_{q=1}^{Q} p_i^{(q)}(n) R_i^{(q)}(n). \tag{4.8}$$

The weighting coefficients $p_i^{(q)}(n)$ can be selected according to various optimality criteria. Among them, the MMSE strategy aims at minimizing the following MSE

$$J_i(n) = \mathrm{E}\left\{\left|\sum_{q=1}^{Q} p_i^{(q)}(n) R_i^{(q)}(n) - c_i(n)\right|^2\right\}. \tag{4.9}$$

Assuming for simplicity that the noise power σ_w^2 is the same at each branch, the optimum weights are found to be

$$p_i^{(q)}(n) = \frac{[H_i^{(q)}(n)]^*}{\rho + \sum_{\ell=1}^{Q}\left|H_i^{(\ell)}(n)\right|^2}, \tag{4.10}$$

where $\rho = \sigma_w^2/C_2$. Interestingly, setting $\rho = 0$ in the above equation results into the well-known maximum-ratio-combining (MRC) strategy, which has the appealing property of maximizing the SNR at the output of the combining/equalization unit.

4.2   Pilot-aided channel estimation

In multicarrier systems the transmission is normally organized in frames, each containing a specified number of OFDM blocks. As mentioned in Chapter 3, some reference blocks carrying known data are usually appended in front of the frame to assist the synchronization process as well as to provide initial estimates of the channel frequency response. If the channel remains static over the frame duration, the estimates obtained from the reference blocks can be used to coherently detect the entire payload. This situation is typical of WLAN systems, where the user terminals are characterized by low mobility and, in consequence, the channel coherence time is expected to be much greater than the packet length. On the other hand, in applications characterized by relatively high mobility as those envisioned by the IEEE 802.16e standard for WMANs, the channel response undergoes significant variations over one frame and must continuously be tracked to maintain reliable data detection. In this case, in addition to initial reference blocks, known symbols called *pilots* are normally inserted into the payload section of the frame at some convenient positions. These pilots are scattered in both the time and frequency directions (*i.e.*, they are positioned over different blocks and different subcarriers), and are used as reference values for channel estimation and tracking. In practice, the channel transfer function is first estimated at the positions where pilots are placed. Interpolation techniques are next employed to obtain the channel response over information-bearing subcarriers. This approach is usually referred to as pilot-aided channel estimation and is the subject of this Section.

4.2.1   *Scattered pilot patterns*

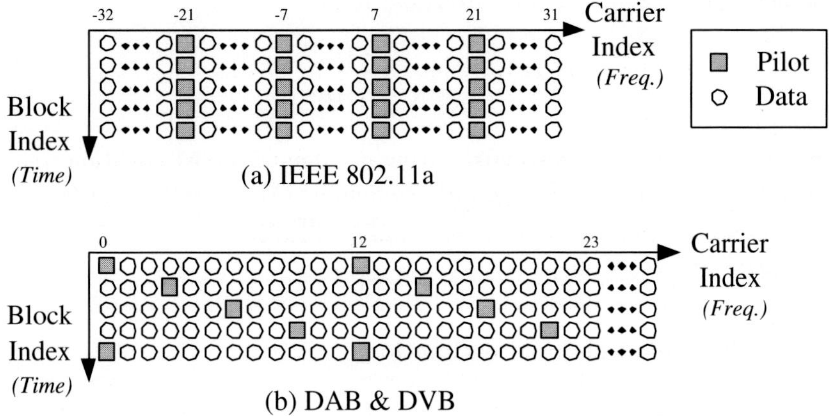

Fig. 4.3   Pilot arrangements in commercial systems: IEEE 802.11a WLAN standard (a); DAB and DVB systems (b).

Figure 4.3 illustrates two major examples of pilot arrangements in the time- and frequency-domains adopted in commercial applications. In particular, Fig. 4.3 (a) refers to the IEEE 802.11a standard for WLANs [41,59], while the pattern of Fig. 4.3 (b) is employed in digital audio broadcasting (DAB) [39] and digital video broadcasting (DVB) systems [40]. The vertical axis represents the time direction and spans over the OFDM blocks, while the horizontal axis indicates the frequency direction and counts the indices of subcarriers in a given block.

As is seen, in the WLAN some specified subcarriers (called *pilot tones*) are exclusively reserved for pilot insertion. In these systems, initial channel acquisition is performed at the beginning of each frame by exploiting two reference blocks (not shown in the figure) carrying known symbols over all subcarriers. During the payload section, pilot tones can be exploited for channel tracking, even though in the IEEE 802.11a standard they are specifically employed to track any residual frequency error that may remain after initial frequency acquisition.

Generally speaking, the arrangement of Fig. 4.3 (a) is advantageous in terms of system complexity because of the fixed positions occupied by pilot tones in the frequency-domain. On the other hand, it is not robust against possible deep fades that might hit some of these pilot tones for the entire frame duration. As shown in Fig. 4.3 (b), in DAB and DVB systems this problem is mitigated by shifting the pilot positions in the frequency-domain at each new OFDM block. Compared to the pilot insertion strategy adopted in the WLAN, this approach offers increased robustness against deep fades and provides the system with improved channel tracking capabilities.

4.2.2 *Pilot distances in time and frequency directions*

A fundamental issue in the design of the pilot grid is the determination of the time and frequency distances between adjacent pilots. These parameters are strictly related to the rapidity of channel fluctuations in both the time- and frequency-domains, and their selection is driven by the two-dimensional sampling theorem.

Let $f_{D,max}$ be the maximum expected Doppler frequency and assume that, at any given frequency $\overline{f}$, the channel response $H(\overline{f}, t)$ can be modeled in the time direction as a narrow-band stochastic process whose power spectral density is confined within the interval $[-f_{D,max}, f_{D,max}]$. Then, from the sampling theorem we know that the distance $\Delta_{p,t}$ (measured in OFDM blocks) between neighboring pilots in the time-domain must satisfy the inequality

$$\Delta_{p,t} \leq \lceil \frac{1}{2f_{D,max}T_B} \rceil, \tag{4.11}$$

where $T_B = N_T T_s$ is the length of the OFDM block (including the cyclic prefix) and $\lceil x \rceil$ is the largest integer not exceeding x.

On the other hand, at any given instant $\overline{t}$, the rate of variation of $H(f, \overline{t})$ with respect to f is related to the channel delay spread or, equivalently, to the length of the channel impulse response (CIR) $h(\tau, \overline{t})$ over the τ-axis. Thus, assuming that $h(\tau, \overline{t})$ has support $[0, \tau_{max}]$, the frequency spacing between pilots is subject to the following constraint

$$\Delta_{p,f} \leq \left\lceil \frac{1}{\tau_{max} f_{cs}} \right\rceil, \tag{4.12}$$

where $\Delta_{p,f}$ is normalized to the subcarrier spacing $f_{cs} = 1/(NT_s)$. A practical criterion for the design of $\Delta_{p,t}$ and $\Delta_{p,f}$ is to fix them to approximately one-half of their maximum allowable values given in Eqs. (4.11) and (4.12). This approach corresponds to two-times oversampling of $H(f,t)$ and helps to relax the requirements of the interpolation filters used for channel estimation.

The optimal arrangement of pilot symbols in both the time and frequency directions has extensively been studied in the literature [36,93,106]. One major result is that in many cases a uniform pilot distribution represents a good choice as it maximizes the channel estimation accuracy for a given number of pilots.

Example 4.1 In this example we evaluate the maximum time and frequency distances among pilots in the DAB system. We consider a typical urban (TU) channel with $\tau_{max} = 5$ μs and $f_{D,max} = 180$ Hz, which corresponds to a mobile speed of approximately 100 km/h if the carrier frequency is fixed to 2 GHz. The subcarrier spacing is $f_{cs} = 992$ Hz while the duration of the OFDM block is $T_B = 1.3$ ms. Substituting these parameters into Eqs. (4.11) and (4.12) produces

$$\Delta_{p,t} \leq \left\lceil \frac{1}{2 \times 180 \times 1.3 \times 10^{-3}} \right\rceil = 2, \tag{4.13}$$

and

$$\Delta_{p,f} = \left\lceil \frac{1}{5 \times 10^{-6} \times 992} \right\rceil = 201. \tag{4.14}$$

Actually, the pilot arrangement specified in the DAB system is characterized by $\Delta_{p,t} = 1$ and $\Delta_{p,f} = 12$, as shown in Fig. 4.3 (b). This means that, in principle, the DAB system can correctly operate in multipath environments with delay spreads much larger than 5 μs and with user terminals moving at speeds greater than 100 km/h.

4.2.3 *Pilot-aided channel estimation*

Channel estimation by means of scattered pilots is normally accomplished in two successive steps. Let i' and n' be the coordinates of the pilot positions in the time/frequency grid of Fig. 4.3 (a) or (b), and denote $\mathcal{P}$ the set of all ordered pairs (i', n'). Then, in the first step an estimate $\widetilde{H}_{i'}(n')$

of the channel transfer function is computed for each pair $(i', n') \in \mathcal{P}$ by exploiting the corresponding DFT output $R_{i'}(n')$. During the second step, the quantities $\widetilde{H}_{i'}(n')$ are interpolated in some way to obtain channel state information over data-bearing subcarriers.

One simple method to compute $\widetilde{H}_{i'}(n')$ results from application of the least-squares (LS) approach to the signal model Eq. (4.1). This produces

$$\widetilde{H}_{i'}(n') = \frac{R_{i'}(n')}{c_{i'}(n')}, \qquad \text{for} \quad (i', n') \in \mathcal{P} \tag{4.15}$$

where $c_{i'}(n')$ is the corresponding pilot symbol. Substituting Eq. (4.1) into Eq. (4.15) yields

$$\widetilde{H}_{i'}(n') = H_{i'}(n') + \frac{W_{i'}(n')}{c_{i'}(n')}, \tag{4.16}$$

from which it follows that $\widetilde{H}_{i'}(n')$ is unbiased with variance σ_w^2/σ_p^2, where $\sigma_p^2 = |c_{i'}(n')|^2$ is the pilot power. If information about the channel covariance matrix and noise power is available, channel estimation at the pilot positions can be performed according to the MMSE optimality criterion. Compared to the LS solution in Eq. (4.15), the MMSE approach is expected to achieve better performance at the price of higher complexity. The latter is somewhat reduced by resorting to low-rank techniques available in the literature [37].

As mentioned previously, channel estimates over information-bearing subcarriers are obtained by suitable interpolation of the quantities $\widetilde{H}_{i'}(n')$. Two alternative approaches can be adopted for this purpose. The first one is based on two-dimensional (2D) filtering in both the time and frequency directions. This technique provides optimum performance at the expense of heavy computational load [54]. A better trade-off between complexity and estimation accuracy is achieved by the second approach, where the 2D interpolator is replaced by the cascade of two one-dimensional (1D) filters working sequentially and performing independent interpolations in the time- and frequency-domains. The design of 2D and 1D interpolating filters is discussed hereafter under some specified optimality criterions.

4.2.4 *2D Wiener interpolation*

With 2D Wiener filtering, the estimated channel frequency response over the nth subcarrier of the ith OFDM block is given by

$$\widehat{H}_i(n) = \sum_{(i', n') \in \mathcal{P}} q(i, n; i', n')\widetilde{H}_{i'}(n'), \tag{4.17}$$

where $\tilde{H}_{i'}(n')$ is the channel estimate at the pilot position $(i', n') \in \mathcal{P}$ as given in Eq. (4.15), while $\{q(i, n; i', n')\}$ are suitable coefficients minimizing the mean-square channel estimation error

$$J_i(n) = \mathrm{E}\left\{ \left| \hat{H}_i(n) - H_i(n) \right|^2 \right\}. \tag{4.18}$$

Equation (4.17) can be rewritten in matrix form as

$$\hat{H}_i(n) = \boldsymbol{q}^T(i, n)\widetilde{\boldsymbol{H}}, \tag{4.19}$$

where $\boldsymbol{q}(i, n)$ and $\widetilde{\boldsymbol{H}}$ are column vectors of dimension N_p equal to the cardinality of $\mathcal{P}$ and collect the quantities $q(i, n; i', n')$ and $\tilde{H}_{i'}(n')$, respectively. From the orthogonality principle [123], we know that $J_i(n)$ achieves its global minimum when the error $\hat{H}_i(n) - H_i(n)$ is orthogonal to the observations $\tilde{H}_{i'}(n')$ for each pair $(i', n') \in \mathcal{P}$, i.e.,

$$\mathrm{E}\left\{ \left[\hat{H}_i(n) - H_i(n) \right] \widetilde{\boldsymbol{H}}^H \right\} = \boldsymbol{0}^T. \tag{4.20}$$

Substituting Eq. (4.19) into Eq. (4.20) leads to the following set of Wiener–Hopf equations

$$\boldsymbol{q}^T(i, n)\boldsymbol{R}_{\tilde{H}} = \boldsymbol{\theta}^T(i, n), \tag{4.21}$$

where $\boldsymbol{R}_{\tilde{H}} = E\{\widetilde{\boldsymbol{H}}\widetilde{\boldsymbol{H}}^H\}$ is the autocorrelation matrix of $\widetilde{\boldsymbol{H}}$ while $\boldsymbol{\theta}^T(i, n) = E\{H_i(n)\widetilde{\boldsymbol{H}}^H\}$. The entries of $\boldsymbol{R}_{\tilde{H}}$ are given by $R_{\tilde{H}}(i'', n''; i', n') = E\{\tilde{H}_{i''}(n'')\tilde{H}_{i'}^*(n')\}$ with both (i'', n'') and (i', n') belonging to $\mathcal{P}$, while $\boldsymbol{\theta}^T(i, n)$ is a row-vector with elements $\theta(i, n; i', n') = E\{H_i(n)\tilde{H}_{i'}^*(n')\}$.

Bearing in mind Eq. (4.16) and assuming that the channel response and thermal noise are statistically independent, we may write

$$R_{\tilde{H}}(i'', n''; i', n') = R_H(i'', n''; i', n') + \frac{\sigma_w^2}{\sigma_p^2} \cdot \delta(i'' - i')\delta(n'' - n'), \quad (4.22)$$

and

$$\theta(i, n; i', n') = R_H(i, n; i', n'), \tag{4.23}$$

where $\delta(\ell)$ is the Kronecker delta function and $R_H(i, n; j, m) = E\{H_i(n)H_j^*(m)\}$ the two-dimensional channel autocorrelation function. In [90] it is shown that for a typical mobile wireless channel $R_H(i, n; j, m)$ can be separated into the multiplication of a time-domain correlation $R_t(\cdot)$ by a frequency-domain correlation $R_f(\cdot)$, i.e.,

$$R_H(i, n; j, m) = R_t(i - j) \cdot R_f(n - m). \tag{4.24}$$

Clearly, $R_f(\cdot)$ depends on the multipath delay spread and power delay profile, while $R_t(\cdot)$ is related to the vehicle speed or, equivalently, to the Doppler frequency.

The optimum interpolating coefficients for the estimation of $H_i(n)$ are computed from Eq. (4.21) and read

$$\boldsymbol{q}^T(i,n) = \boldsymbol{\theta}^T(i,n)\boldsymbol{R}_{\tilde{\boldsymbol{H}}}^{-1}. \qquad (4.25)$$

A critical issue in 2D Wiener filtering is the inversion of the N_p - dimensional matrix $\boldsymbol{R}_{\tilde{\boldsymbol{H}}}$, which may be prohibitively complex for large N_p values. Also, computing $\boldsymbol{R}_{\tilde{\boldsymbol{H}}}$ and $\boldsymbol{\theta}(i,n)$ requires information about the channel statistics and noise power, which are typically unknown at the receiver. One possible strategy is to derive suitable estimates of these parameters, which are then used in Eqs. (4.22) and (4.23) in place of their true values. In general, this approach provides good results but requires the on-time inversion of $\boldsymbol{R}_{\tilde{\boldsymbol{H}}}$.

An alternative method relies on some a-priori assumptions about the channel statistics and optimizes the filter coefficients for specified values of the noise power and channel correlation functions. In practice, the Wiener coefficients are often designed for a uniform Doppler spectrum and power delay profile [90]. This amounts to assuming a wireless channel with the following time- and frequency-correlation functions

$$R_t(i) = \operatorname{sinc}\left(2\overline{f}_D i T_B\right), \qquad (4.26)$$

and

$$R_f(n) = \operatorname{sinc}\left(n f_{cs}\overline{\tau}\right) e^{-j\pi n f_{cs}\overline{\tau}}, \qquad (4.27)$$

in which $\overline{f}_D$ and $\overline{\tau}$ are conservatively chosen a bit larger than the maximum expected Doppler frequency and multipath delay spread, respectively.

This approach leads to a significant reduction of complexity because the filter coefficients are now pre-computed and stored in the receiver. Clearly, the price for this simplification is a certain degradation of the system performance due to a possible mismatch between the assumed operating parameters and their actual values. However, theoretical analysis and numerical results indicate that the mismatching effect is tolerable if the interpolating coefficients are designed on the basis of the autocorrelation functions given in Eqs. (4.26) and (4.27).

4.2.5 *Cascaded 1D interpolation filters*

A simple method to avoid the complexity of 2D Wiener filtering is based on the use of two cascaded 1D filters which perform independent interpolation

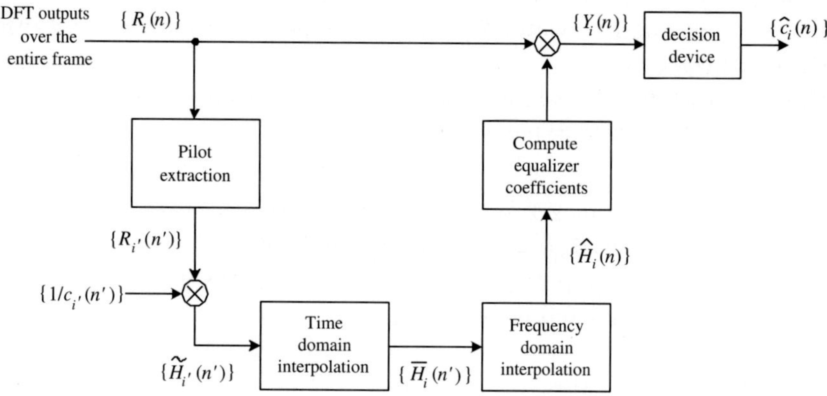

Fig. 4.4   A typical equalizer structure with two-cascaded 1D interpolation filters.

in the time and frequency directions. This idea is illustrated in Fig. 4.4, where interpolation in the time-domain precedes that in the frequency-domain, even though the opposite ordering could be used as well due to the linearity of the filters. Regardless of the actual filtering order, the essence of the first interpolation is to compute channel estimates over some specific data subcarriers that are subsequently used as additional pilots for the second interpolation stage.

Consider a specific subcarrier n' (represented by a column in the time-frequency grids of Fig. 4.3) and assume that the latter conveys pilot symbols over a number $N_{p,t}$ of OFDM blocks specified by the indices $i' \in \mathcal{P}_t(n')$. For example, the WLAN pilot arrangement of Fig. 4.3 (a) results into $\mathcal{P}_t(n') = \{1, 2, 3, \ldots\}$ for $n' = \pm 7$ or ± 21 and $\mathcal{P}_t(n') = \varnothing$ for the remaining subcarriers. In the DAB/DVB system of Fig. 4.3 (b) we have $\mathcal{P}_t(n') = \varnothing$ if n' is not multiple of three while $\mathcal{P}_t(3m') = \{|m'|_4 + 4\ell\}$, where m' and ℓ are non-negative integers and $|m'|_4$ denotes the remainder of the ratio $m'/4$.

As indicated in Fig. 4.4, pilot tones are extracted from the DFT output and used to compute the quantities $\{\widetilde{H}_{i'}(n')\}$ specified in Eq. (4.15). The latter are then interpolated by the time-domain filter to obtain the following channel estimates over the n'th subcarrier of *each* OFDM block $(i = 1, 2, \ldots)$

$$\overline{H}_i(n') = \sum_{i' \in \mathcal{P}_t(n')} q_t(i; i', n')\widetilde{H}_{i'}(n'), \qquad n' \in \mathcal{P}_f \qquad (4.28)$$

where $q_t(i; i', n')$ are suitable coefficients designed according to some optimality criterion while the set $\mathcal{P}_f$ collects the indices of pilot-bearing subcarriers and has cardinality $N_{p,f}$. Clearly, $\mathcal{P}_f = \{\pm 7, \pm 21\}$ in Fig. 4.3 (a) while $\mathcal{P}_f = \{0, 3, 6, \ldots\}$ in Fig. 4.3 (b).

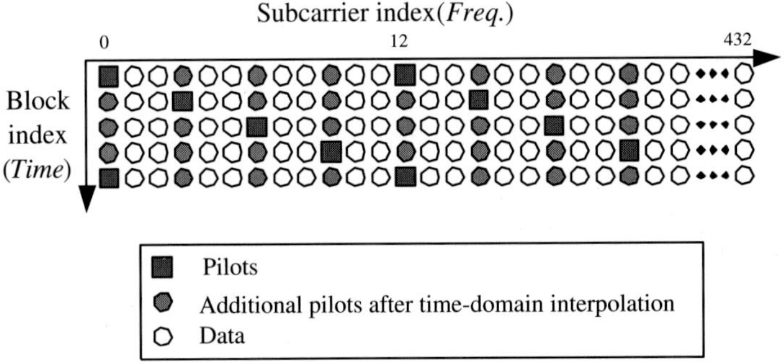

Fig. 4.5   Increase of effective pilots after time-domain interpolation.

Figure 4.5 illustrates the position of the time-interpolated channel estimates $\overline{H}_i(n')$ in the DAB frame. As mentioned previously, these quantities are viewed by the second interpolation filter as additional pilots, and used to obtain the channel transfer function over the entire time-frequency grid. In particular, the estimate of $H_i(n)$ is computed as

$$\widehat{H}_i(n) = \sum_{n' \in \mathcal{P}_f} q_f(n; n')\overline{H}_i(n'), \qquad (4.29)$$

where the weights $q_f(n; n')$ are independent of the time index i and, accordingly, are the same over all OFDM blocks. Popular approaches for designing the filtering coefficients $q_t(i; i', n')$ and $q_f(n; n')$ are discussed hereafter.

4.2.5.1 *Cascaded 1D Wiener interpolators*

Wiener interpolators are based on the MMSE optimality criterion. Specifically, for a given n' the coefficients $\boldsymbol{q}_t(i, n') = \{q_t(i; i', n'); \ i' \in \mathcal{P}_t(n')\}$ of the time-domain Wiener filter are designed so as to minimize the following MSE:

$$J_i(n') = \mathrm{E}\left\{\left|\overline{H}_i(n') - H_i(n')\right|^2\right\}, \qquad (4.30)$$

with $\overline{H}_i(n')$ as given in Eq. (4.28). After invoking the orthogonality principle, we find that

$$\boldsymbol{q}_t^T(i, n') = \boldsymbol{\theta}_t^T(i, n')\boldsymbol{R}_t^{-1}, \tag{4.31}$$

where $\boldsymbol{\theta}_t(i, n')$ is a column vector of length $N_{p,t}$ whose entries are related to the time-domain channel correlation function $R_t(\cdot)$ by

$$[\boldsymbol{\theta}_t(i, n')]_{i'} = R_t(i - i'), \qquad i' \in \mathcal{P}_t(n') \tag{4.32}$$

while $\boldsymbol{R}_t$ is a matrix of order $N_{p,t}$ with elements

$$[\boldsymbol{R}_t]_{i'',i'} = R_t(i'' - i') + \frac{\sigma_w^2}{\sigma_p^2} \cdot \delta(i'' - i'), \qquad i'', i' \in \mathcal{P}_t(n'). \tag{4.33}$$

It is worth noting that $\boldsymbol{R}_t$ is independent of n' and i, whereas $\boldsymbol{\theta}_t(i, n')$ may depend on n' through $i' \in \mathcal{P}_t(n')$. However, if the pilot arrangement is such that the same set $\mathcal{P}_t(n')$ is used for each $n' \in \mathcal{P}_f$ as in Fig. 4.3 (a), vector $\boldsymbol{\theta}_t(i, n')$ becomes independent of n' and the same occurs to the filter coefficients in Eq. (4.31). This property is clearly appealing because in such a case the same set of time-interpolation coefficients are used over all subcarriers $n' \in \mathcal{P}_f$, thereby reducing the computational effort and storage requirement of the channel estimation unit.

The orthogonality principle is also used to obtain the interpolation coefficients $\boldsymbol{q}_f(n) = \{q_f(n; n'); \; n' \in \mathcal{P}_f\}$ of the frequency-domain Wiener filter. This yields

$$\boldsymbol{q}_f^T(n) = \boldsymbol{\theta}_f^T(n)\boldsymbol{R}_f^{-1}, \tag{4.34}$$

where $\boldsymbol{\theta}_f(n)$ is a vector of length $N_{p,f}$ and $\boldsymbol{R}_f$ a matrix of the same order. Their entries are related to the frequency-domain channel correlation function $R_f(\cdot)$ by

$$\left[\boldsymbol{\theta}_f^T(n)\right]_{n'} = R_f(n - n'), \qquad n' \in \mathcal{P}_f \tag{4.35}$$

and

$$[\boldsymbol{R}_f]_{n'',n'} = R_f(n'' - n') + \frac{\sigma_w^2}{\sigma_p^2} \cdot \delta(n'' - n'), \qquad n'', n' \in \mathcal{P}_f. \tag{4.36}$$

Although much simpler than 2D Wiener filtering, the use of two-cascaded 1D Wiener interpolators may still be impractical for a couple of reasons. The first one is the dependence of the filtering coefficients on the channel statistics and noise power. As discussed previously, a robust filter design based on the sinc-shaped autocorrelation functions in Eqs. (4.26) and (4.27) can mitigate this problem to some extent. The second difficulty is that time-domain Wiener interpolation cannot be started until *all* blocks carrying pilot symbols have been received. This results into a significant filtering delay, which may be intolerable in many practical applications. A possible solution to this problem is offered by piecewise polynomial interpolation, as it is now discussed.

4.2.5.2 *Cascaded 1D polynomial-based interpolators*

The concept of piecewise polynomial interpolation is extensively covered in the digital signal processing literature [28, 136]. One of the main conclusions is that excellent interpolators can be implemented with a small number of taps, say either two or three. The limited amount of complexity associated with polynomial-based filters makes them particularly attractive in a number of applications. In the ensuing discussion, they are applied to OFDM systems in order to find practical schemes for interpolating channel estimates in both the time- and frequency-domains [132].

For illustration purposes, we concentrate on the DAB pilot arrangement of Fig. 4.3 (b) and observe that, for any given pilot-bearing subcarrier with index $n' \in \mathcal{P}_f = \{0, 3, 6, \ldots\}$, two neighboring pilots are separated in the time direction by three OFDM blocks. In other words, if a pilot is present on the n'th subcarrier of the i'th block, the next pilot on the same subcarrier will not be available until reception of the $(i' + 4)$th block.

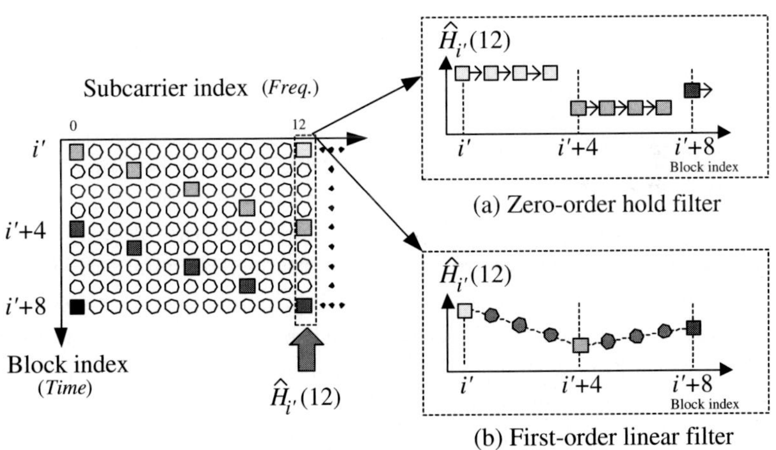

Fig. 4.6   Time-domain interpolation by means of (a) zero-order and (b) first-order polynomial filters.

The simplest form of piecewise polynomial interpolation is represented by the zero-order hold filter. When applied in the time direction over the n'th subcarrier, this filter receives a channel estimate $\widetilde{H}_{i'}(n')$ and keeps it fixed until the arrival of the next pilot. Mathematically, we have

$$\overline{H}_i(n') = \widetilde{H}_{i'}(n'), \quad \text{for } i' \le i \le i' + p_t - 1 \ \text{ and } \ n' \in \mathcal{P}_f \qquad (4.37)$$

where $p_t = 4$ is the time-distance between adjacent pilots. The concept of time-domain zero-order interpolation is illustrated in Fig. 4.6 (a) for $n' = 12$. This technique does not introduce any filtering delay but can only be used in those applications where the channel transfer function $H_i(n)$ keeps almost unchanged between adjacent pilots. Channel variations occurring in high-mobility systems are better handled by first-order interpolation. In this case $\overline{H}_i(n')$ varies in a piecewise-linear fashion as depicted in Fig. 4.6 (b), and is computed as

$$\overline{H}_i(n') = \frac{1}{p_t}\left[(p_t + i' - i)\,\widetilde{H}_{i'}(n') + (i' - i)\,\widetilde{H}_{i'+p_t}(n')\right], \qquad (4.38)$$

for $i' \le i \le i' + p_t - 1$ and $n' \in \mathcal{P}_f$.

Intuitively speaking, first-order interpolation is expected to provide more accurate estimates than zero-order filtering. However, it results into an inherent filtering delay since the estimate $\overline{H}_i(n')$ in Eq. (4.38) cannot be computed before reception of the $(i' + p_t)$th OFDM block. Polynomial filters based on second or higher order interpolation provide even better performance at the price of increased delays. For this reason, they are rarely used in practice.

The idea of piecewise polynomial filtering can also be applied in the frequency direction to obtain final channel estimates $\widehat{H}_i(n)$. Contrarily to time-domain interpolation, however, in this case the filtering delay is not a critical issue. The reason is that the frequency-domain interpolator operates on a block-by-block basis, so that in principle the quantities $\overline{H}_i(n')$ are filtered as soon as the ith OFDM block has been received. It follows that low-order filters with a small number of taps are not strictly necessary for frequency-domain interpolation. More sophisticated schemes based on LS reasoning can be resorted to as it is now illustrated.

4.2.5.3 *LS-based interpolation in frequency domain*

The quantity $\overline{H}_i(n')$ produced by the time-domain interpolation filter are modeled as

$$\overline{H}_i(n') = H_i(n') + \overline{W}_i(n'), \qquad n' \in \mathcal{P}_f \qquad (4.39)$$

where $\overline{W}_i(n')$ is a disturbance term that accounts for thermal noise and possible interpolation errors. We denote $\boldsymbol{h}_i = [h_i(0), h_i(1), \ldots, h_i(L-1)]^T$

the T_s-spaced samples of the CIR during the ith OFDM block, and recall that the channel transfer function is obtained by taking the DFT of $\boldsymbol{h}_i$, *i.e.*,

$$H_i(n) = \sum_{\ell=0}^{L-1} h_i(\ell)\, e^{-j2\pi n\ell/N}. \tag{4.40}$$

Substituting Eq. (4.40) into Eq. (4.39) produces

$$\overline{\boldsymbol{H}}_i = \overline{\boldsymbol{F}}\boldsymbol{h}_i + \overline{\boldsymbol{W}}_i, \tag{4.41}$$

where $\overline{\boldsymbol{H}}_i$ and $\overline{\boldsymbol{W}}_i$ are $N_{p,f}$-dimensional vectors with elements $\overline{H}_i(n')$ and $\overline{W}_i(n')$, respectively, while $\overline{\boldsymbol{F}} \in \mathbb{C}^{N_{p,f} \times L}$ is a matrix with entries $e^{-j2\pi n'\ell/N}$ for $0 \le \ell \le L - 1$ and $n' \in \mathcal{P}_f$. The quantities $\overline{\boldsymbol{H}}_i$ in Eq. (4.41) are now exploited to derive an estimate of $\boldsymbol{h}_i$. For this purpose, we adopt a LS approach and obtain

$$\widehat{\boldsymbol{h}}_i = (\overline{\boldsymbol{F}}^H \overline{\boldsymbol{F}})^{-1} \overline{\boldsymbol{F}}^H \overline{\boldsymbol{H}}_i. \tag{4.42}$$

Note that a *necessary* condition for the invertibility of $\overline{\boldsymbol{F}}^H \overline{\boldsymbol{F}}$ in Eq. (4.42) is that $N_{p,f} \ge L$. This amounts to saying that the number of pilots in the frequency direction cannot be less than the number of channel taps, otherwise the observations $\{\overline{H}_i(n')\}$ are not sufficient to estimate all unknown parameters $\{h_i(\ell)\}$.

From Eq. (4.40), an estimate of the channel transfer function is obtained as

$$\widehat{H}_i(n) = \sum_{\ell=0}^{L-1} \widehat{h}_i(\ell)\, e^{-j2\pi n\ell/N}, \quad 0 \le n \le N - 1. \tag{4.43}$$

After substituting Eq. (4.42) into Eq. (4.43), we get the final channel estimate in the form

$$\widehat{H}_i(n) = \sum_{n' \in \mathcal{P}_f} q_f^{LS}(n; n')\overline{H}_i(n'), \tag{4.44}$$

where the LS coefficients $q_f^{LS}(n; n')$ are given by

$$q_f^{LS}(n; n') = \sum_{\ell_1=0}^{L-1}\sum_{\ell_2=0}^{L-1} \left[(\boldsymbol{F}^H \boldsymbol{F})^{-1}\right]_{\ell_1,\ell_2} e^{j2\pi(n'\ell_2 - n\ell_1)/N}. \tag{4.45}$$

In [101] it is shown that the accuracy of the estimator Eq. (4.44) is optimized when the pilot symbols are uniformly spaced in the frequency-domain with a separation interval $\Delta_{p,f} = N/N_{p,f}$. In this case $\overline{\boldsymbol{F}}^H \overline{\boldsymbol{F}} = N_{p,f} \cdot \boldsymbol{I}_L$ and the filtering coefficients in Eq. (4.45) take the form

$$q_f^{LS}(n; n') = \frac{1}{N_{p,f}} e^{j\pi(L-1)(n'-n)/N} \frac{\sin[\pi L (n' - n) N]}{\sin[\pi (n' - n)/N]}. \tag{4.46}$$

It is worth noting that in many commercial systems a specified number of subcarriers at both edges of the signal spectrum are left unmodulated (virtual or null subcarriers) so as to reduce out-of-band emission. If this number is greater than $N/N_{p,f}$, a uniform distribution of pilots in the frequency-domain is not possible. In this case, the optimum pilots' positions can only be determined through a numerical search. Simulation results reported in [101] indicate that in the presence of virtual subcarriers (VCs) it is convenient to adopt a non-uniform pilot arrangement with a smaller separation distance in the neighborhood of the spectrum edges. An alternative method is depicted in Fig. 4.7. Here, the transmitter inserts uniformly spaced pilots only within the signal spectrum while leaving the suppressed bandwidth empty. At the receiving terminal, the pilot symbols closest to the spectrum boundaries are artificially duplicated over the suppressed bandwidth and used by the interpolation filters as if they were regular pilots. Clearly, this approach is more practical then using non-uniformly spaced pilots, even though channel estimates in the vicinity of the suppressed bandwidth are expected to be less accurate than those in the middle of the signal spectrum.

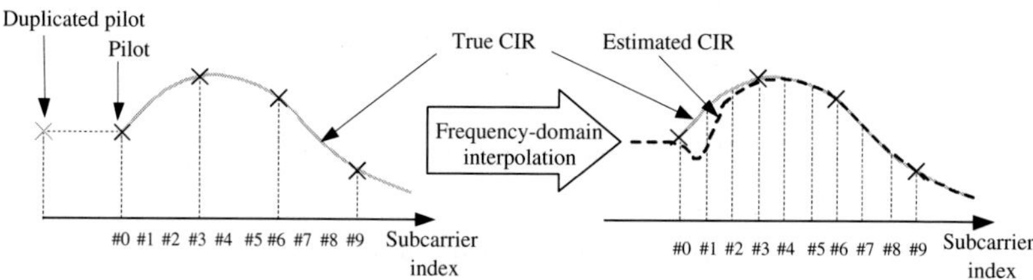

Fig. 4.7   Channel estimation in the vicinity of suppressed carriers.

In a sparse multipath environment where only a few multipath components are present with relatively large differential delays, most of the CIR coefficients $h_i(\ell)$ are expected to be vanishingly small. In such a scenario, the accuracy of the LS estimator can be improved by adopting a parametric channel model characterized by a reduced number of unknown parameters. This approach is suggested in [179], where the minimum description length (MDL) criterion [169] is employed to detect the number of paths in the channel. After recovering the path delays through rotational invariant techniques (ESPRIT) [135], estimates of the path gains are eventually obtained using LS or MMSE methods.

4.3 Advanced techniques for blind and semi-blind channel estimation

The insertion of pilot symbols into the transmitted data stream simplifies the channel estimation task to a large extent, but inevitably reduces the spectral efficiency of the communication system. This problem has inspired considerable interest in blind or semi-blind channel estimation techniques where only a few pilots are required. These schemes are largely categorized into subspace-based or decision-directed (DD) methods. In the former case, the intrinsic redundancy provided by the cyclic prefix (CP) or by VCs is exploited as a source of channel state information. A good sample of the results obtained in this area are found in [86, 103, 167] and references therein. Although attractive because of the considerable saving in training overhead, the subspace approach is effective as long as a large amount of data is available for channel estimation. This is clearly a disadvantage in high-mobility applications, since in this case the time-varying channel might preclude accumulation of a large data record.

In DD methods, tentative data decisions are exploited in addition to a few pilots to improve the channel estimation accuracy. An example of this idea is presented in [91], where trellis decoding is employed for joint equalization and data detection of differentially-encoded PSK signals. Differential encoding is performed in the frequency direction while trellis decoding is efficiently implemented through a standard Viterbi processor. The latter operates in a per-survivor fashion [128] wherein a separate channel estimate is computed for each surviving path.

The idea of exploiting data decisions to improve the channel estimation accuracy is also the rationale behind EM-based methods [102, 176]. These schemes operate in an iterative mode with channel estimates at a given step being derived from symbol decisions obtained at the previous step. In this way, data detection and channel estimation are no longer viewed as separate tasks but, rather, are coupled together and accomplished in a joint fashion. Other blind approaches for channel estimation in OFDM systems exploit either the cyclostationarity property induced by the CP on the received time-domain samples [70] or the fact that the information-bearing symbols belong to a finite alphabet set [183].

It is fair to say that strictly blind channel estimation techniques exploiting no pilots at all are hardly usable in practice as they are plagued by an inherent scalar ambiguity. This amounts to saying that, even in the absence of noise and/or interference, the channel response can only be estimated up to a complex-valued factor. The only way to solve the ambiguity is to insert a few pilot symbols into the transmitted blocks in order to provide a phase reference for the receiving terminal. The use of pilots in combination with blind algorithms results into semi-blind schemes with improved estimation accuracy. Compared to the pilot-aided methods discussed previously, the semi-blind approach suffers from some drawbacks in terms of computational complexity and prolonged acquisition time.

4.3.1 *Subspace-based methods*

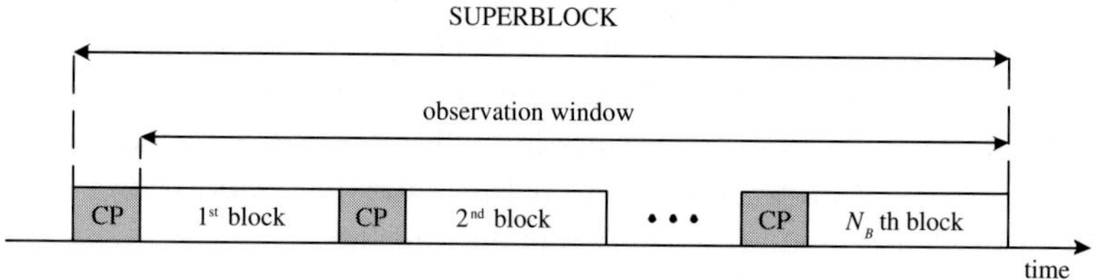

Fig. 4.8   Observation of a superblock for subspace-based channel estimation.

Subspace-based methods derive channel information from the inherent redundancy introduced in the transmitted signal by the use of the CP and/or VCs. To explain the basic idea behind this class of blind estimation techniques, we define a *superblock* as the concatenation of N_B successive OFDM blocks, where N_B is a suitably chosen design parameter. As depicted in Fig. 4.8, at the receiver side the observation window spans an entire superblock, except for the CP of the first OFDM block which is intentionally discarded to avoid IBI from the previously transmitted superblock. The total number of time-domain samples falling within the kth observation window is thus $M_T = N_B N_T - N_g$. These samples are arranged into a vector

$$\boldsymbol{r}(k) = \boldsymbol{s}_R(k) + \boldsymbol{w}(k), \tag{4.47}$$

where $\boldsymbol{s}_R(k)$ is the signal component while $\boldsymbol{w}(k)$ accounts for thermal noise. We assume that some VCs are present in the signal spectrum, so that only P subcarriers out of a total of N are actually employed for data transmission. This means that each superblock conveys $N_B P$ data symbols, which are collected into a vector $c(k)$. Hence, we can rewrite $s_R(k)$ in the form

$$s_R(k) = G(h)c(k), \qquad (4.48)$$

where $h = [h(0), h(1), \ldots, h(L-1)]^T$ is the CIR vector (assumed static for simplicity) while $G(h) \in \mathbb{C}^{M_T \times N_B P}$ is a *tall* matrix whose entries depend on the indices of the modulated subcarriers and are also *linearly* related to h. It is worth noting that the mapping $c(k) \longrightarrow s_R(k)$ in Eq. (4.48) can be interpreted as a sort of coding scheme wherein $G(h)$ is the code generator matrix and the introduced redundancy is proportional to the difference between the dimensions of $s_R(k)$ and $c(k)$, say $N_r = M_T - N_B P$. This redundancy originates from the use of VCs and CPs, and can be exploited for the purpose of channel estimation as it is now explained.

Returning to Eq. (4.48), we observe that $s_R(k)$ is a linear combination of the columns of $G(h)$, each weighted by a given transmitted symbol. As a result, $s_R(k)$ belongs to the subspace of $\mathbb{C}^{M_T}$ spanned by the columns of $G(h)$, which is referred to as the *signal subspace*. If $G(h)$ is full-rank (an event which occurs with unit probability), the signal subspace has dimension $N_B P$. Its orthogonal complement in $\mathbb{C}^{M_T}$ is called the *noise subspace* and has dimension N_r. To proceed further, we consider the correlation matrix R_{rr} of the received vector $r(k)$. After substituting Eq. (4.48) into Eq. (4.47) we obtain

$$R_{rr} = V(h) + \sigma_w^2 I_{M_T}, \qquad (4.49)$$

where σ_w^2 is the noise power and $V(h) = G(h)R_{cc}G^H(h)$, with $R_{cc} = \mathrm{E}\{c(k)c^H(k)\}$ denoting the correlation matrix of the data vector. At this stage we observe that rank $\{V(h)\} = \min\{M_T, N_B P\} = N_B P$. This means that $V(h)$ has only $N_B P$ non-zero eigenvalues μ_j $(1 \le j \le N_B P)$ out of a total of M_T. Thus, from Eq. (4.49) it follows that the eigenvalues of R_{rr} (arranged in a decreasing order of magnitude) are given by

$$\lambda_j = \begin{cases} \mu_j + \sigma_w^2, & 1 \le j \le N_B P, \\ \sigma_w^2, & N_B P + 1 \le j \le M_T. \end{cases} \qquad (4.50)$$

A fundamental property of R_{rr} is that the set $U = \{u_1, u_2, \ldots, u_{N_r}\}$ of N_r eigenvectors associated to the smallest eigenvalues σ_w^2 constitute a basis for the noise subspace, while the remaining $N_B P$ eigenvectors lie in the signal subspace. Since the latter is spanned by the columns of $G(h)$ and is also orthogonal to the noise subspace (hence, to each vector u_j in the basis U), we may write

$$u_j^H G(h) = 0_{N_B P}^T, \quad 1 \le j \le N_r \qquad (4.51)$$

where $\mathbf{0}_{N_B P}$ is a column vector of $N_B P$ zeros.

Recalling that the entries of $\boldsymbol{G}(\boldsymbol{h})$ are related to the unknown channel vector $\boldsymbol{h}$ in a linear fashion, we may interpret the constraints Eq. (4.51) as a set of $N_r N_B P$ linear homogeneous equations in the variables $\{h(\ell)\}$. Hence, they can equivalently be rewritten as

$$\boldsymbol{h}^H \boldsymbol{B}(\boldsymbol{U}) = \mathbf{0}_{N_r N_B P}^T, \tag{4.52}$$

where $\boldsymbol{B}(\boldsymbol{U})$ is a suitable matrix of dimensions $L \times N_B P N_r$ whose entries depend on the basis $\boldsymbol{U}$ of the noise subspace. Solving the set of equations in Eq. (4.52) and discarding the trivial solution $\boldsymbol{h} = \mathbf{0}_L$ provides an estimate of the CIR vector up to a complex scaling factor.

From the above discussion it turns out that subspace-based methods rely on the decomposition of the observation space $\mathbb{C}^{M_T}$ into a signal subspace plus a noise subspace, and determine the channel estimate by exploiting the reciprocal orthogonality among them. This decomposition is performed over the correlation matrix $\boldsymbol{R}_{rr}$ which, however, is typically unknown. In practice, $\boldsymbol{R}_{rr}$ is replaced by the so-called *sample-correlation matrix*, which is obtained by averaging the received time-domain samples over a specified number K_B of superblocks, *i.e.*,

$$\widehat{\boldsymbol{R}}_{rr} = \frac{1}{K_B} \sum_{k=1}^{K_B} \boldsymbol{r}(k) \boldsymbol{r}^H(k). \tag{4.53}$$

The eigenvectors of $\widehat{\boldsymbol{R}}_{rr}$ associated with the N_r smallest eigenvalues are taken as an estimate $\widehat{\boldsymbol{U}}$ of the noise subspace, which is then used in Eq. (4.52) in place of the true $\boldsymbol{U}$. Under normal operating conditions, the set of linear equations $\boldsymbol{h}^H \boldsymbol{B}(\widehat{\boldsymbol{U}}) = \mathbf{0}_{N_r N_B P}^T$ has $\boldsymbol{h} = \mathbf{0}_L$ as unique solution. To overcome this problem, the equations are solved in the LS sense under an amplitude constraint $\|\boldsymbol{h}\| = 1$. This leads to the following minimization problem

$$\widehat{\boldsymbol{h}} = \underset{\|\tilde{\boldsymbol{h}}\|=1}{\arg\min} \left\{ \tilde{\boldsymbol{h}}^H \boldsymbol{B}(\widehat{\boldsymbol{U}}) \boldsymbol{B}^H(\widehat{\boldsymbol{U}}) \tilde{\boldsymbol{h}} \right\}, \tag{4.54}$$

where $\tilde{\boldsymbol{h}}$ represents a trial value of $\boldsymbol{h}$. The solution is well known and is attained by choosing $\widehat{\boldsymbol{h}}$ as the unit-norm eigenvector associated to the smallest eigenvalue of $\boldsymbol{B}(\widehat{\boldsymbol{U}}) \boldsymbol{B}^H(\widehat{\boldsymbol{U}})$.

In conclusion, we can summarize the subspace-based procedure as follows:

(1) observe a specified number K_B of superblocks and compute the sample correlation matrix $\widehat{\boldsymbol{R}}_{rr}$ as indicated in Eq. (4.53);

(2) determine the noise subspace by computing the N_r smallest eigenvalues of $\widehat{\boldsymbol{R}}_{rr}$. Arrange the corresponding eigenvectors into a set $\widehat{\boldsymbol{U}} = \{\widehat{\boldsymbol{u}}_1, \widehat{\boldsymbol{u}}_2, \ldots, \widehat{\boldsymbol{u}}_{N_r}\}$;

(3) use $\widehat{\boldsymbol{U}}$ to construct matrix $\boldsymbol{B}(\widehat{\boldsymbol{U}})$;

(4) compute the smallest eigenvalue of $\boldsymbol{B}(\widehat{\boldsymbol{U}})\boldsymbol{B}^H(\widehat{\boldsymbol{U}})$ and take the corresponding unit-norm eigenvector as an estimate $\widehat{\boldsymbol{h}}$ of the CIR vector.

For a given observation window, the accuracy of subspace-based methods increases with the amount of redundancy introduced by the use of CPs and/or VCs. In particular, simulation results shown in [86] indicate that enlarging the CP is more beneficial than increasing the number of VCs. As mentioned previously, a major drawback of this class of schemes is represented by the large number of blocks that are normally required to achieve the desired estimation accuracy.

4.3.2 *EM-based channel estimation*

In conventional OFDM systems with coherent detection, channel estimation and data decoding are normally kept as separate tasks. Albeit reasonable and easy to implement, this approach is not based over any optimality criterion. Better results are expected if the channel response and data symbols are jointly estimated under a maximum likelihood (ML) framework. Unfortunately, using this strategy over an entire OFDM frame is computationally unfeasible due to lack of efficient ways for maximizing the likelihood function over all candidate data sequences. This problem is alleviated if the receiver only exploits channel correlation in the frequency direction while neglecting any time correlation over adjacent OFDM blocks. In this way the equalization algorithm can operate on a block-by-block basis, with a substantial reduction of the number of candidate sequences. However, even with the adoption of this simplified approach, joint ML estimation of channel response and data symbols remains a challenging task as it is now shown.

4.3.2.1 *Likelihood function for joint data detection and channel estimation*

In the following derivations we focus on a single OFDM block and neglect the time index i for notational simplicity. The DFT output is given by

$$R(n) = H(n)c(n) + W(n), \quad 0 \le n \le N - 1 \tag{4.55}$$

where $H(n) = \sum_{\ell=0}^{L-1} h(\ell)\, e^{-j2\pi n\ell/N}$ and $\boldsymbol{h} = [h(0), h(1), \ldots, h(L-1)]^T$ collects the CIR coefficients. Denoting $\boldsymbol{R} = [R(0), R(1), \ldots, R(N-1)]^T$ the observation vector, we may rewrite Eq. (4.55) in matrix form as

$$\boldsymbol{R} = \boldsymbol{A}(\boldsymbol{c})\boldsymbol{F}\boldsymbol{h} + \boldsymbol{W}, \tag{4.56}$$

where $\boldsymbol{c} = [c(0), c(1), \ldots, c(N-1)]^T$ is the transmitted data sequence, $\boldsymbol{A}(\boldsymbol{c})$ is a diagonal matrix with $\boldsymbol{c}$ along its main diagonal and $\boldsymbol{F}$ is an $N \times L$ matrix with entries

$$[\boldsymbol{F}]_{n,\ell} = e^{-j2\pi n\ell/N}, \quad 0 \leq n \leq N-1, \quad 0 \leq \ell \leq L-1. \tag{4.57}$$

Vector $\boldsymbol{W}$ represents the noise contribution and is Gaussian distributed with zero-mean and covariance matrix $\sigma_w^2 \boldsymbol{I}_N$.

From Eq. (4.56), the likelihood function for the joint estimation of $\boldsymbol{c}$ and $\boldsymbol{h}$ is found to be

$$\Lambda(\widetilde{\boldsymbol{c}}, \widetilde{\boldsymbol{h}}) = \frac{1}{(\pi\sigma_w^2)^N} \exp\left\{ -\frac{1}{\sigma_w^2} \left\| \boldsymbol{R} - \boldsymbol{A}(\widetilde{\boldsymbol{c}})\boldsymbol{F}\widetilde{\boldsymbol{h}} \right\|^2 \right\}, \tag{4.58}$$

where $\widetilde{\boldsymbol{c}}$ and $\widetilde{\boldsymbol{h}}$ are trial values of $\boldsymbol{c}$ and $\boldsymbol{h}$, respectively. The ML estimates of the unknown vectors are eventually obtained looking for the location where $\Lambda(\widetilde{\boldsymbol{c}}, \widetilde{\boldsymbol{h}})$ achieves its global maximum, *i.e.*,

$$(\widehat{\boldsymbol{c}}, \widehat{\boldsymbol{h}}) = \arg\max_{(\widetilde{\boldsymbol{c}}, \widetilde{\boldsymbol{h}})} \left\{ \Lambda(\widetilde{\boldsymbol{c}}, \widetilde{\boldsymbol{h}}) \right\}. \tag{4.59}$$

4.3.2.2 *Likelihood function maximization by EM algorithm*

The maximum of $\Lambda(\widetilde{\boldsymbol{c}}, \widetilde{\boldsymbol{h}})$ in Eq. (4.58) can be found in two successive steps. First, we keep $\widetilde{\boldsymbol{c}}$ fixed and maximize with respect to $\widetilde{\boldsymbol{h}}$. This produces

$$\widehat{\boldsymbol{h}}(\widetilde{\boldsymbol{c}}) = [\boldsymbol{A}(\widetilde{\boldsymbol{c}})\boldsymbol{F}]^\dagger \boldsymbol{R}, \tag{4.60}$$

where $[\boldsymbol{A}(\widetilde{\boldsymbol{c}})\boldsymbol{F}]^\dagger = \left[\boldsymbol{F}^H \boldsymbol{A}^H(\widetilde{\boldsymbol{c}})\boldsymbol{A}(\widetilde{\boldsymbol{c}})\boldsymbol{F} \right]^{-1} \boldsymbol{F}^H \boldsymbol{A}^H(\widetilde{\boldsymbol{c}})$ is the Moore-Penrose generalized inverse of $\boldsymbol{A}(\widetilde{\boldsymbol{c}})\boldsymbol{F}$. After substituting Eq. (4.60) into Eq. (4.58) and letting $\widetilde{\boldsymbol{c}}$ vary, we see that maximizing Eq. (4.58) is equivalent to maximizing the following metric

$$g(\widetilde{\boldsymbol{c}}) = \Re\left\{ \boldsymbol{R}^H \boldsymbol{A}(\widetilde{\boldsymbol{c}})\boldsymbol{F} [\boldsymbol{A}(\widetilde{\boldsymbol{c}})\boldsymbol{F}]^\dagger \boldsymbol{R} \right\}. \tag{4.61}$$

Inspection of Eqs. (4.60) and (4.61) indicates that the estimates of $\boldsymbol{c}$ and $\boldsymbol{h}$ are decoupled in that the former can be computed first and is then exploited to get the latter. However, maximizing $g(\widetilde{\boldsymbol{c}})$ in Eq. (4.61) appears a formidable task. A certain simplification is possible if the data symbols belong to a PSK constellation. In this case we have $\boldsymbol{A}^H(\widetilde{\boldsymbol{c}})\boldsymbol{A}(\widetilde{\boldsymbol{c}}) = \boldsymbol{I}_N$, so that $[A(\widetilde{c})F]^{\dagger}$ reduces to $\left[F^H F\right]^{-1} F^H A^H(\widetilde{c})$. Observing that $F^H F = N \cdot I_N$, Eqs. (4.60) and (4.61) become

$$\widehat{h}(\widetilde{c}) = \frac{1}{N} F^H A^H(\widetilde{c}) R, \tag{4.62}$$

$$g(\widetilde{c}) = \frac{1}{N} \left\| R^H A(\widetilde{c}) F \right\|^2. \tag{4.63}$$

Unfortunately, the direct maximization of $g(\widetilde{c})$ in Eq. (4.63) is still intractable as it requires an exhaustive search over all possible data sequences $\widetilde{c}$, whose number grows exponentially with N. A possible way to overcome this obstacle is the use of the EM algorithm. Under some mild conditions, the latter can locate the global maximum of the likelihood function through an iterative procedure which is much simpler than the exhaustive search [34]. In the EM parlance, the observed measurements are replaced with some *complete data* from which the original measurements are obtained through a many-to-one mapping. At each iteration, the algorithm computes the expectation of the log-likelihood function for the complete data (E-step), which is next maximized with respect to the unknown parameters (M-step). Here, we follow the guidelines suggested in [102] and view the DFT output R as the incomplete data, whereas the complete data set is defined as the pair $\{R, h\}$. Under these assumptions, during the jth iteration the EM algorithm proceeds as follows [102]:

EM-based joint channel estimation and data detection

- **E-step**
  Compute

  $$Q\left(\widetilde{c} \middle| \widehat{c}^{(j-1)}\right) = \mathrm{E}_h \left\{ p\left(R \middle| h, \widehat{c}^{(j-1)}\right) \cdot \ln p\left(R \middle| h, \widetilde{c}\right) \right\}, \tag{4.64}$$

  where $\widehat{c}^{(j-1)}$ is the estimate of c at the $(j-1)$th step, $p(\cdot)$ is the probability density function (pdf) of the enclosed quantities and $\mathrm{E}_h\{\cdot\}$ indicates statistical expectation over the pdf of h.

- **M-step**
  Maximize $Q\left(\widetilde{c} \middle| \widehat{c}^{(j-1)}\right)$ over the set spanned by $\widetilde{c}$ to obtain data decisions in the form

  $$\widehat{c}^{(j)} = \arg\max_{\widetilde{c}} \left\{ Q\left(\widetilde{c} \middle| \widehat{c}^{(j-1)}\right) \right\}. \tag{4.65}$$

Assuming that h is Gaussian distributed with zero-mean (Rayleigh fading) and covariance matrix $C_h = \mathrm{E}\{hh^H\}$, after some manipulations it is found that Eq. (4.65) can equivalently be rewritten as [102]

$$\widehat{\boldsymbol{c}}^{(j)} = \arg\max_{\widetilde{\boldsymbol{c}}} \left\{ \Re \left[\boldsymbol{R}^H \boldsymbol{A}(\widetilde{\boldsymbol{c}}) \boldsymbol{F} \widehat{\boldsymbol{h}}_{MMSE}(\widehat{\boldsymbol{c}}^{(j-1)}) \right] \right\}, \qquad (4.66)$$

where

$$\widehat{\boldsymbol{h}}_{MMSE}(\widehat{\boldsymbol{c}}^{(j-1)}) = (N \cdot \boldsymbol{I}_N + \sigma_w^2 \boldsymbol{C}_h^{-1})^{-1} \boldsymbol{F}^H \boldsymbol{A}^H(\widehat{\boldsymbol{c}}^{(j-1)}) \boldsymbol{R} \qquad (4.67)$$

is the MMSE estimator of $\boldsymbol{h}$ as derived from the model Eq. (4.56) after replacing the true data vector $\boldsymbol{c}$ by its corresponding estimate $\widehat{\boldsymbol{c}}^{(j-1)}$. Denoting $\{\widehat{H}_{MMSE}(n, \widehat{\boldsymbol{c}}^{(j-1)})\}$ the N-point DFT of $\widehat{\boldsymbol{h}}_{MMSE}(\widehat{\boldsymbol{c}}^{(j-1)})$, we may rewrite Eq. (4.66) in the following way

$$\widehat{\boldsymbol{c}}^{(j)} = \arg\max_{\widetilde{\boldsymbol{c}}} \left\{ \sum_{n=0}^{N-1} \Re \left[R^*(n)\widetilde{c}(n)\widehat{H}_{MMSE}(n, \widehat{\boldsymbol{c}}^{(j-1)}) \right] \right\}. \qquad (4.68)$$

With uncoded transmissions, the above maximization is equivalent to maximizing each individual term in the sum, *i.e.*, making symbol-by-symbol decisions

$$\widehat{c}^{(j)}(n) = \arg\max_{\widetilde{c}(n)} \left\{ \Re \left[R^*(n)\widetilde{c}(n)\widehat{H}_{MMSE}(n, \widehat{\boldsymbol{c}}^{(j-1)}) \right] \right\}, \quad 0 \le n \le N-1$$

$$(4.69)$$

where $\widehat{c}^{(j)}(n)$ is the nth entry of $\widehat{\boldsymbol{c}}^{(j)}$.

Inspection of Eq. (4.69) reveals the physical rationale behind the EM algorithm. As is seen, at the jth iteration the estimate of $\boldsymbol{c}$ is computed through conventional frequency-domain detection/equalization techniques, where channel state information is achieved by means of the MMSE criterion using data decisions $\widehat{\boldsymbol{c}}^{(j-1)}$ from the previous iteration. Clearly, an initial estimate $\widehat{\boldsymbol{h}}^{(0)}$ of the channel vector is needed to initialize the iterative procedure. One possibility is to insert some pilots within each OFDM block and use them to compute $\widehat{\boldsymbol{h}}^{(0)}$ according to Eq. (4.42). Alternatively, the channel estimate obtained during the current OFDM block can be used in the next block for initialization purposes.

As indicated in Eq. (4.67), the MMSE channel estimator requires knowledge of the channel statistics and noise power. These quantities can be estimated on-time from the received samples as suggested in [102]. A simpler solution is found assuming high SNR values. In this case σ_w^2 is vanishingly small and $\widehat{\boldsymbol{h}}_{MMSE}(\widehat{\boldsymbol{c}}^{(j-1)})$ in Eq. (4.66) is thus replaced by the following LS estimate

$$\widehat{\boldsymbol{h}}_{LS}(\widehat{\boldsymbol{c}}^{(j-1)}) = \frac{1}{N} \boldsymbol{F}^H \boldsymbol{A}^H(\widehat{\boldsymbol{c}}^{(j-1)}) \boldsymbol{R}. \qquad (4.70)$$

Albeit simple, this approach is expected to incur some performance penalty with respect to the optimal solution Eq. (4.66).

4.4 Performance comparison

In this section we use computer simulations to compare the performance of some of the channel estimation techniques described throughout the chapter. In doing so we consider an OFDM system with $N = 256$ subcarriers and QPSK data symbols. The DAB/DVB pilot pattern of Fig. 4.3 (b) is employed to multiplex 16 scattered pilots in each OFDM block. The transmission channel is characterized by $N_p = 4$ multipath components. The path delays are kept fixed at $\tau_1 = 0$, $\tau_2 = 1.4T_s$, $\tau_3 = 4.8T_s$ and $\tau_4 = 9.7T_s$, while the path gains $\alpha_m(t)$ $(m = 1, 2, 3, 4)$ are modeled as statistically independent Gaussian random processes with zero-mean and autocorrelation function

$$R_m(\tau) = \sigma_m^2 J_0(2\pi f_D \tau). \qquad (4.71)$$

In the above equation, $J_0(x)$ denotes the zero-order Bessel function of the first kind, f_D is the Doppler frequency and $\sigma_m^2 = \mathrm{E}\{|\alpha_m(t)|^2\}$ the statistical power of $\alpha_m(t)$. We assume an exponentially-decaying power delay profile where

$$\sigma_m^2 = \beta e^{-m}, \qquad m = 1, 2, 3, 4 \qquad (4.72)$$

and parameter β is chosen so as to normalize the received signal power to unity.

The channel taps $h_i(\ell)$ are expressed by

$$h_i(\ell) = \sum_{m=1}^{4} \alpha_m(iT_B) g(\ell T_s - \tau_m), \qquad \ell = 0, 1, \ldots, L - 1 \qquad (4.73)$$

where $g(t)$ accounts for the signal shaping operated by the transmit and receive filters, and has a raised-cosine Fourier transform with roll-off 0.22. The Doppler frequency is $f_D = 10^{-2}/T_B$, while the channel length is $L = 16$. To prevent IBI, a CP of length $N_g = 16$ is appended to each block.

Figure 4.9 shows the BER performance as a function of E_s/N_0 for an uncoded QPSK transmission. The curve labeled "Ideal" refers to a system with perfect channel state information while the curves labeled "Two-cascaded 1D EQ" are obtained by performing zero-order or first-order 1D polynomial interpolation in the time-domain followed by 1D LS interpolation in the frequency-direction as indicated by Eq. (4.42). The EM-based equalizer is initialized with channel estimates provided by the two-cascaded 1D filters with first-order polynomial interpolation. We see that the first-order filter provides much better performance than zero-order interpolation

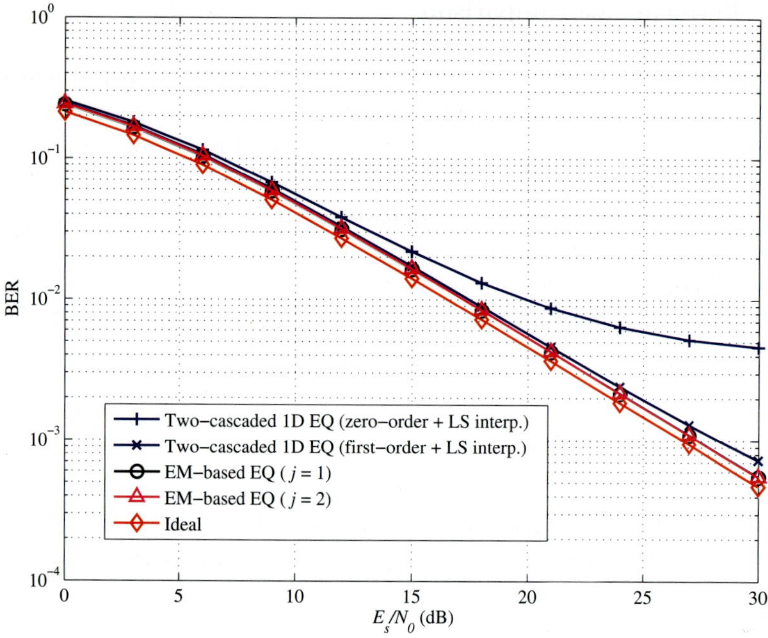

Fig. 4.9   BER comparison between two-cascaded 1D interpolation filters and EM-based equalization as a function of E_s/N_0.

due to its enhanced tracking capability. The BER slightly improves if the channel estimation and data detection tasks are coupled together by means of the EM algorithm. Figure 4.9 indicates that in this way the error-rate performance approaches that of the ideal system after only one iteration ($j = 1$), while marginal improvements are observed with more iterations.

Chapter 5

Joint Synchronization, Channel Estimation and Data Symbol Detection in OFDMA Uplink

A frequency offset estimator based on the space-alternating generalized expectation-maximization (SAGE) algorithm has been presented in Chapter 3 for OFDMA uplink transmissions with generalized CAS. This scheme computes estimates of all users' carrier frequency offsets (CFOs) by exploiting a training block transmitted at the beginning of the uplink frame. The frequency estimates are then employed during the payload section to restore orthogonality among the uplink signals by means of interference cancellation or linear multiuser detection techniques.

In a high-mobility environment such as air traffic control and management [50], the users' CFOs and channel responses may vary with time and their variations must continuously be tracked for reliable data detection. Hence, a robust scheme where data decisions are exploited in addition to pilot symbols for the purpose of frequency and channel tracking is highly desirable.

In this chapter we investigate the issue of joint frequency synchronization, channel estimation and data detection for all active users in the uplink of a quasi-synchronous OFDMA system. As we shall see, the exact maximum likelihood (ML) solution to this problem turns out to be too complex for practical purposes as it involves a search over a multidimensional domain. The complexity requirement is greatly reduced by resorting to the EM principle. This leads to an iterative scheme where the superimposed signals arriving at the base station (BS) are first separated by means of the SAGE algorithm. The separated signals are subsequently passed to an expectation-conditional maximization (ECM)-based processor, which updates frequency estimates while performing channel estimation and data detection for each user. The resulting architecture is reminiscent of the parallel interference cancellation (PIC) receiver, where at each step inter-

135

ference is generated and removed from the received signal to improve the reliability of data decisions.

Simulations indicate that the joint synchronization, channel estimation and data detection scheme provides an effective means to track possible frequency variations that may occur in high-mobility applications. In particular, it turns out that large CFOs can be corrected without incurring severe performance degradation with respect to a perfectly synchronized system where neither interchannel interference (ICI) nor multiple-access interference (MAI) is present. It is nevertheless fair to say that these advantages come at the price of a higher computational load compared to other existing methods as those presented in [12, 18, 55, 158].

5.1   Uncoded OFDMA uplink

5.1.1   *Signal model*

We consider the uplink of a quasi-synchronous OFDMA system in which the cyclic prefix (CP) is sufficiently long to accommodate both the channel delay spreads and timing offsets of all active terminals. The channel impulse responses (CIRs) are assumed static over one OFDMA block, even though they can vary from block to block. We denote $\boldsymbol{h}_{m,i} = [h_{m,i}(0), h_{m,i}(1), \ldots, h_{m,i}(L_m - 1)]^T$ the discrete-time CIR of the mth user during the ith block and assume that the channel length L_m keeps constant over an entire frame. For convenience, we also define the mth extended channel vector as

$$\boldsymbol{h}'_{m,i} = \begin{bmatrix} \mathbf{0}_{\theta_m}^T & \boldsymbol{h}_{m,i}^T & \mathbf{0}_{L-\theta_m-L_m}^T \end{bmatrix}^T, \tag{5.1}$$

where θ_m is the mth timing error (normalized to the sampling interval T_s) and $L = \max_{m} \{L_m + \theta_m\}$. As explained in Chapter 3, the fractional part of the timing error can be absorbed into the CIR and, accordingly, is not considered in the following derivations.

At the BS receiver, the samples of the superimposed uplink signals that fall within the ith DFT window are given by

$$r_i(k) = \sum_{m=1}^{M} r_{m,i}(k) + w_i(k), \quad 0 \leq k \leq N - 1 \tag{5.2}$$

in which M is the number of active terminals, $w_i(k)$ represents Gaussian noise with zero-mean and power σ_w^2 and, finally, $r_{m,i}(k)$ is the signal from the mth user. Apart from an irrelevant phase shift that can be incorporated as part of the channel response, from (3.70) we have

$$r_{m,i}(k) = \frac{1}{\sqrt{N}} e^{j2\pi k\varepsilon_{m,i}/N} \sum_{n\in\mathcal{I}_m} H'_{m,i}(n)c_{m,i}(n)\, e^{j2\pi nk/N}, \quad 0 \le k \le N-1 \tag{5.3}$$

where $\varepsilon_{m,i}$ is the CFO of the mth user (possibly varying from block to block), $\{c_{m,i}(n)\}$ are uncoded information symbols and $H'_{m,i}(n)$ denotes the mth channel frequency response over the nth subcarrier, which reads

$$H'_{m,i}(n) = \sum_{\ell=0}^{L-1} h'_{m,i}(\ell)\, e^{-j2\pi n\ell/N}, \qquad 0 \le n \le N-1. \tag{5.4}$$

Without loss of generality, in the ensuing discussion we concentrate on the ith received block and omit the time index i for notational simplicity. Then, collecting the received samples into a vector $\boldsymbol{r} = [r(0), r(1), \ldots, r(N-1)]^T$, after substituting Eqs. (5.3) and (5.4) into Eq. (5.2) we obtain

$$\boldsymbol{r} = \sum_{m=1}^{M} \boldsymbol{\Gamma}(\varepsilon_m)\boldsymbol{F}^H \boldsymbol{D}(\boldsymbol{c}_m)\boldsymbol{U}\boldsymbol{h}'_m + \boldsymbol{w}, \tag{5.5}$$

where

- $\boldsymbol{\Gamma}(\varepsilon_m) = \text{diag}\left\{1, e^{j2\pi\varepsilon_m/N}, \ldots, e^{j2\pi(N-1)\varepsilon_m/N}\right\}$;
- $\boldsymbol{F}$ is the N-point DFT matrix with entries

$$[\boldsymbol{F}]_{p,q} = \frac{1}{\sqrt{N}} \exp\left(-j2\pi pq/N\right), \tag{5.6}$$

  for $0 \le p,q \le N-1$;
- $\boldsymbol{c}_m$ is an N-dimensional vector with entries $c_m(n)$ for $n \in \mathcal{I}_m$ and zero otherwise;
- $\boldsymbol{D}(\boldsymbol{c}_m)$ is a diagonal matrix with $\boldsymbol{c}_m$ on its main diagonal;
- $\boldsymbol{U}$ is an $N \times L$ matrix with elements $[\boldsymbol{U}]_{p,q} = \exp\left(-j2\pi pq/N\right)$ for $0 \le p \le N-1$ and $0 \le q \le L-1$. In practice, the columns of $\boldsymbol{U}$ are scaled versions of the first L columns of $\boldsymbol{F}$;
- $\boldsymbol{w}$ is circularly symmetric white Gaussian noise with zero-mean and covariance matrix $\sigma_w^2 \boldsymbol{I}_N$.

5.1.2 *Iterative detection and frequency synchronization*

Since timing errors θ_m do not explicitly appear in the signal model Eq. (5.5), timing estimation is not strictly necessary in the considered system. Hence, we only investigate the joint estimation of $\boldsymbol{\varepsilon} = [\varepsilon_1, \varepsilon_2, \ldots, \varepsilon_M]^T$, $\boldsymbol{h}' = [\boldsymbol{h}_1'^T, \boldsymbol{h}_2'^T, \ldots, \boldsymbol{h}_M'^T]^T$ and $\boldsymbol{c} = [\boldsymbol{c}_1^T, \boldsymbol{c}_2^T, \ldots, \boldsymbol{c}_M^T]^T$ based on received vector $\boldsymbol{r}$. In doing so, we follow an ML approach. Recalling that the entries of $\boldsymbol{w}$ are independent Gaussian random variables with zero-mean and variance σ_w^2, the log-likelihood function for the unknown parameters $\boldsymbol{\varepsilon}$, $\boldsymbol{h}'$ and $\boldsymbol{c}$ takes the form

$$\Lambda(\tilde{\boldsymbol{\varepsilon}}, \tilde{\boldsymbol{h}}', \tilde{\boldsymbol{c}}) = -N \ln\left(\pi \sigma_w^2\right) - \frac{1}{\sigma_w^2} \left\| \boldsymbol{r} - \sum_{m=1}^{M} \boldsymbol{\Gamma}(\tilde{\varepsilon}_m) \boldsymbol{F}^H \boldsymbol{D}(\tilde{\boldsymbol{c}}_m) \boldsymbol{U} \tilde{\boldsymbol{h}}'_m \right\|^2 , \quad (5.7)$$

where the notation $\tilde{\lambda}$ is used to indicate a trial value of an unknown parameter λ.

The joint ML estimates of $\boldsymbol{\varepsilon}$, $\boldsymbol{h}'$ and $\boldsymbol{c}$ are found by searching for the maximum of $\Lambda(\tilde{\boldsymbol{\varepsilon}}, \tilde{\boldsymbol{h}}', \tilde{\boldsymbol{c}})$ with respect to $\tilde{\boldsymbol{\varepsilon}}$, $\tilde{\boldsymbol{h}}'$ and $\tilde{\boldsymbol{c}}$. Unfortunately, this operation requires an exhaustive search over the multidimensional space spanned by $\tilde{\boldsymbol{\varepsilon}}$, $\tilde{\boldsymbol{h}}'$ and $\tilde{\boldsymbol{c}}$, which is prohibitively complex for practical implementation. To circumvent this obstacle, we consider the iterative scheme proposed in [126] and depicted in Fig. 5.1. As is seen, a SAGE-based processor [45] is first used to extract the contribution of each user, say $\hat{\boldsymbol{r}}_p$ $(p = 1, 2, \ldots, M)$, from the received vector $\boldsymbol{r}$. Each $\hat{\boldsymbol{r}}_p$ is then exploited to jointly estimate ε_p, $\boldsymbol{h}_p'$ and $\boldsymbol{c}_p$ following an ECM approach [94].

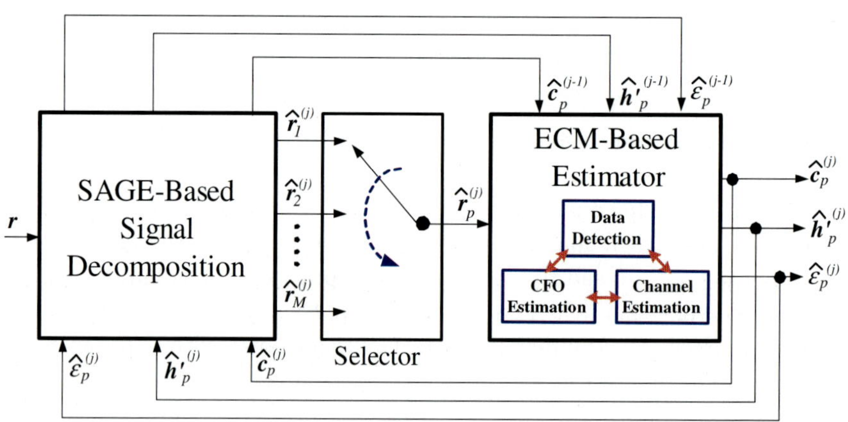

Fig. 5.1   Block diagram of the EM-based iterative receiver.

5.1.2.1 *SAGE-based signal decomposition*

In a variety of ML problems, direct maximization of the likelihood function is analytically challenging. In such a case, the EM algorithm proves to be effective as it achieves the same final result with a comparatively simpler iterative procedure. In the EM formulation, the observed measurements are replaced with some *complete data* from which original measurements are obtained through a many-to-one mapping [94]. At each iteration, the EM algorithm calculates the expectation of the log-likelihood function of the complete data set (E-step), which is then maximized with respect to the unknown parameters (M-step). The process is terminated as soon as no significant changes are observed in the estimated parameters.

As mentioned in Chapter 3, the SAGE algorithm improves upon EM in that it has a faster convergence rate. The reason is that maximization in the EM algorithm is *simultaneously* performed with respect to all unknown parameters, which results in a slow process requiring searches over a space with many dimensions. In contrast, the maximization in the SAGE is performed by updating a smaller group of parameters at a time. The SAGE algorithm was first proposed in [45] and provides a practical solution to parameter estimation from superimposed signals [43]. In particular, it is now exploited to decompose the maximization of $\Lambda(\tilde{\boldsymbol{\varepsilon}}, \tilde{\boldsymbol{h}}', \tilde{\boldsymbol{c}})$ in Eq. (5.7) into M simpler maximization problems.

For this purpose, we view the received vector $\boldsymbol{r}$ as the observed data and take $\{\boldsymbol{r}_m; m = 1, 2, \ldots, M\}$ as the complete data, where $\boldsymbol{r}_m$ is the contribution of the mth user to $\boldsymbol{r}$ in form of

$$\boldsymbol{r}_m = \boldsymbol{\Gamma}(\varepsilon_m)\boldsymbol{F}^H \boldsymbol{D}(\boldsymbol{c}_m)\boldsymbol{U}\boldsymbol{h}'_m + \boldsymbol{w}_m, \quad m = 1, 2, \ldots, M. \tag{5.8}$$

and $\boldsymbol{w}_m$ $(m = 1, 2, \ldots, M)$ are circularly symmetric and statistically independent Gaussian vectors satisfying the identity $\boldsymbol{w} = \sum_{m=1}^{M} \boldsymbol{w}_m$ [43].

The SAGE algorithm is applied in such a way that the parameters of a single user are updated at a time. This leads to a procedure consisting of *iterations* and *cycles*, where M cycles make an iteration and each cycle updates the parameters of a given user. To see how this comes about, we call $\hat{\varepsilon}_m^{(j)}$, $\hat{\boldsymbol{h}}'^{(j)}_m$ and $\hat{\boldsymbol{c}}_m^{(j)}$ estimates of ε_m, $\boldsymbol{h}'_m$ and $\boldsymbol{c}_m$ after the jth iteration, respectively. Given initial estimates $\hat{\varepsilon}_m^{(0)}$, $\hat{\boldsymbol{h}}'^{(0)}_m$ and $\hat{\boldsymbol{c}}_m^{(0)}$, we compute

$$\hat{\boldsymbol{z}}_m^{(0)} = \boldsymbol{\Gamma}(\hat{\varepsilon}_m^{(0)})\boldsymbol{F}^H \boldsymbol{D}(\hat{\boldsymbol{c}}_m^{(0)})\boldsymbol{U}\hat{\boldsymbol{h}}'^{(0)}_m, \quad m = 1, 2, \ldots, M. \tag{5.9}$$

Then, during the pth cycle of the jth iteration (with $p = 1, 2, \ldots, M$), the SAGE proceeds as follows [45].

E-Step:

Compute

$$\hat{r}_p^{(j)} = r - \sum_{m=1}^{p-1} \hat{z}_m^{(j)} - \sum_{m=p+1}^{M} \hat{z}_m^{(j-1)} \tag{5.10}$$

where $\sum_l^u$ is zero if $u < l$.

M-Step:

Compute

$$\left[\hat{\varepsilon}_p^{(j)}, \hat{h}'^{(j)}_p, \hat{c}_p^{(j)}\right] = \arg\min_{\tilde{\varepsilon}_p, \tilde{h}_p, \tilde{c}_p} \left\{ \left\| \hat{r}_p^{(j)} - \Gamma(\tilde{\varepsilon}_p) F^H D(\tilde{c}_p) U \tilde{h}'_p \right\|^2 \right\}, \tag{5.11}$$

and then use updated parameters to obtain the following vector

$$\hat{z}_p^{(j)} = \Gamma(\hat{\varepsilon}_p^{(j)}) F^H D(\hat{c}_p^{(j)}) U \hat{h}'^{(j)}_p. \tag{5.12}$$

We see from Eq. (5.11) that the SAGE algorithm splits the maximization of $\Lambda(\tilde{\varepsilon}, \tilde{h}', \tilde{c})$ in Eq. (5.7) into a series of M simpler optimization problems. However, the multidimensional minimization in Eq. (5.11) still remains a formidable task. An iterative solution to this problem is presented in the next subsection by resorting to the ECM algorithm.

5.1.2.2 *ECM-based iterative estimator*

Substituting Eq. (5.5) into Eq. (5.10) yields

$$\hat{r}_p^{(j)} = \Gamma(\varepsilon_p) F^H D(c_p) U h'_p + \eta_p^{(j)}, \tag{5.13}$$

where

$$\eta_p^{(j)} = w + \sum_{m=1}^{p-1} [z_m - \hat{z}_m^{(j)}] + \sum_{m=p+1}^{M} [z_m - \hat{z}_m^{(j-1)}], \tag{5.14}$$

and $z_m = \Gamma(\varepsilon_m) F^H D(c_m) U h'_m$ is the signal received from the mth user. Note that $\eta_p^{(j)}$ is a disturbance term that accounts for thermal noise and residual MAI after the jth SAGE iteration, and is linearly related to the data symbols of all interfering users. Then, assuming that these symbols are independent and identically distributed with zero-mean, it follows from the central limit theorem that the entries of $\eta_p^{(j)}$ are nearly Gaussian distributed with zero-mean and some variance $\sigma_\eta^2(j)$. Under this assumption, it turns out that the minimization problem in Eq. (5.11) is equivalent to the ML estimation of ε_p , $\boldsymbol{h}'_p$ and $\boldsymbol{c}_p$ starting from the observation of $\hat{\boldsymbol{r}}_p^{(j)}$.

The ECM algorithm offers a practical solution to this problem. The only difference between this technique and the conventional EM algorithm is that the maximization step in the ECM algorithm is divided into several stages, where at each stage only one parameter is updated while all the others are kept constant at their most updated values. This makes the ECM algorithm suitable for *multidimensional* ML estimation problems, where the likelihood function has to be optimized over several parameters [94].

In the following, the ECM algorithm is employed to solve the optimization problem stated in Eq. (5.11). In doing so, we view $\hat{\boldsymbol{r}}_p^{(j)}$ as the observed data and $[\hat{\boldsymbol{r}}_p^{(j)T} \ \boldsymbol{h}'^T_p]^T$ as the complete set of data. Also, we denote $\boldsymbol{\xi}_p \overset{\text{def}}{=} [\boldsymbol{c}_p^T \ \varepsilon_p]^T$ the parameters to be estimated and $\hat{\boldsymbol{\xi}}_p^{(j,u)} = [\hat{\boldsymbol{c}}_p^{(j,u)T} \ \hat{\varepsilon}_p^{(j,u)}]^T$ the estimate of $\boldsymbol{\xi}_p$ at the uth ECM and jth SAGE iterations. Then, after initializing $\hat{\boldsymbol{c}}_p^{(j,0)} = \hat{\boldsymbol{c}}_p^{(j-1)}$ and $\hat{\varepsilon}_p^{(j,0)} = \hat{\varepsilon}_p^{(j-1)}$, the ECM algorithm alternates between an E-step and an M-step as follows.

E-Step:

We define

$$\Upsilon\left(\tilde{\boldsymbol{\xi}}_p \,\Big|\, \hat{\boldsymbol{\xi}}_p^{(j,u)}\right) = E_{\boldsymbol{h}'_p}\left\{\ln\left[p\left(\hat{\boldsymbol{r}}_p^{(j)} \,\Big|\, \boldsymbol{h}'_p, \tilde{\boldsymbol{\xi}}_p\right)\right] p\left(\hat{\boldsymbol{r}}_p^{(j)} \,\Big|\, \boldsymbol{h}'_p, \hat{\boldsymbol{\xi}}_p^{(j,u)}\right)\right\}, \quad (5.15)$$

where $p\left(\hat{\boldsymbol{r}}_p^{(j)} \,\Big|\, \boldsymbol{h}'_p, \tilde{\boldsymbol{\xi}}_p\right)$ and $p\left(\hat{\boldsymbol{r}}_p^{(j)} \,\Big|\, \boldsymbol{h}'_p, \hat{\boldsymbol{\xi}}_p^{(j,u)}\right)$ are conditional probability density functions (pdf), $E_{\boldsymbol{h}'_p}\{\cdot\}$ denotes the statistical expectation over the pdf of $\boldsymbol{h}'_p$ and $\tilde{\boldsymbol{\xi}}_p = [\tilde{\boldsymbol{c}}_p^T \ \tilde{\varepsilon}_p]^T$ is a trial value of $\boldsymbol{\xi}_p$.

Function Υ defined in Eq. (5.15) can be rewritten as

$$\Upsilon\left(\tilde{\boldsymbol{\xi}}_p \,\Big|\, \hat{\boldsymbol{\xi}}_p^{(j,u)}\right) = \int_\Omega \ln\left[p(\hat{\boldsymbol{r}}_p^{(j)} \,\Big|\, \boldsymbol{h}'_p, \tilde{\boldsymbol{\xi}}_p)\right] \cdot p\left(\hat{\boldsymbol{r}}_p^{(j)} \,\Big|\, \boldsymbol{h}'_p, \hat{\boldsymbol{\xi}}_p^{(j,u)}\right) p(\boldsymbol{h}'_p) \, d\boldsymbol{h}'_p,$$

$$(5.16)$$

where $p(\boldsymbol{h}'_p)$ is the a-priori pdf of $\boldsymbol{h}'_p$.

To proceed further, we make the following assumptions:

(1) $\boldsymbol{h}'_p$ is a circularly symmetric Gaussian vector with zero-mean (Rayleigh fading) and covariance matrix $\boldsymbol{C}_p = E\{\boldsymbol{h}'_p \boldsymbol{h}'^H_p\}$;

(2) the disturbance $\boldsymbol{\eta}_p^{(j)}$ in Eq. (5.13) is nearly Gaussian distributed with zero-mean and covariance matrix $\sigma_\eta^2(j)\boldsymbol{I}_N$.

Thus, bearing in mind Eq. (5.13), we may write

$$p(\boldsymbol{h}'_p) = \frac{1}{\pi^L \det(\boldsymbol{C}_p)} \exp\{-\boldsymbol{h}'^H_p \boldsymbol{C}_p^{-1} \boldsymbol{h}'_p\}, \tag{5.17}$$

$$p\left(\hat{\boldsymbol{r}}_p^{(j)} \,\middle|\, \boldsymbol{h}'_p, \hat{\boldsymbol{\xi}}_p^{(j,u)}\right) \approx \frac{1}{[\pi\sigma_\eta^2(j)]^N} \exp\left\{-\frac{1}{\sigma_\eta^2(j)} \left\|\hat{\boldsymbol{r}}_p^{(j)} - \hat{\boldsymbol{z}}_p^{(j,u)}\right\|^2\right\}, \tag{5.18}$$

$$\ln\left[p(\hat{\boldsymbol{r}}_p^{(j)} \,\middle|\, \boldsymbol{h}'_p, \tilde{\boldsymbol{\xi}}_p)\right] \approx -N\ln[\pi\sigma_\eta^2(j)] - \frac{1}{\sigma_\eta^2(j)} \left\|\hat{\boldsymbol{r}}_p^{(j)} - \tilde{\boldsymbol{z}}_p\right\|^2, \tag{5.19}$$

with

$$\hat{\boldsymbol{z}}_p^{(j,u)} = \boldsymbol{\Gamma}(\hat{\varepsilon}_p^{(j,u)})\boldsymbol{F}^H \boldsymbol{D}(\hat{\boldsymbol{c}}_p^{(j,u)})\boldsymbol{U}\boldsymbol{h}'_p, \tag{5.20}$$

and

$$\tilde{\boldsymbol{z}}_p = \boldsymbol{\Gamma}(\tilde{\varepsilon}_p)\boldsymbol{F}^H \boldsymbol{D}(\tilde{\boldsymbol{c}}_p)\boldsymbol{U}\boldsymbol{h}'_p. \tag{5.21}$$

Substituting Eqs. (5.17)-(5.19) into Eq. (5.16) and skipping additive and multiplicative terms independent of $\tilde{\boldsymbol{\xi}}_p$, we may replace $\Upsilon\left(\tilde{\boldsymbol{\xi}}_p \,\middle|\, \hat{\boldsymbol{\xi}}_p^{(j,u)}\right)$ with the equivalent function

$$\Phi\left(\tilde{\boldsymbol{\xi}}_p \,\middle|\, \hat{\boldsymbol{\xi}}_p^{(j,u)}\right) = -\left\|\hat{\boldsymbol{r}}_p^{(j)} - \boldsymbol{\Gamma}(\tilde{\varepsilon}_p)\boldsymbol{F}^H \boldsymbol{D}(\tilde{\boldsymbol{c}}_p)\boldsymbol{U}\hat{\boldsymbol{h}}'_{p,\mathrm{MMSE}}(\hat{\boldsymbol{\xi}}_p^{(j,u)})\right\|^2 -$$
$$\sigma_\eta^2(j) \cdot \mathrm{tr}\{\boldsymbol{D}(\tilde{\boldsymbol{c}}_p)\boldsymbol{U}[\boldsymbol{P}(\hat{\boldsymbol{c}}_p^{(j,u)})]^{-1}\boldsymbol{U}^H \boldsymbol{D}^H(\tilde{\boldsymbol{c}}_p)\}, \tag{5.22}$$

where

$$\hat{\boldsymbol{h}}'_{p,\mathrm{MMSE}}(\hat{\boldsymbol{\xi}}_p^{(j,u)}) = [\boldsymbol{P}(\hat{\boldsymbol{c}}_p^{(j,u)})]^{-1}\boldsymbol{U}^H \boldsymbol{D}^H(\hat{\boldsymbol{c}}_p^{(j,u)})\boldsymbol{F}\boldsymbol{\Gamma}^H(\hat{\varepsilon}_p^{(j,u)})\hat{\boldsymbol{r}}_p^{(j)} \tag{5.23}$$

is the MMSE estimate of $\boldsymbol{h}'_p$ obtained with $\boldsymbol{\xi}_p = \hat{\boldsymbol{\xi}}_p^{(j,u)}$, while

$$\boldsymbol{P}(\hat{\boldsymbol{c}}_p^{(j,u)}) = \boldsymbol{U}^H \boldsymbol{E}_p(\hat{\boldsymbol{c}}_p^{(j,u)})\boldsymbol{U} + \sigma_\eta^2(j)\boldsymbol{C}_p^{-1} \tag{5.24}$$

with

$$\boldsymbol{E}_p(\hat{\boldsymbol{c}}_p^{(j,u)}) = \mathrm{diag}\left\{\left|\hat{c}_p^{(j,u)}(n)\right|^2 ; n = 0, 1, \ldots, N-1\right\}. \tag{5.25}$$

We see from Eqs. (5.22)-(5.24) that evaluating $\Phi\left(\tilde{\boldsymbol{\xi}}_p \,\middle|\, \hat{\boldsymbol{\xi}}_p^{(j,u)}\right)$ requires knowledge of $\boldsymbol{C}_p$ and $\sigma_\eta^2(j)$. Thus, suitable schemes must be devised to estimate these parameters. A practical solution to this problem is found by assuming high SNR values. In this case we expect that $\sigma_\eta^2(j)$ becomes vanishingly small and $\Phi\left(\tilde{\boldsymbol{\xi}}_p \,\middle|\, \hat{\boldsymbol{\xi}}_p^{(j,u)}\right)$ can reasonably be approximated by

$$\bar{\Phi}\left(\tilde{\boldsymbol{\xi}}_p \,\middle|\, \hat{\boldsymbol{\xi}}_p^{(j,u)}\right) = -\left\|\hat{\boldsymbol{r}}_p^{(j)} - \boldsymbol{\Gamma}(\tilde{\varepsilon}_p)\boldsymbol{F}^H \boldsymbol{D}(\tilde{\boldsymbol{c}}_p)\boldsymbol{U}\hat{\boldsymbol{h}}'_{p,\mathrm{LS}}(\hat{\boldsymbol{\xi}}_p^{(j,u)})\right\|^2, \tag{5.26}$$

where $\hat{\boldsymbol{h}}'_{p,\mathrm{LS}}(\hat{\boldsymbol{\xi}}_p^{(j,u)})$ is the least-squares (LS) estimate of $\boldsymbol{h}'_p$ and takes the form

$$\hat{\boldsymbol{h}}'_{p,\mathrm{LS}}(\hat{\boldsymbol{\xi}}_p^{(j,u)}) = [\boldsymbol{U}^H \boldsymbol{E}_p(\hat{\boldsymbol{c}}_p^{(j,u)})\boldsymbol{U}]^{-1}\boldsymbol{U}^H \boldsymbol{D}^H(\hat{\boldsymbol{c}}_p^{(j,u)})\boldsymbol{F}\boldsymbol{\Gamma}^H(\hat{\varepsilon}_p^{(j,u)})\hat{\boldsymbol{r}}_p^{(j)}. \tag{5.27}$$

In the sequel, function $\bar{\Phi}\left(\tilde{\boldsymbol{\xi}}_p \middle| \hat{\boldsymbol{\xi}}_p^{(j,u)}\right)$ is used in place of $\Phi\left(\tilde{\boldsymbol{\xi}}_p \middle| \hat{\boldsymbol{\xi}}_p^{(j,u)}\right)$. Although this approach may entail some performance penalty at low and medium SNRs, it has the advantage of being practically implementable, while computing $\Phi\left(\tilde{\boldsymbol{\xi}}_p \middle| \hat{\boldsymbol{\xi}}_p^{(j,u)}\right)$ seems hardly viable in practice.

M-Step:

The M-step aims at maximizing the right-hand-side of Eq. (5.26) with respect to $\tilde{\boldsymbol{\xi}}_p$. This goal is achieved using a two-stage procedure. Following the notation of [94], we denote $\hat{\boldsymbol{\xi}}_p^{(j,u+g/2)}$ the estimate of $\boldsymbol{\xi}_p$ at the gth stage of the uth ECM iteration, where $g = 1, 2$. Then, the maximum of $\bar{\Phi}\left(\tilde{\boldsymbol{\xi}}_p \middle| \hat{\boldsymbol{\xi}}_p^{(j,u)}\right)$ is found as follows.

- **Step 1:**

$$\hat{\boldsymbol{\xi}}_p^{(j,u+1/2)} = \left[(\hat{\boldsymbol{c}}_p^{(j,u)})^T \; \hat{\varepsilon}_p^{(j,u+1)} \right]^T, \tag{5.28}$$

  where

$$\hat{\varepsilon}_p^{(j,u+1)} = \arg\max_{\tilde{\varepsilon}_p} \left\{ - \left\| \hat{\boldsymbol{r}}_p^{(j)} - \boldsymbol{\Gamma}(\tilde{\varepsilon}_p)\boldsymbol{F}^H \boldsymbol{D}(\hat{\boldsymbol{c}}_p^{(j,u)})\boldsymbol{U}\hat{\boldsymbol{h}}'_{p,\mathrm{LS}}(\hat{\boldsymbol{\xi}}_p^{(j,u)}) \right\|^2 \right\}. \tag{5.29}$$

  Note that the quantity $\left\| \boldsymbol{\Gamma}(\tilde{\varepsilon}_p)\boldsymbol{F}^H \boldsymbol{D}(\hat{\boldsymbol{c}}_p^{(j,u)})\boldsymbol{U}\hat{\boldsymbol{h}}'_{p,\mathrm{LS}}(\hat{\boldsymbol{\xi}}_p^{(j,u)}) \right\|^2$ is independent of $\tilde{\varepsilon}_p$ since $\boldsymbol{\Gamma}^H(\tilde{\varepsilon}_p)\boldsymbol{\Gamma}(\tilde{\varepsilon}_p) = \boldsymbol{I}_N$. Thus, Eq. (5.29) can equivalently be replaced by

$$\hat{\varepsilon}_p^{(j,u+1)} = \arg\max_{\tilde{\varepsilon}_p} \left\{ \Re e\left[\hat{\boldsymbol{r}}_p^{(j)H}\boldsymbol{\Gamma}(\tilde{\varepsilon}_p)\boldsymbol{F}^H \boldsymbol{D}(\hat{\boldsymbol{c}}_p^{(j,u)})\boldsymbol{U}\hat{\boldsymbol{h}}'_{p,\mathrm{LS}}(\hat{\boldsymbol{\xi}}_p^{(j,u)}) \right] \right\}. \tag{5.30}$$

- **Step 2:**

$$\hat{\boldsymbol{\xi}}_p^{(j,u+1)} = \left[(\hat{\boldsymbol{c}}_p^{(j,u+1)})^T \; \hat{\varepsilon}_p^{(j,u+1)} \right]^T, \tag{5.31}$$

  where

$$\hat{\boldsymbol{c}}_p^{(j,u+1)} = \arg\min_{\tilde{\boldsymbol{c}}_p} \left\{ \sum_{n=0}^{N-1} \left| \widehat{R}_p^{(j)}(n, \hat{\varepsilon}_p^{(j,u+1)}) - \tilde{c}_p(n)\hat{H}'^{(j,u)}_{p,\mathrm{LS}}(n) \right|^2 \right\}, \tag{5.32}$$

  with $\{\widehat{R}_p^{(j)}(n, \hat{\varepsilon}_p^{(j,u+1)}); n = 0, 1, \ldots, N - 1\}$ and $\{\hat{H}'^{(j,u)}_{p,\mathrm{LS}}(n); n = 0, 1, \ldots, N - 1\}$ being the N-point DFTs of $\boldsymbol{\Gamma}^H(\hat{\varepsilon}_p^{(j,u+1)})\hat{\boldsymbol{r}}_p^{(j)}$ and $\hat{\boldsymbol{h}}'_{p,\mathrm{LS}}(\hat{\boldsymbol{\xi}}_p^{(j,u)})$, respectively.

An approximation of the CFO estimate in Eq. (5.30) can be obtained in closed-form after replacing $\mathbf{\Gamma}(\tilde{\varepsilon}_p)$ with its Taylor series expansion truncated to the second order term and using $\hat{\varepsilon}_p^{(j,u)}$ as starting point, *i.e.,*

$$\mathbf{\Gamma}(\tilde{\varepsilon}_p) \approx \mathbf{\Gamma}(\hat{\varepsilon}_p^{(j,u)}) + j(\tilde{\varepsilon}_p - \hat{\varepsilon}_p^{(j,u)})\mathbf{\Gamma}'(\hat{\varepsilon}_p^{(j,u)}) - \frac{1}{2}(\tilde{\varepsilon}_p - \hat{\varepsilon}_p^{(j,u)})^2\mathbf{\Gamma}''(\hat{\varepsilon}_p^{(j,u)}),$$

(5.33)

where $\mathbf{\Gamma}'(\hat{\varepsilon}_p^{(j,u)}) = \mathbf{\Psi}\mathbf{\Gamma}(\hat{\varepsilon}_p^{(j,u)})$, $\mathbf{\Gamma}''(\hat{\varepsilon}_p^{(j,u)}) = \mathbf{\Psi}^2\mathbf{\Gamma}(\hat{\varepsilon}_p^{(j,u)})$ and $\mathbf{\Psi} = (2\pi/N) \cdot \mathrm{diag}\{0, 1, \ldots, N-1\}$. Substituting Eq. (5.33) into Eq. (5.30) and setting the derivative with respect to $\tilde{\varepsilon}_p$ to zero yields

$$\hat{\varepsilon}_p^{(j,u+1)} = \hat{\varepsilon}_p^{(j,u)} + \frac{\Im m\left\{\hat{\boldsymbol{r}}_p^{(j)H}\mathbf{\Gamma}'(\hat{\varepsilon}_p^{(j,u)})\boldsymbol{F}^H\boldsymbol{D}(\hat{\boldsymbol{c}}_p^{(j,u)})\boldsymbol{U}\hat{\boldsymbol{h}}'_{p,\mathrm{LS}}(\hat{\boldsymbol{\xi}}_p^{(j,u)})\right\}}{\Re e\left\{\hat{\boldsymbol{r}}_p^{(j)H}\mathbf{\Gamma}''(\hat{\varepsilon}_p^{(j,u)})\boldsymbol{F}^H\boldsymbol{D}(\hat{\boldsymbol{c}}_p^{(j,u)})\boldsymbol{U}\hat{\boldsymbol{h}}'_{p,\mathrm{LS}}(\hat{\boldsymbol{\xi}}_p^{(j,u)})\right\}}.$$

(5.34)

After a specified number N_U of iterations, we terminate the ECM process and replace Eq. (5.11) with

$$[\hat{\varepsilon}_p^{(j)}, \hat{\boldsymbol{h}}'^{(j)}_p, \hat{\boldsymbol{c}}_p^{(j)}] = [\hat{\varepsilon}_p^{(j,N_U)}, \hat{\boldsymbol{h}}'_{p,\mathrm{LS}}(\hat{\boldsymbol{\xi}}_p^{(j,N_U)}), \hat{\boldsymbol{c}}_p^{(j,N_U)}].$$

(5.35)

In the sequel, the iterative scheme relying on Eqs. (5.27), (5.32) and (5.34) is referred to as the EM-based receiver (EMBR).

5.1.3 *Practical adjustments*

The following guidelines may be helpful for a practical implementation of EMBR:

(1) It is well known that a good initialization is essential for EM-type algorithms. Hence, the problem arises of how to obtain initial estimates $\hat{\varepsilon}_m^{(0)}$, $\hat{\boldsymbol{h}}'^{(0)}_m$ and $\hat{\boldsymbol{c}}_m^{(0)}$ to start the SAGE procedure. If $\boldsymbol{\varepsilon}$ and $\boldsymbol{h}'$ vary slowly in time, frequency and channel estimates obtained in a given block can be used to initialize the iterative process in the next block. Estimates for the first data block may be obtained in a data-aided fashion by exploiting a training sequence placed at the beginning of the uplink frame [124, 127].
The initial CFO estimates $\hat{\boldsymbol{\varepsilon}}^{(0)} = [\hat{\varepsilon}_1^{(0)}, \hat{\varepsilon}_2^{(0)}, \ldots, \hat{\varepsilon}_M^{(0)}]^T$ are next exploited to accomplish frequency correction using one of the methods discussed in [12,18,55,158]. This operation aims at restoring orthogonality among subcarriers and produces the following N-dimensional vectors (one for

each user)

$$\boldsymbol{\psi}_m = \boldsymbol{D}\left(\boldsymbol{c}_m\right)\boldsymbol{U}\boldsymbol{h}'_m + \boldsymbol{\gamma}_m, \quad m = 1, 2, \ldots, M. \tag{5.36}$$

where $\boldsymbol{\gamma}_m$ is a disturbance term that accounts for thermal noise and residual MAI caused by imperfect separation of the users' signals. Finally, initial data decisions are obtained as in conventional OFDM transmission, *i.e.*,

$$\hat{\boldsymbol{c}}_m^{(0)} = \underset{\tilde{\boldsymbol{c}}_m}{\arg\min} \left\{ \sum_{n=0}^{N-1} \left| \psi_m(n) - \tilde{c}_m(n)\hat{H}'^{(0)}_m(n) \right|^2 \right\}, \tag{5.37}$$

where $\psi_m(n)$ is the nth entry of $\boldsymbol{\psi}_m$ and $\{\hat{H}'^{(0)}_m(n); n = 0, 1, \ldots, N-1\}$ is the N-point DFT of $\hat{\boldsymbol{h}}'^{(0)}_m$.

In applications characterized by high user mobility, initializing the SAGE iterations with channel estimates from the previous block may result in poor performance due to fast fading. In these circumstances, a possible solution is to insert scattered pilots in each OFDMA block and compute $\hat{\boldsymbol{h}}'^{(0)}_m$ through conventional pilot-aided estimation techniques [101]. Albeit robust against rapidly varying channels, this approach inevitably results into a reduction of the overall data throughput due to the increased overhead.

(2) For PSK transmissions, matrix $\boldsymbol{E}_p(\hat{\boldsymbol{c}}_p^{(j,u)})$ defined in Eq. (5.25) becomes independent of $\hat{c}_p^{(j,u)}$ since $\left|\hat{c}_p^{(j,u)}\right|^2$ is either unitary or zero depending on whether the nth subcarrier is assigned to the mth user or not. In such a case, evaluating $\hat{\boldsymbol{h}}'_{p,\mathrm{LS}}(\hat{\boldsymbol{\xi}}_p^{(j,u)})$ in Eq. (5.27) does not require any on-line matrix inversion since $[\boldsymbol{U}^H\boldsymbol{E}_p(\hat{\boldsymbol{c}}_p^{(j,u)})\boldsymbol{U}]^{-1}$ can be pre-computed and stored in the receiver. A further simplification is possible if the subcarriers of the pth user are uniformly distributed over the signal bandwidth with separation interval N/P, where P is the number of subcarriers in each subchannel. In this hypothesis, $\boldsymbol{U}^H\boldsymbol{E}_p(\hat{\boldsymbol{c}}_p^{(j,u)})\boldsymbol{U}$ reduces to $P \cdot \boldsymbol{I}_L$ and Eq. (5.27) becomes

$$\hat{\boldsymbol{h}}'_{p,\mathrm{LS}}(\hat{\boldsymbol{\xi}}_p^{(j,u)}) = \frac{1}{P}\boldsymbol{U}^H\boldsymbol{D}^H(\hat{\boldsymbol{c}}_p^{(j,u)})\boldsymbol{F}\boldsymbol{\Gamma}^H(\hat{\varepsilon}_p^{(j,u)})\hat{\boldsymbol{r}}_p^{(j)}. \tag{5.38}$$

(3) Intuitively speaking, the SAGE procedure should be stopped when no significant variations are observed in the log-likelihood function, *i.e.*,

$$\Lambda(\hat{\boldsymbol{\varepsilon}}^{(j)}, \hat{\boldsymbol{h}}'^{(j)}, \hat{\boldsymbol{c}}^{(j)}) - \Lambda(\hat{\boldsymbol{\varepsilon}}^{(j-1)}, \hat{\boldsymbol{h}}'^{(j-1)}, \hat{\boldsymbol{c}}^{(j-1)}) < \lambda_{th},$$

for some threshold λ_{th}. A simpler stopping criterion is to terminate the SAGE procedure after a preassigned number of iterations.

5.1.4 *Performance assessment*

The performance of EMBR has been assessed by computer simulation in an OFDMA scenario inspired by the IEEE 802.16 standard for Wireless Metropolitan Area Networks [177]. Without loss of generality, we only provide results for user #1.

The simulated system has $N = 128$ subcarriers and a signal bandwidth of 1.429 MHz, which corresponds to a sampling period of $T_s = 0.7$ μs. The useful part of each OFDMA block has length $T = NT_s = 89.6$ μs while the subcarrier spacing is $1/T = 11.16$ kHz. We consider an interleaved CAS where each user is provided with a set of $P = 32$ subcarriers uniformly spaced over the signal bandwidth. In this way, the maximum number of active users in each OFDMA block is $R = 4$. We assume a fully-loaded system with $M = 4$ active terminals and let the users' CFOs be $\boldsymbol{\varepsilon} = \rho \cdot [1, -1, 1, -1]^T$, where the attenuation factor ρ is modeled as a deterministic parameter belonging to interval $[0, 0.5]$ [55]. Information bits are mapped onto uncoded QPSK symbols using a Gray map. The channel responses $\boldsymbol{h}_{m,i}$ have length $L = 5$ while the timing errors θ_m are independently generated at the beginning of each frame and take values in the set $\{0, 1, 2, 3\}$. A CP of length $N_g = 8$ is used to avoid interblock interference (IBI). In this way, the duration of the extended OFDMA block (including the CP) is $T_B = (N + N_g)T_s = 95.2$ μs.

The channel taps $\{h_{m,i}(\ell)\}$ are modeled as statistically independent narrow-band Gaussian processes with zero-mean and autocorrelation function

$$E\left\{h_{m,i}(\ell)h_{m,i+n}^*(\ell)\right\} = \sigma_\ell^2 J_0\left(2\pi n f_D T_B\right), \quad \ell = 0, 1, 2, 3, 4 \qquad (5.39)$$

where f_D is the Doppler bandwidth, $J_0(x)$ is the zero-order Bessel function of the first kind and

$$\sigma_\ell^2 = E\{|h_{m,i}(\ell)|^2\} = \beta_m \cdot \exp(-\ell). \qquad (5.40)$$

In Eq. (5.40), β_1 is chosen such that the signal power of user #1 is normalized to unity, *i.e.*, $E\{\|\boldsymbol{h}_1\|^2\} = 1$, while parameters β_m $(m \geq 2)$ affect the signal-to-interference ratio. The Doppler bandwidth is related to the carrier frequency f_c and mobile velocity v by $f_D = f_c v/c$. Letting $f_c = 2$ GHz and $v = 60$ km/h, we obtain $f_D \approx 110$ Hz, corresponding to 1% of subcarrier spacing.

The uplink frame is composed by 10 OFDMA blocks. Frequency and channel estimates obtained in a given block are used to initialize the iterative process in the next block, while initialization for the first block is achieved using a training sequence placed at the beginning of the frame [127]. For each block, initial CFO estimates $\hat{\varepsilon}^{(0)}$ are employed to restore orthogonality among subcarriers by resorting to the scheme proposed by Cao, Tureli, Yao and Honan (CTYH) in [12], where a linear transformation is applied to the DFT output to obtain vectors $\boldsymbol{\psi}_m$ $(m = 1, 2, \ldots, M)$ in Eq. (5.36). The latter are exploited to get initial channel estimates $\hat{\boldsymbol{h}}'^{(0)}_m$. For this task we employ the pilot-aided estimator described in [101] and assume that 8 pilots are uniformly placed in each subchannel. Initial data decisions are eventually obtained according to Eq. (5.37).

The number N_U of ECM iterations is set to 1 while the number N_i of SAGE iterations is varied throughout simulations to assess its impact on the system performance.

Performance with ideal frequency and channel information

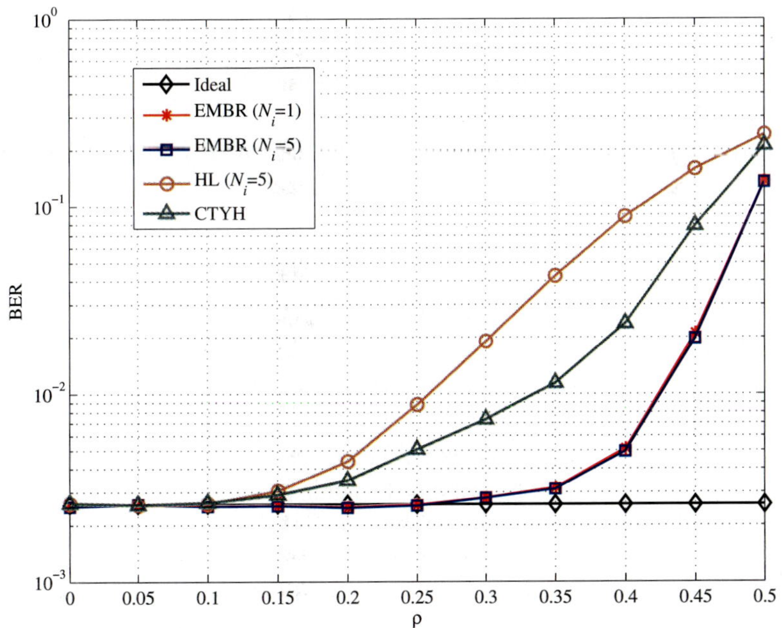

Fig. 5.2 BER performance vs. ρ for uncoded QPSK with $E_b/N_0 = 20$ dB and perfect knowledge of the CFOs and channel responses.

Figure 5.2 shows the BER performance as a function of ρ in case of perfect knowledge of CFOs and channel responses, *i.e.*, $\hat{\varepsilon}_m = \varepsilon_m$ and $\hat{h}'_{m,\mathrm{LS}} = h'_m$ for $m = 1, 2, 3, 4$. This scenario was also considered in [12,18,55,158] and is used here to assess the ability of the system to mitigate ICI and MAI produced by frequency offsets. Users have equal power with $E_b/N_0 = 20$ dB. Comparisons are made with both CTYH [12] and the iterative scheme proposed by Huang and Letaief (HL) in [55], where frequency correction is accomplished at the output of the receive DFT by means of interference cancellation techniques and windowing functions. Five iterations are employed with HL while the number of SAGE iterations is either $N_i = 1$ or 5.

The curve labeled "ideal" is obtained by assuming that all CFOs have perfectly been corrected at the mobile terminals (MTs), *i.e.*, $\varepsilon_m = 0$ for $m = 1, 2, 3, 4$. This provides a benchmark for the BER performance since in this case users' signals at the DFT output are orthogonal and no interference is thus present. As expected, the BER of all considered schemes degrades with ρ due to the increased amount of ICI and MAI. Interestingly, EMBR provides similar results with either $N_i = 1$ or $N_i = 5$, meaning that convergence is achieved after one single iteration. Also, this scheme largely outperforms the other methods. A possible explanation is that CTYH operates similarly to a linear multiuser detector where interference is mitigated at the price of non-negligible noise enhancement. As to the HL scheme, the windowing functions applied to the DFT output may lead to a significant loss of signal energy in the presence of relatively large CFOs.

Performance with estimated frequency offsets and channel responses

We now assess the performance of EMBR when the frequency and channel estimation tasks are coupled with the decision making process. Figure 5.3 shows the BER of the considered schemes as a function of E_b/N_0 with $\rho = 0.3$. Users have equal power and the number of iterations is $N_i = 5$ with both EMBR and HL. For comparison, we also illustrate the performance of the ideal system with perfect frequency and channel information, where all CFOs have been corrected at the MTs. Again, the best performance is achieved by EMBR. In particular, at an error rate of 10^{-2}, the gain over CTYH is approximately 4 dB while a loss of 3 dB is incurred with respect to the ideal system. As for HL, it performs poorly and exhibits an error floor at high SNRs.

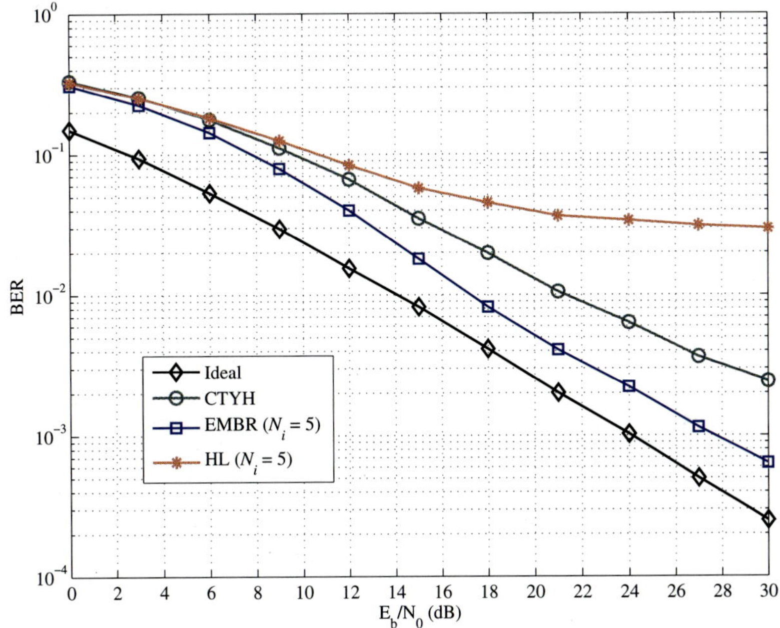

Fig. 5.3   BER performance vs. E_b/N_0 for uncoded QPSK and $\rho = 0.3$.

Resistance to near-far effect

In practical systems, power control is employed to mitigate the near-far problem arising from the different path losses incurred by uplink signals. However, power control cannot be assumed when a new user is entering the system as its power level is still to be measured. Therefore, it is of interest to assess the performance of the considered schemes in the presence of a strong interferer. For this purpose, we consider a scenario in which the power of user #2 is larger than that of the others by a factor $\alpha \geq 1$. This condition is obtained setting $\beta_2 = \sqrt{\alpha} \cdot \beta_1$ in Eq. (5.40), while keeping $\beta_m = \beta_1$ for $m = 3, 4$. Simulation results illustrating the BER of user #1 are shown in Fig. 5.4 as a function of α (expressed in dB) for $\rho = 0.3$ and $E_b/N_0 = 20$ dB. As expected, the system performance degrades with α. In particular, the BER of EMBR and CTYH increases by a factor of two when α passes from 0 to 5 dB, while larger degradations occur with HL.

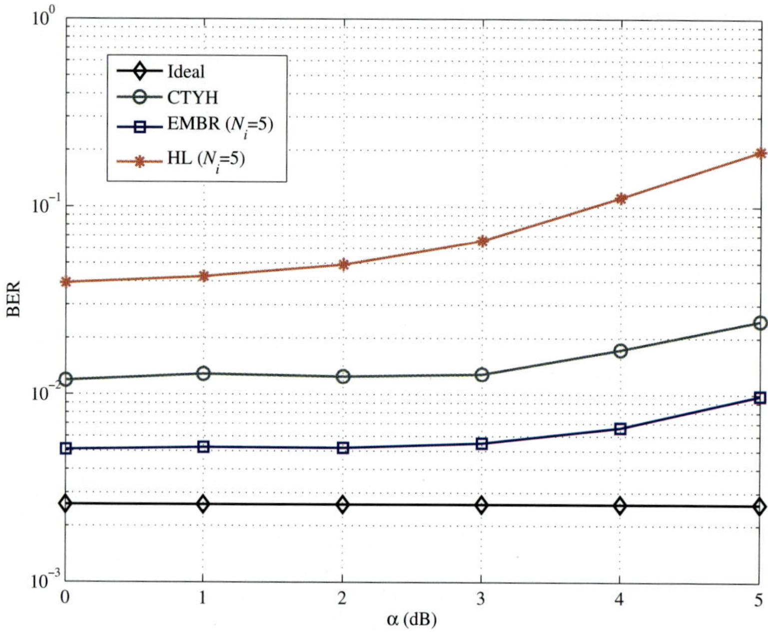

Fig. 5.4   BER performance in the presence of a strong interferer for uncoded QPSK with $E_b/N_0 = 20$ dB and $\rho = 0.3$.

5.2   Trellis-coded OFDMA uplink

The receiver structures discussed in the previous subsection are specifically designed for uncoded transmissions. On the other hand, we know that channel coding is a fundamental part of any multicarrier system as it provides a natural way for exploiting the frequency diversity offered by the multipath channel. For this reason, it is of practical interest to extend the EMBR to *coded* systems.

5.2.1   *Signal model for coded transmissions*

Figure 5.5 illustrates the basic block diagram of the mth MT transmitter in a coded OFDMA uplink. Here, a block of binary information data $\boldsymbol{a}_m$ is trellis-encoded into a vector $\boldsymbol{b}_m$ of coded bits. The latter are then fed to a block interleaver, which helps to break up error bursts. After dividing the interleaved bits $\boldsymbol{x}_m$ into adjacent segments of length ϑ, each segment

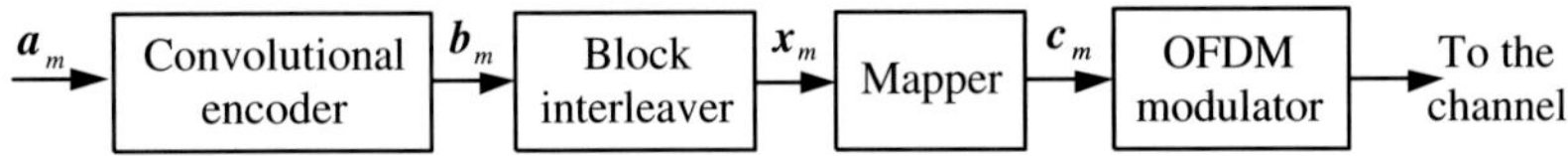

Fig. 5.5   Block diagram of the mth MT transmitter in a coded OFDMA system.

is mapped onto a modulation symbol taken from a constellation with 2^ϑ points. This produces a vector $\boldsymbol{c}_m$ of N symbols which is finally passed to the OFDM modulator and launched over the channel. At the BS receiver, the observation vector $\boldsymbol{r}$ is still expressed as in Eq. (5.5), where the entries of $\boldsymbol{c}_m$ are now coded symbols obtained as illustrated in Fig. 5.5.

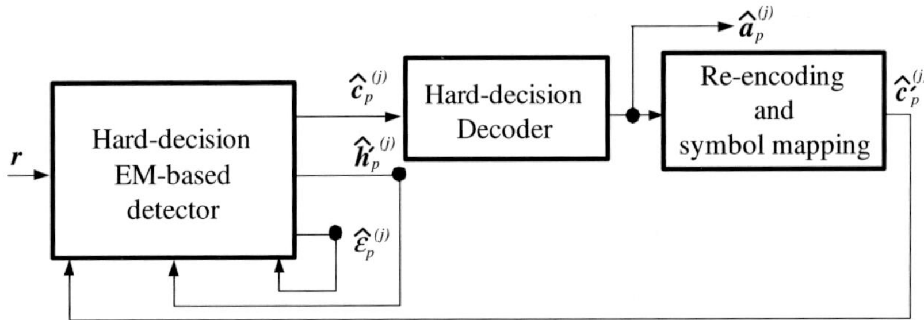

Fig. 5.6   Block diagram of an EM-based receiver employing a hard-decoding strategy.

One possible way for applying the EMBR to a coded OFDMA system is depicted in Fig. 5.6. As is seen, at each iteration the EM-based detector provides decisions about the coded symbols of all users, which are then passed to the hard-decoding unit. The retrieved information bits are re-encoded and re-mapped before being returned to the EM detector for the next iteration. This approach is relatively simple, but cannot provide optimum performance as it does not exploit any information regarding the likelihood of the detected symbols (also referred to as *soft information*). Inspired by the turbo decoding principle, a number of turbo processing techniques have recently been developed to improve the channel estimation [116] or interference suppression tasks [47] by taking advantage of the soft information associated with the decoded data. In the ensuing discussion, the turbo principle is applied to a coded OFDMA uplink. In particular, we exploit soft-decision feedback from a maximum a posteriori (MAP) decoder to jointly perform frequency synchronization, channel estimation and interference cancellation.

5.2.2 *Iterative detection and frequency synchronization with coded transmissions*

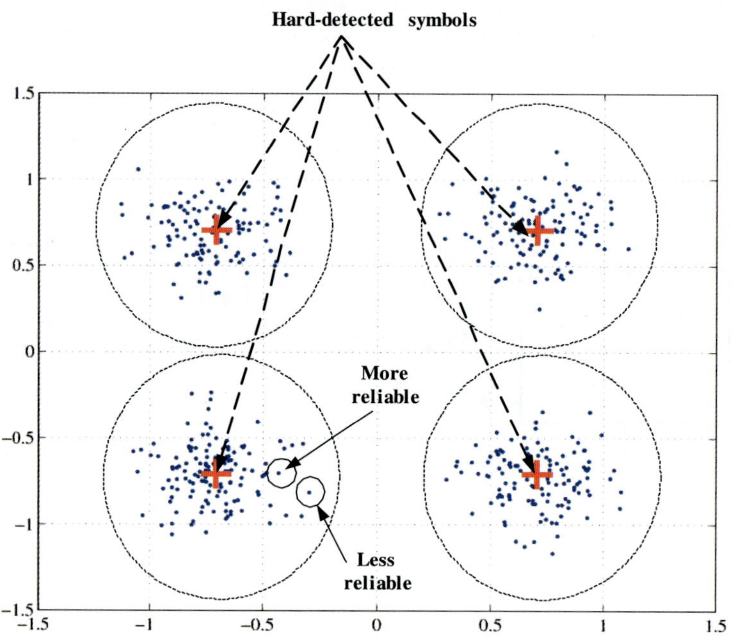

Fig. 5.7   Hard-decision detection in a QPSK transmission.

Figure 5.7 shows the classical concept of hard data detection of QPSK symbols. The noisy points in the I/Q diagram represent the output of the channel equalizer and are classified into one out of four possible constellation symbols. Although some of these points may be more reliable than others, the hard-decision process masks out this reliability since points lying in the same decision region are treated exactly in the same way, regardless of their distances from the corresponding constellation symbol. In coded systems, reliability information can be exploited by representing the tentative decoded symbols through their *statistical expectation*. In this way

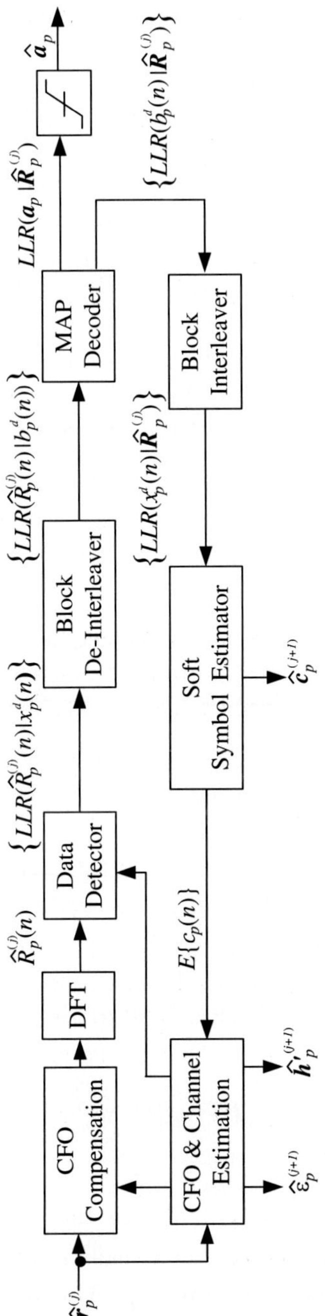

Fig. 5.8 Block diagram of the ECM-based MAP decoder.

the system performance is greatly improved as compared to hard-decision decoding.

We follow the same approach employed with uncoded transmissions and consider an iterative receiver structure in which a SAGE-based processor is first used to extract the contribution of each user, say $\hat{r}_p$ $(p = 1, 2, \ldots, M)$, from the received vector r. Each $\hat{r}_p$ is then exploited to estimate ε_p, h'_p and c_p in a joint fashion according to the ECM principle. The overall receiver architecture is depicted in Fig. 5.8. The main difference with respect to the uncoded case is that now the receiver can effectively exploit information about the reliability of the detected symbols.

The SAGE algorithm is applied in the same way as in uncoded systems. In particular, during the pth cycle of the jth iteration (with $p = 1, 2, \ldots, M$), the contribution of the pth user to the received vector r is estimated as

$$\hat{r}_p^{(j)} = r - \sum_{m=1}^{p-1} \hat{z}_m^{(j)} - \sum_{m=p+1}^{M} \hat{z}_m^{(j-1)}, \tag{5.41}$$

where $\hat{z}_m^{(j)}$ is given in Eq. (5.12) and represents an estimate of the signal $z_m = \Gamma(\varepsilon_m) F^H D(c_m) U h'_m$ received from the mth user.

Following the same steps outlined in Sec. 5.1.2.2, we substitute Eq. (5.5) into Eq. (5.41) and obtain

$$\hat{r}_p^{(j)} = \Gamma(\varepsilon_p) F^H D(c_p) U h'_p + \eta_p^{(j)}, \tag{5.42}$$

where $\eta_p^{(j)}$ is defined in Eq. (5.14).

The ML estimates of ε_p, h'_p and c_p are derived from $\hat{r}_p^{(j)}$ using the ECM algorithm. After initializing $\hat{c}_p^{(j,0)} = \hat{c}_p^{(j-1)}$ and $\hat{\varepsilon}_p^{(j,0)} = \hat{\varepsilon}_p^{(j-1)}$, the uth iteration of the ECM-based MAP decoder proceeds in the following way [116]. The estimated CFO $\hat{\varepsilon}_p^{(j,u)}$ is first used to compute the N-dimensional vector

$$\hat{R}_p^{(j)} = F \Gamma^H(\hat{\varepsilon}_p^{(j,u)}) \hat{r}_p^{(j)}, \tag{5.43}$$

with entries $\hat{R}_p^{(j)}(n)$ for $n = 0, 1, \ldots, N - 1$.

Next, we call $\{x_p^d(n); d = 0, 1, \ldots, \vartheta - 1\}$ the nth segment of ϑ interleaved bits that are mapped onto $c_p(n)$. Recalling that $\eta_p^{(j)}$ is nearly Gaussian distributed, the log-likelihood ratio (LLR) of $\hat{R}_p^{(j)}(n)$ conditioned on $x_p^d(n)$ is given by

$$LLR\left(\hat{R}_p^{(j)}(n)\,\big|x_p^d(n)\right) = \log\frac{\Pr\left(\hat{R}_p^{(j)}(n)\,\big|x_p^d(n)=+1\right)}{\Pr\left(\hat{R}_p^{(j)}(n)\,\big|x_p^d(n)=-1\right)}$$

$$= \log\frac{\displaystyle\sum_{\tilde{c}_p(n)\in\mathcal{S}_{+1}^d}\exp\left\{-\frac{|\hat{R}_p^{(j)}(n)-\hat{H}_{p,LS}^{\prime(j,u)}(n)\,\tilde{c}_p(n)|^2}{\sigma_\eta^2(j)}\right\}}{\displaystyle\sum_{\tilde{c}_p(n)\in\mathcal{S}_{-1}^d}\exp\left\{-\frac{|\hat{R}_p^{(j)}(n)-\hat{H}_{p,LS}^{\prime(j,u)}(n)\,\tilde{c}_p(n)|^2}{\sigma_\eta^2(j)}\right\}},$$

$$(5.44)$$

where $\mathcal{S}_\alpha^d$ (with $\alpha=\pm1$) is the set of constellation symbols for which $x^d=\alpha$, while $\hat{H}_{p,LS}^{\prime(j,u)}(n)$ represents the nth entry of $\hat{\boldsymbol{H}}_{p,LS}^{\prime}(\hat{\boldsymbol{\xi}}_p^{(j,u)})$. The latter is the LS estimate of the channel frequency response for a given $\hat{\boldsymbol{\xi}}_p^{(j,u)}=[\,\hat{\boldsymbol{c}}_p^{(j,u)T}\,\,\hat{\varepsilon}_p^{(j,u)}\,]^T$, and reads

$$\hat{\boldsymbol{H}}_{p,LS}^{\prime}(\hat{\boldsymbol{\xi}}_p^{(j,u)}) = \boldsymbol{U}\hat{\boldsymbol{h}}_{p,\mathrm{LS}}^{\prime}(\hat{\boldsymbol{\xi}}_p^{(j,u)}), \qquad (5.45)$$

where $\hat{\boldsymbol{h}}_{p,\mathrm{LS}}^{\prime}(\hat{\boldsymbol{\xi}}_p^{(j,u)})$ is defined in Eq. (5.27).

In an attempt of reducing the computational complexity, one can use the max-log approximation in Eq. (5.44) to obtain [116]

$$LLR\left(\hat{R}_p^{(j)}(n)\,\big|x_p^d(n)\right) \approx \max_{\tilde{c}_p(n)\in\mathcal{S}_{+1}^d}\left\{-|\hat{R}_p^{(j)}(n)-\hat{H}_{p,LS}^{\prime(j,u)}\tilde{c}_p(n)|^2\right\}$$

$$-\max_{\tilde{c}_p(n)\in\mathcal{S}_{-1}^d}\left\{-|\hat{R}_p^{(j)}(n)-\hat{H}_{p,LS}^{\prime(j,u)}\tilde{c}_p(n)|^2\right\},$$

$$(5.46)$$

where the quantity $\sigma_\eta^2(j)$ has been dropped since the frequent renormalization process during MAP decoding removes in practice the effect of any common factors.

The sequence $\left\{LLR\left(\hat{R}_p^{(j)}(n)\,\big|x_p^d(n)\right)\right\}$ at the output of the data detector is then de-interleaved to yield $\left\{LLR\left(\hat{R}_p^{(j)}(n)\,\big|b_p^d(n)\right)\right\}$. These quantities are employed by the MAP decoder to generate the sequence $\left\{LLR\left(b_p^d(n)\,\big|\hat{\boldsymbol{R}}_p^{(j)}\right)\right\}$ and $LLR\left(\boldsymbol{a}_p\,\big|\hat{\boldsymbol{R}}_p^{(j)}\right)$ using the BCJR algorithm [5]. Readers are referred to [79] and references therein for a formal treatment of the BCJR algorithm. Finally, the stream $\left\{LLR\left(b_p^d(n)\,\big|\hat{\boldsymbol{R}}_p^{(j)}\right)\right\}$ is interleaved and employed to evaluate the expected values of the coded channel symbols $\boldsymbol{c}_p$.

Letting $\hat{c}_p^{(j,u+1)}(n) = E\{c_p(n)\}$ and assuming for simplicity a QPSK constellation ($d = 0, 1$), it can be shown that [116]

$$\hat{c}_p^{(j,u+1)}(n) = \frac{1}{\sqrt{2}}\left[\frac{e^{LLR\left(x_p^0(n)\mid\hat{R}_p^{(j)}\right)} - 1}{e^{LLR\left(x_p^0(n)\mid\hat{R}_p^{(j)}\right)} + 1} + j\frac{e^{LLR\left(x_p^1(n)\mid\hat{R}_p^{(j)}\right)} - 1}{e^{LLR\left(x_p^1(n)\mid\hat{R}_p^{(j)}\right)} + 1}\right].$$

$$(5.47)$$

The detected symbols $\{\hat{c}_p^{(j,u+1)}(n)\}$ are grouped to form a vector $\hat{\boldsymbol{c}}_p^{(j,u+1)}$ defined as

$$\hat{\boldsymbol{c}}_p^{(j,u+1)} \stackrel{\text{def}}{=} [\hat{c}_p^{(j,u+1)}(0), \hat{c}_p^{(j,u+1)}(1), \ldots, \hat{c}_p^{(j,u+1)}(N-1)]^T, \qquad (5.48)$$

which is next employed to update the CFO estimate according to Eq. (5.34).

Finally, $\hat{\varepsilon}_p^{(j,u+1)}$ and $\hat{\boldsymbol{c}}_p^{(j,u+1)}$ are substituted into Eq. (5.45) to update the channel estimates. After N_U iterations, we terminate the ECM process and update the SAGE processor with

$$[\hat{\varepsilon}_p^{(j)}, \hat{\boldsymbol{h}}'_p^{(j)}, \hat{\boldsymbol{c}}_p^{(j)}] = [\hat{\varepsilon}_p^{(j,N_U)}, \hat{\boldsymbol{h}}'_{p,\text{LS}}(\hat{\boldsymbol{\xi}}_p^{(j,N_U)}), \hat{\boldsymbol{c}}_p^{(j,N_U)}]. \qquad (5.49)$$

In summary, during the pth cycle of the jth iteration (with $p = 1, 2, \ldots, M$), the iterative algorithm proceeds as follows.

E-Step:

Compute $\hat{\boldsymbol{r}}_p^{(j)}$ according to Eq. (5.41);

M-Step:

- Update $\hat{\boldsymbol{H}}'_{p,LS}(\hat{\boldsymbol{\xi}}_p^{(j,u)})$ based on Eq. (5.45) and compute

$$LLR\left(\hat{R}_p^{(j)}(n)\mid x_p^d(n)\right) \approx \max_{\tilde{c}_p(n)\in\mathcal{S}_{+1}^d}\left\{-|\hat{R}_p^{(j)}(n) - \hat{H}'^{(j,u)}_{p,LS}\tilde{c}_p(n)|^2\right\}$$

$$- \max_{\tilde{c}_p(n)\in\mathcal{S}_{-1}^d}\left\{-|\hat{R}_p^{(j)}(n) - \hat{H}'^{(j,u)}_{p,LS}\tilde{c}_p(n)|^2\right\}.$$

$$(5.50)$$

- Generate $\left\{LLR\left(b_p^d(n)\mid\hat{R}_p^{(j)}\right)\right\}$ and $LLR\left(a_p\mid\hat{R}_p^{(j)}\right)$ by exploiting $\left\{LLR\left(\hat{R}_p^{(j)}(n)\mid b_p^d(n)\right)\right\}$ using the BCJR algorithm;
- Update $\hat{\boldsymbol{c}}_p^{(j,u+1)}$ and the estimation parameters based on Eqs. (5.47) and (5.49), respectively;
- Finally, use updated parameters to obtain the following vector

$$\hat{\boldsymbol{z}}_p^{(j)} = \boldsymbol{\Gamma}(\hat{\varepsilon}_p^{(j)})\boldsymbol{F}^H \boldsymbol{D}(\hat{\boldsymbol{c}}_p^{(j)})\boldsymbol{U}\hat{\boldsymbol{h}}'_p^{(j)}. \qquad (5.51)$$

5.2.3 *Performance assessment*

The performance of EMBR when applied to a coded OFDMA uplink is assessed by computer simulations under the same operating conditions of Fig. 5.3. The only difference is that the information bits are now encoded by a rate-1/2 convolutional encoder with generator polynomials $(5, 7)$ (in octal) and an 8×8 block interleaver is employed to scramble the coded bits within the OFDM block. The interleaved bits are then mapped onto QPSK symbols using a Gray map. The number N_U of ECM iterations is set to 3 while the number of SAGE iterations is $N_i = 5$. The CTYH scheme is used to initialize the EMBR. Again, results are only provided for user #1.

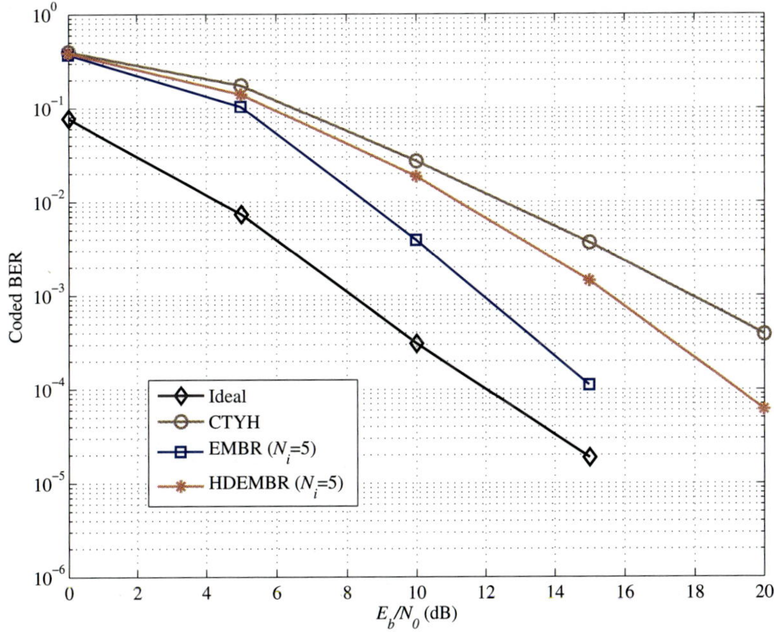

Fig. 5.9   BER performance vs. E_b/N_0 for a coded QPSK transmission.

Figure 5.9 illustrates BER results as a function of E_b/N_0 in case of users with equal average power and $\rho = 0.3$. The curve labeled "ideal" corresponds to perfect knowledge of CFOs and channel responses and provides a benchmark for the BER performance. At an error rate of 10^{-3}, the gain of EMBR with respect to CTYH is nearly 6 dB after five iterations, while a loss of 4 dB is incurred with respect to the ideal system. For comparison, we also show the performance of a hard-decision EM-based receiver (HDEMBR) which operates as illustrated in Fig. 5.6 using a hard-decoding Viterbi processor. As is seen, HDEMBR performs poorly since hard-decoding does not allow to exploit any reliability information.

Chapter 6

Dynamic Resource Allocation

One attractive feature of multicarrier transmissions is the possibility of dynamically allocating system resources according to the changing environmental conditions. Many studies have demonstrated that significant performance improvement is achieved in single-user OFDM systems if transmission power and data rate are properly adjusted over each subcarrier to take advantage of the channel frequency selectivity. This idea is usually referred to as *adaptive modulation* while the set of algorithms and protocols governing it is known as *link adaptation* [13, 75].

The goal of any link adaptation algorithm is to ensure that the most efficient set of modulation parameters (or transmission mode) is always used over varying channel conditions. Different mode selection criteria can be envisaged depending on whether the system is attempting to maximize the overall data throughput under a total power constraint or to minimize the overall transmit power given a fixed throughput. In any case, the adaptation algorithm tends to allocate more information bits onto better quality subcarriers, *i.e.*, those exhibiting the highest signal-to-noise ratios (SNRs), whereas small-size constellations are normally employed over severely faded subcarriers in order to increase their robustness against thermal noise. In some extreme situations a number of subcarriers may even be left unused if the corresponding SNR is too poor for reliable data transmission. In the related literature, the problem of efficiently mapping information bits over the available carriers is referred to as *bit loading*.

The concept of link adaptation has also been extended to OFDMA systems. In this case the base station (BS) not only has the opportunity of optimally allocating power and data rate over different subchannels, but can also exploit instantaneous channel state information for dynamically distributing subcarriers to the active users. The adoption of a dynamic

159

carrier assignment scheme allows a more effective use of the available system resources, even though it complicates the link adaptation problem to a large extent as compared to point-to-point communications.

The aim of this chapter is to present the basic concept of link adaptation in multicarrier systems. Section I investigates adaptive bit and power loading in single-user OFDM applications. Here, we revisit the classical water-filling power allocation policy and formulate the *rate-maximization* and *margin-maximization* problems. Practical bit loading schemes based on greedy techniques are illustrated for either uniform or non-uniform power allocation. We also present the concept of subband adaptation and discuss some signaling schemes enabling exchange of side information between the transmit and receive ends of an adaptive modulation system.

Section II is devoted to link adaptation in a multiuser OFDM network. After discussing the multiaccess water-filling principle, we extend the rate-maximization and margin-maximization concepts to a typical OFDMA downlink scenario. As we shall see, in such a case optimum assignment of system resources results into a multidimensional optimization problem which does not lend itself to any practical solution. To overcome this difficulty, we present some suboptimum schemes in which the subcarrier allocation and bit loading tasks are performed separately and with affordable complexity.

6.1   Resource allocation in single-user OFDM systems

The research on resource allocation in multicarrier systems was fueled by the success of the asymmetric digital subscriber line (ADSL) service in the early nineties [1, 8]. This technology employs a Digital Multitone (DMT) modulation for high-speed wireline data transmissions. Due to crosstalk from adjacent copper twisted pairs, the ADSL channel is characterized by remarkable frequency-selectivity. The latter can usefully be exploited as a source of diversity by applying suitable link adaptation techniques.

In this Section we review the main concepts behind bit and power loading in point-to-point OFDM transmissions. Although originally devised for ADSL applications, the investigated methods apply to multicarrier wireless services as well. The only requirement is that the fading rate is not too fast, as dynamic resource allocation is hardly usable in the presence of rapidly-varying transmission channels.

6.1.1 *Classic water-filling principle*

We start discussing the water-filling power allocation principle, which allows one to achieve the theoretical capacity offered by a frequency-selective channel. Capacity is operationally defined as the maximum data rate that the channel can support with an arbitrarily low error-rate probability. From an information theoretic perspective, it represents the maximum mutual information between the transmitted data symbols and the received signal vector, where maximization is performed over the probability density function (pdf) of the transmitted data [27]. In the ensuing discussion, these concepts are applied to an OFDM communication system.

Assuming perfect timing and frequency synchronization, the output from the receive DFT is expressed by

$$R(n) = H(n)S(n) + W(n), \qquad 0 \le n \le N - 1. \tag{6.1}$$

where $H(n)$ is the channel frequency response over the nth subcarrier, $S(n)$ the corresponding input symbol with power $P_n = \mathrm{E}\{|S(n)|^2\}$ and $W(n)$ is white Gaussian noise with zero-mean and variance σ_w^2. Inspection of Eq. (6.1) indicates that the OFDM channel can be viewed as a collection of parallel independent AWGN subchannels, one for each subcarrier.

In a practical system, the transmitted power is normally constrained to some value P_{budget}. Mathematically, this amounts to setting

$$\sum_{n=0}^{N-1} P_n \le P_{\text{budget}}, \tag{6.2}$$

with $P_n \ge 0$ for $n = 0, 1, \ldots, N - 1$. It is known that among all input vectors $\boldsymbol{S} = [S(0), S(1), \ldots, S(N - 1)]^T$ satisfying the overall power constraint Eq. (6.2), the mutual information $I(\boldsymbol{S}, \boldsymbol{R})$ between $\boldsymbol{S}$ and the observation vector $\boldsymbol{R} = [R(0), R(1), \ldots, R(N - 1)]^T$ is maximized when the data symbols $\{S(n)\}$ are statistically independent and Gaussian distributed with zero-mean [105]. In this case we have

$$I(\boldsymbol{S}, \boldsymbol{R}) = \sum_{n=0}^{N-1} \log_2 \left(1 + \frac{P_n |H(n)|^2}{\sigma_w^2} \right). \tag{6.3}$$

The channel capacity C is obtained by maximizing the right-hand-side of Eq. (6.3) with respect to $\boldsymbol{P} = [P_0, P_1, \ldots, P_{N-1}]^T$, *i.e.*,

$$C = \max_{\boldsymbol{P}} \left\{ \sum_{n=0}^{N-1} \log_2 \left(1 + \frac{P_n |H(n)|^2}{\sigma_w^2} \right) \right\}. \tag{6.4}$$

Since the objective function in Eq. (6.4) is convex in the variables $\{P_n\}$, the optimum power allocation under the convex constraints Eq. (6.2) can be found using Lagrangian methods. For this purpose, we consider the augmented cost function

$$J = \sum_{n=0}^{N-1} \log_2\left(1 + \frac{P_n\,|H(n)|^2}{\sigma_w^2}\right) + \lambda\left(P_{\text{budget}} - \sum_{n=0}^{N-1} P_n\right), \qquad (6.5)$$

where λ is the Lagrangian multiplier. The Kuhn–Tucker (KT) optimality conditions are given by

$$\text{KT conditions:} \quad \begin{cases} \frac{\partial J}{\partial P_n} = 0 & \text{if } P_n > 0 \\[2mm] \frac{\partial J}{\partial P_n} \leq 0 & \text{if } P_n = 0 \end{cases} \qquad (6.6)$$

where $\partial J/\partial P_n$ is the derivative of J with respect to P_n, which reads

$$\frac{\partial J}{\partial P_n} = \frac{1}{\left[P_n + \sigma_w^2/\,|H(n)|^2\right]\ln 2} - \lambda. \qquad (6.7)$$

The optimum power allocation satisfying the KT conditions is found to be

$$P_n^{(opt)} = \left(\mu - \frac{1}{\gamma_n}\right)^+, \qquad (6.8)$$

where $(x)^+ = \max\{x, 0\}$, $\gamma_n = |H(n)|^2/\sigma_w^2$ is the so-called *channel SNR* and $\mu = 1/(\lambda\ln 2)$ is a parameter that must be chosen so as to meet the total transmit power constraint

$$\sum_{n=0}^{N-1}\left(\mu - \frac{1}{\gamma_n}\right)^+ = P_{\text{budget}}. \qquad (6.9)$$

This solution lends itself to an interesting physical interpretation. As depicted in Fig. 6.1, the quantities $1/\gamma_n$ can be thought of as the bottom of a vessel in which the transmit power P_{budget} is poured similarly to water. In particular, the quantity μ represents the height of the water surface, while $P_n^{(opt)}$ is the depth of the water at subcarrier n. Since the power allocation process resembles the way by which water distributes itself in a vessel, this optimal strategy is referred to as *water-filling* or *water-pouring*. It is worth noting that the bottom level may occasionally become higher than the water surface. When this happens, no power is allocated over the corresponding subcarriers since the latter are too faded for supporting reliable data transmission. In general, the water-filling strategy takes advantage of the channel frequency-selectivity by giving more power to high-quality subcarriers while those characterized by the worst channel SNRs are used to a lesser extent or avoided altogether. Once the power has been optimally distributed over the signal spectrum according to Eq. (6.8), specific coding techniques should be employed over each subcarrier to attain the data rate promised by the channel capacity.

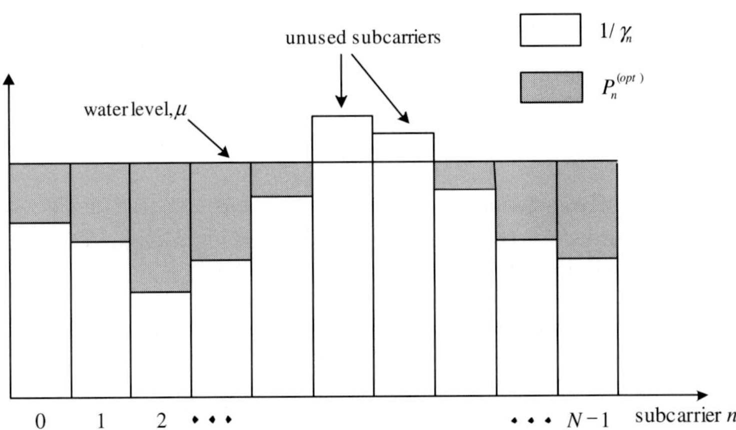

Fig. 6.1   Water-filling power allocation over the available subcarriers.

Inspection of Eq. (6.9) reveals that the water level μ is related to the quantities $1/\gamma_n$ and P_{budget} but, unfortunately, the presence of the non-linear operator $(\cdot)^+$ prevents the possibility of computing it in closed-form. As a consequence, the optimum power allocation specified by Eq. (6.8) can only be found through iterative procedures. Two prominent schemes have been suggested in the literature. In the first one, a tentative level μ is re-calculated at each new iteration after discarding the subcarrier that exhibits the lowest channel SNR. Specifically, denote $\mathcal{N}^{(i)}$ the set of subcarrier indices that are considered for power allocation during the ith iteration, where $\mathcal{N}^{(0)} = \{0, 1, 2, \ldots, N-1\}$ is used for initialization purposes. Then, the water level is first computed from Eq. (6.9) as

$$\mu^{(i)} = \frac{1}{\text{card}\left\{\mathcal{N}^{(i)}\right\}} \left[P_{\text{budget}} + \sum_{n \in \mathcal{N}^{(i)}} 1/\gamma_n \right], \qquad (6.10)$$

where card$\{\cdot\}$ represents the cardinality of the enclosed set. This value is next inserted into Eq. (6.8) to obtain the tentative power allocated over the

nth subchannel in the form

$$P_n^{(i)} = \begin{cases} \mu^{(i)} - 1/\gamma_n, & \text{if } n \in \mathcal{N}^{(i)}, \\ 0 & \text{otherwise.} \end{cases} \tag{6.11}$$

At the end of each iteration, if the subcarrier with the lowest channel gain has a negative power assignment (*i.e.*, $P_n^{(i)} < 0$), we discard this subcarrier from the iterative process by setting the corresponding power level to zero and removing its index from $\mathcal{N}^{(i)}$. The remaining subcarriers are then used to form the set $\mathcal{N}^{(i+1)}$ which is employed in the next iteration. The algorithm stops as soon as all power assignments are non-negative. In the sequel, this method is referred to as the *iterative subcarrier-removal* algorithm.

An alternative scheme to solve the non-linear Eq. (6.8) with respect to μ relies on the use of the well-known bisection algorithm. To explain this method, we denote

$$P(\widetilde{\mu}) = \sum_{n=0}^{N-1} \left(\widetilde{\mu} - \frac{1}{\gamma_n} \right)^+ \tag{6.12}$$

the total required power for a given water level $\widetilde{\mu}$, and assume that during the ith iteration the desired water level μ lies in a coarsely estimated interval $I^{(i)} = [\mu_\ell^{(i)}, \mu_u^{(i)}]$. Then, we take the middle point of $I^{(i)}$ as a rough estimate of μ, say $\widehat{\mu}^{(i)} = (\mu_\ell^{(i)} + \mu_u^{(i)})/2$, and evaluate the corresponding required power $P(\widehat{\mu}^{(i)})$ based on Eq. (6.12). A refined estimate of μ is thus obtained by comparing $P(\widehat{\mu}^{(i)})$ with P_{budget}. Specifically, if $P(\widehat{\mu}^{(i)}) < P_{\text{budget}}$ the interval $I^{(i+1)} = [\mu_\ell^{(i+1)}, \mu_u^{(i+1)}]$ to be used in the next iteration is such that $\mu_\ell^{(i+1)} = \widehat{\mu}^{(i)}$ and $\mu_u^{(i+1)} = \mu_u^{(i)}$, otherwise we set $\mu_\ell^{(i+1)} = \mu_\ell^{(i)}$ and $\mu_u^{(i+1)} = \widehat{\mu}^{(i)}$. In this way the interval width is halved at each new iteration, thereby improving the accuracy of the estimated water level. The algorithm is stopped as soon as $\mu_u^{(i)} - \mu_\ell^{(i)} < \epsilon$, where ϵ is a specified positive parameter. Clearly, smaller values of ϵ result into more accurate estimates of μ.

Example 6.1 For illustration purposes, in this example we consider an OFDM system with only eight subcarriers. The channel is frequency-selective and characterized by the SNR values given in Table 6.1.

The goal is to distribute an overall power $P_{\text{budget}} = 1$ over the available subcarriers using either the iterative subcarrier-removal method or the bisection algorithm. The latter is initialized with $\mu_\ell^{(0)} = 0.1$ and $\mu_u^{(0)} = 0.6$, while the stopping criterion is $\mu_u^{(i)} - \mu_\ell^{(i)} < 10^{-4}$. Although both schemes achieve the same final power distribution depicted in Fig. 6.2, the

Table 6.1   Channel SNRs in Example 6.1.

Subcarrier index, n	Channel SNR, γ_n (dB)
1	-0.7791
2	6.1063
3	19.7239
4	36.8800
5	41.3190
6	23.1618
7	31.4632
8	26.6705

subcarrier-removal method stops after just one iteration whereas it takes 13 iterations for the bisection algorithm to reach the same result. Clearly, the convergence speed of the bisection procedure is largely determined by the width of the initialization interval $I^{(0)}$. As a final remark, we observe that the first two subcarriers in Fig. 6.2 are left unused due to their poor channel quality.

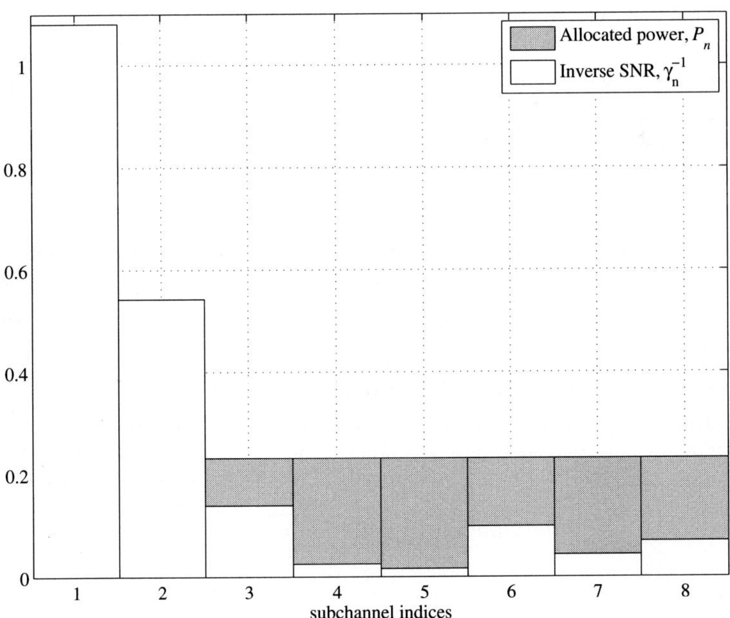

Fig. 6.2   Water-filling power distribution in Example 6.1.

6.1.2 *Rate maximization and margin maximization*

Although the water-filling solution represents the optimal power assignment strategy for maximizing the data rate, its practical relevance is limited by the fact that it does not provide any clear indication about the kind of signaling and coding schemes that must be used over each subcarrier to approach the theoretical channel capacity. In addition, it tacitly assumes an arbitrarily low error-rate probability, whereas practical communication systems are normally designed for a non-zero target error-rate which is specified by the requested quality-of-service. These inherent drawbacks of the water-filling principle have motivated an intense research activity toward the development of efficient bit and power loading schemes operating under a variety of error probability constraints. For instance, in [10, 20, 71] transmission power and data rate are assigned such that the bit-error-rate (BER) across tones does not exceed a given threshold $p_{e,\max}$. This results into the following *uniform* BER constraint

$$p_{e,n} \le p_{e,\max}, \quad n \in \mathcal{N} \tag{6.13}$$

where $p_{e,n}$ is the BER over the nth subcarrier and $\mathcal{N}$ the set of modulated subcarriers. A less stringent requirement is adopted in [171] and [173] by specifying the *average* error probability over the entire OFDM block. If b_n is the number of bits allocated over the nth subcarrier, the corresponding constraint is stated as

$$\bar{p}_e = \frac{\sum_{n=0}^{N-1} b_n\, p_{e,n}}{\sum_{n=0}^{N-1} b_n} \le \bar{p}_{e,\max}, \tag{6.14}$$

and results into a *non-uniform* error probability across subcarriers.

Whatever the adopted BER constraint, practical loading algorithms are normally derived on the basis of two main optimization criterions. A first possibility is to distribute a given amount of power P_{budget} over the available subcarriers such that the number of bits per transmitted block is maximized. This results into the following *rate-maximization* concept (RMC)

$$\text{maximize} \quad R_b = \sum_{n=0}^{N-1} b_n \tag{6.15}$$

subject to

$$\sum_{n=0}^{N-1} P_n = P_{\mathrm{budget}}, \quad \text{with} \quad b_n, P_n \ge 0 \tag{6.16}$$

where P_n is the power allocated over the nth subcarrier.

The second approach is known as the *margin-maximization* concept (MMC), which aims at minimizing the overall transmission power for a given target data rate R_{target}. Mathematically, we have

$$\text{minimize} \quad P_T = \sum_{n=0}^{N-1} P_n \tag{6.17}$$

subject to

$$\sum_{n=0}^{N-1} b_n = R_{\text{target}}, \quad \text{with} \quad b_n, P_n \geq 0. \tag{6.18}$$

Although RMC and MMC represent the most popular approaches for the design of loading algorithms, in some applications there might be the desire to employ a given power P_{budget} to transmit at a target data rate with the lowest possible error probability. A practical scheme based on this concept is found in [3].

6.1.3 *Rate-power function*

The uniform BER constraint Eq. (6.13) establishes a strict relationship between the number b_n of bits allocated over the nth subcarrier and the corresponding transmission power P_n. The functional dependence between these quantities is dictated by the specified BER $p_{e,n}$ and by the available coding and modulation schemes. For instance, with an uncoded BPSK transmission ($b_n = 1$) we have [123]

$$p_{e,n} = Q\left(\sqrt{2P_n\gamma_n}\right), \tag{6.19}$$

where $\gamma_n = |H(n)|^2/\sigma_w^2$ is the channel SNR over the nth subcarrier while the Q-function is defined as

$$Q(x) = \frac{1}{\sqrt{2\pi}} \int_x^\infty e^{-t^2/2} dt. \tag{6.20}$$

For QPSK ($b_n = 2$), 16-QAM ($b_n = 4$) and 64-QAM ($b_n = 6$) constellations with Gray mapping the uncoded BER is reasonably approximated as [123]

$$p_{e,n} \approx \frac{4}{b_n}\left(1 - \frac{1}{2^{b_n/2}}\right) Q\left(\sqrt{\frac{3P_n\gamma_n}{2^{b_n} - 1}}\right). \tag{6.21}$$

In some works [10, 114] the *gap-approximation* analysis is adopted to establish a more general relationship between P_n and b_n in the form [23]

$$b_n = \log_2\left(1 + \frac{P_n\gamma_n}{\Gamma_n}\right), \tag{6.22}$$

where Γ_n is the so-called *SNR gap*, which is calculated on the basis of the target BER, the selected coding scheme and the system performance margin. Unfortunately, the gap approximation provides accurate results only when the size of the employed constellation is adequately large, a situation that is typical of ADSL applications but rarely occurs in wireless communications. Some useful comments on the validity of Eq. (6.22) are given in [3].

Solving Eqs. (6.19), (6.21) or (6.22) with respect to $P_n\gamma_n$ yields

$$P_n\gamma_n = f(b_n, p_{e,n}), \qquad (6.23)$$

where $f(b,p)$ is referred to as the *rate-power* function. The latter is normally viewed as a function of the variable b with p as a parameter. In practice, it represents the *received* SNR that is required on a given sub-carrier for reception of b information bits at a target BER p. Figure 6.3 illustrates $f(b,p)$ vs. b for $p = 10^{-5}$ and some popular coding and modulation schemes. The continuous function approximation is derived from Eq. (6.22) and is expressed by

$$\widetilde{f}(b,p) = \widetilde{\Gamma}\left(2^b - 1\right), \qquad (6.24)$$

where $\widetilde{\Gamma}$ is selected so as to fit the points corresponding to the considered coding/modulation schemes in a least-squares sense.

6.1.4 *Optimal power allocation and bit loading under BER constraint*

The optimal solutions to the RMC and MMC problems are not available in closed-form and can only be approached through iterative methods. To see how this comes about, in what follows we restrict our attention to the RMC criterion (similar reasonings also apply to the MMC case). We begin by considering an average error rate constraint and state the optimization problem as

$$\text{maximize} \quad R_b = \sum_{n=0}^{N-1} b_n \qquad (6.25)$$

with respect to $\boldsymbol{b} = [b_0, b_1, \ldots, b_{N-1}]^T$ and $\boldsymbol{P} = [P_0, P_1, \ldots, P_{N-1}]^T$, subject to

$$\sum_{n=0}^{N-1} P_n = P_{\text{budget}}, \qquad (6.26)$$

$$\overline{p}_e(\boldsymbol{b}, \boldsymbol{P}) = \frac{\sum_{n=0}^{N-1} b_n\, p_{e,n}(b_n, \gamma_n P_n)}{\sum_{n=0}^{N-1} b_n} = \overline{p}_{e,\max}, \qquad (6.27)$$

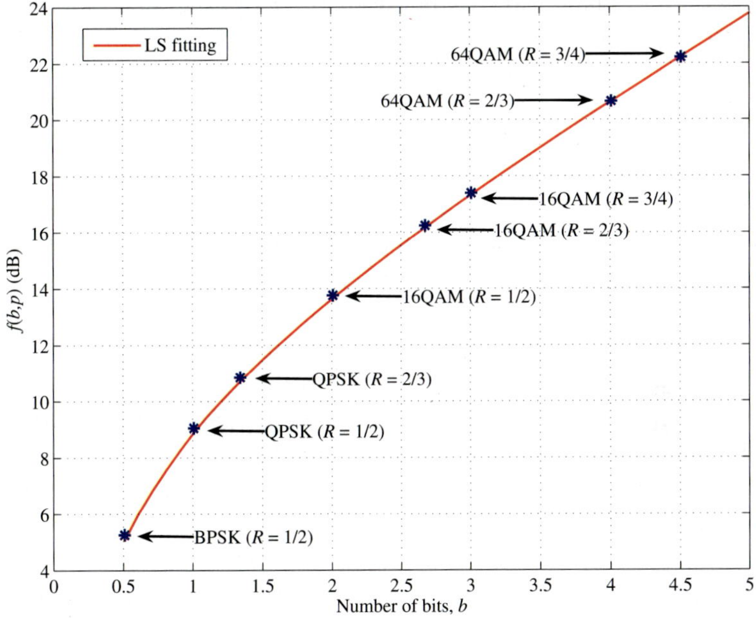

Fig. 6.3   Rate-power function.

with $b_n, P_n \geq 0$. Here, we treat each b_n as a continuous variable and assume that the functional dependence of $p_{e,n}$ on the quantities b_n and $\gamma_n P_n$ is specified in some way using the rate-power function.

The Lagrangian function for the constrained maximization problem Eq. (6.25) is defined as

$$L(\boldsymbol{b}, \boldsymbol{P}, \boldsymbol{\lambda}) = \sum_{n=0}^{N-1} b_n + \lambda_1 \left[P_{\text{budget}} - \sum_{n=0}^{N-1} P_n \right] + \lambda_2 \left[\overline{p}_{e,\max} - \overline{p}_e(\boldsymbol{b}, \boldsymbol{P}) \right],$$

$$(6.28)$$

where $\boldsymbol{\lambda} = [\lambda_1, \lambda_2]^T$ is the set of Lagrangian multipliers. Conditions for optimum bit and power loading are derived by setting to zero the derivatives of $L(\boldsymbol{b}, \boldsymbol{P}, \boldsymbol{\lambda})$ with respect to $\boldsymbol{b}$ and $\boldsymbol{P}$. This produces the following set of $2N$ equations

$$\begin{cases} \dfrac{\partial L}{\partial b_n} = 1 - \lambda_2 \dfrac{\partial \overline{p}_e}{\partial b_n} = 0, \\[2mm] \dfrac{\partial L}{\partial P_n} = -\lambda_1 - \lambda_2 \dfrac{\partial \overline{p}_e}{\partial P_n} = 0, \end{cases} \qquad (6.29)$$

for $n = 0, 1, \ldots, N - 1$.

After appropriate definition of the constant terms ξ_1 and ξ_2, Eq. (6.29) can also be rewritten as

$$\begin{cases} \dfrac{\partial \overline{p}_e}{\partial b_n} = \xi_1, \\[2ex] \dfrac{\partial \overline{p}_e}{\partial P_n} = \xi_2, \end{cases} \tag{6.30}$$

for $n = 0, 1, \ldots, N - 1$.

Unfortunately, there is no explicit solution to the conditions Eq. (6.30). An iterative algorithm for approaching the optimal vectors $\boldsymbol{b}$ and $\boldsymbol{P}$ is proposed in [8] using convex simplex techniques. This scheme requires a search over a multidimensional parameter space and exhibits a long convergence time which makes it unsuited for practical implementation.

A certain reduction of complexity is possible if we replace the average error probability constraint in Eq. (6.27) with a uniform BER constraint in which the same BER $p_{e,\max}$ is imposed over all subcarriers, *i.e.*,

$$p_{e,n}(b_n, \gamma_n P_n) = p_{e,\max}, \tag{6.31}$$

for $n = 0, 1, \ldots, N - 1$.

In this way the optimization process has only to be performed with respect to $\boldsymbol{b}$ rather than over the set $(\boldsymbol{b}, \boldsymbol{P})$ since the power P_n is univocally determined by the constraint Eq. (6.31) once b_n has been specified. Indeed, using the rate-power function defined in Eq. (6.23), we have

$$P_n = \frac{f(b_n, p_{e,\max})}{\gamma_n}. \tag{6.32}$$

The cost function for the new optimization problem takes the form

$$L(\boldsymbol{b}, \lambda) = \sum_{n=0}^{N-1} b_n + \lambda \left[P_{\text{budget}} - \sum_{n=0}^{N-1} \frac{f(b_n, p_{e,\max})}{\gamma_n} \right], \tag{6.33}$$

and conditions for optimal bit allocation are found by setting to zero the derivative of $L(\boldsymbol{b}, \lambda)$ with respect to $\boldsymbol{b}$. This yields

$$\frac{\partial f(b_n, p_{e,\max})}{\partial b_n} = \gamma_n / \lambda, \tag{6.34}$$

for $n = 0, 1, \ldots, N - 1$, from which it follows that the data rate is maximized when each subcarrier operates at a slope γ_n / λ over the rate-power function. An iterative algorithm to approach the solution Eq. (6.34) has been proposed by Campello in [10]. Compared to [171], this scheme is much simpler to implement and also exhibits faster convergence thanks to the reduced number of optimization parameters. The price for these advantages is a slight reduction of the achievable data rate as a consequence of the uniform BER constraint in Eq. (6.31). The latter is more stringent than the average constraint in Eq. (6.27) and inevitably reduces the number of degrees of freedom that are exploited by the optimization process.

6.1.5 *Greedy algorithm for power allocation and bit loading*

The RMC and MMC problems as stated in the previous subsections assume a constellation size with infinite granularity and their optimum solution will invariably lead to noninteger bit allocation across tones. A more practical approach is to specify a finite set of allowable PSK or QAM constellations, which are then selected on a subcarrier-by-subcarrier basis according to the relevant channel gains. Hence, it is of interest to look for efficient bit and power loading schemes that result into the assignment of an integer number of bits over each subcarrier. For this purpose, we still concentrate on the RMC problem which is now restated as

$$\text{maximize} \quad R_b = \sum_{n=0}^{N-1} b_n \tag{6.35}$$

with respect to $\boldsymbol{b} = [b_0, b_1, \ldots, b_{N-1}]^T$ and $\boldsymbol{P} = [P_0, P_1, \ldots, P_{N-1}]^T$ under either a uniform or average BER constraint and subject to

$$\sum_{n=0}^{N-1} P_n \leq P_{\text{budget}}, \tag{6.36}$$

$$b_n \in \{0, 1, \ldots, b_{\max}\}, \tag{6.37}$$

where $P_n \geq 0$ and $2^{b_{\max}}$ is the maximum size of the employed constellations.

The optimization problem formulated in Eqs. (6.35)-(6.37) has been extensively studied by many authors (see for example, [20, 56, 82, 123]). Its solution is found through iterative *greedy* techniques in which bit loading across tones is performed incrementally or decrementally one bit at a time. From an operational point of view, we distinguish between *bit-filling* and *bit-removal* schemes. In the former case we start from an initial all-zero bit allocation and add one bit at a time to the subcarrier requiring the least additional power to meet the specified BER constraint. Vice versa, the bit-removal approach starts with an initial maximum bit allocation $b_n = b_{\max}$ for $n = 0, 1, \ldots, N - 1$ and removes one bit at a time from the subcarrier that guarantees the maximum power saving for operation at the target BER. Both algorithms are stopped as soon as the required transmission power P_T approaches the maximum admissible value P_{budget}.

To better illustrate these iterative procedures, we assume a uniform BER constraint across subcarriers. This allows us to use the rate-power function $f(b, p_{e,\max})$ defined in Eq. (6.23), where $p_{e,\max}$ is the maximum BER that can be tolerated by the system. Then, the bit-filling and bit-removal algorithms are summarized as follows:

Bit-filling algorithm

- *Initialization*
  1) let $b_n = 0$ and $P_T = 0$;
  2) $\Delta P_n^+ = f(1, p_{e,\max})/\gamma_n$ for each $n \in \mathcal{N} = \{0, 1, \ldots, N-1\}$;

- *Bit assignment iterations:*
  repeat the following procedure:
  1) $\widetilde{n} = \arg \min_{n \in \mathcal{N}} \{\Delta P_n^+\}$;
  2) $P_T = P_T + \Delta P_{\widetilde{n}}^+$
  3) if $P_T > P_{\text{budget}}$ then stop the algorithm;
  4) $b_{\widetilde{n}} = b_{\widetilde{n}} + 1$;
  5) $\Delta P_{\widetilde{n}}^+ = \left[f(b_{\widetilde{n}} + 1, p_{e,\max}) - f(b_{\widetilde{n}}, p_{e,\max}) \right]/\gamma_{\widetilde{n}}$;
  6) if $b_{\widetilde{n}} = b_{\max}$, then remove $\widetilde{n}$ from $\mathcal{N}$; end.

Bit-removal algorithm

- *Initialization:*
  1) let $b_n = b_{\max}$ and
  2) initialize ΔP_n^- for each $n \in \mathcal{N} = \{0, 1, \ldots, N-1\}$ as follows

  $$\Delta P_n^- = \left[f(b_{\max}, p_{e,\max}) - f(b_{\max} - 1, p_{e,\max}) \right]/\gamma_n;$$

  3) let $P_T = \sum_{n=0}^{N-1} f(b_{\max}, p_{e,\max})/\gamma_n$.

- *Bit removal iterations:*
  repeat the following procedure until $P_T \leq P_{\text{budget}}$:
  1) $\widetilde{n} = \arg \max_{n \in \mathcal{N}} \{\Delta P_n^-\}$;
  2) $b_{\widetilde{n}} = b_{\widetilde{n}} - 1$;
  3) $P_T = P_T - \Delta P_{\widetilde{n}}^-$;
  4) If $b_{\widetilde{n}} = 0$, then remove $\widetilde{n}$ from $\mathcal{N}$, otherwise compute

  $$\Delta P_{\widetilde{n}}^- = \left[f(b_{\widetilde{n}}, p_{e,\max}) - f(b_{\widetilde{n}} - 1, p_{e,\max}) \right]/\gamma_{\widetilde{n}};$$

  end.

For the bit-filling algorithm, during initialization, the power needed to transmit one bit is calculated for each subcarrier. At each iteration, the subcarrier requiring the minimum additional power $\Delta P_{\tilde{n}}^{+}$ is assigned one more bit and the new additional power for that subcarrier is updated together with the overall transmission power P_T. If the number of bits has achieved its maximum allowable value $b_{\max}$, then the selected subcarrier is excluded from any further assignment by removing its index from $\mathcal{N}$. The stopping criterion is governed by P_T, which cannot overcome the assigned power P_{budget}.

On the other hand, the initialization for the bit-removal algorithm is performed by allocating the maximum number of bits over all subcarriers. At each iteration, one single bit is subtracted from the subcarrier that provides the maximum power saving $\Delta P_{\tilde{n}}^{-}$ for operation at the target BER, and the transmit power P_T is correspondingly updated. If no more bits are left on the selected subcarrier, the latter is excluded from further iterations, otherwise the new amount of power saving is calculated. The optimum bit allocation is obtained as soon as P_T becomes smaller than or equal to P_{budget}.

Although bit-filling and bit-removal procedures converge to the same bit allocation across tones, the computational load involved with these algorithms is typically different and depends on the achieved data rate R_b. In particular, bit-removal is to be preferred when $R_b > Nb_{\max}/2$ since in this case the convergence is faster than with bit-filling. It is also important to note that the resulting bit allocation is optimal only in relation to the considered function $f(b,p)$. Actually, the selection of different modulation schemes as possible transmission modes will lead to the consideration of different rate-power functions, which may result into possibly different bit allocations for the same set of channel SNRs.

6.1.6 *Bit loading with uniform power allocation*

Greedy techniques based on bit-filling or bit-removal strategies provide optimum joint distribution of power and data rate in practical situations where finite-granularity constellations have to be employed. The main difficulty of these methods is the extensive requirement of sorting and searching operations, which may prevent their applicability when the number of bits per OFDM block is relatively large. A simpler approach relies on the observation that in general only negligible throughput penalties occur if the optimal power assignment is replaced by a uniform allocation of power across subcarriers [180]. This simplified strategy has the advantage of reducing the dimensionality of the optimization problem in that the quantities P_n are kept fixed at some specified value $\overline{P}$ and only bit loading is performed adaptively. A scheme based on this suboptimal approach is derived in [9] under an average BER constraint. In this case the RMC problem is reformulated as

$$\text{maximize} \quad R_b = \sum_{n=0}^{N-1} b_n \tag{6.38}$$

under a uniform power allocation and subject to

$$\overline{p}_e(\boldsymbol{b}) = \frac{\sum_{n=0}^{N-1} b_n \, p_{e,n}(b_n, \gamma_n \overline{P})}{\sum_{n=0}^{N-1} b_n} \leq \overline{p}_{e,\max}, \tag{6.39}$$

and

$$b_n \in \{0, 1, \ldots, b_{\max}\}, \tag{6.40}$$

where the BER $p_{e,n}(b_n, \gamma_n \overline{P})$ over the nth subcarrier is univocally determined by the number b_n of allocated bits and by the received SNR $\gamma_n \overline{P}$. Note that the maximization of the objective function R_b is only performed with respect to $\boldsymbol{b} = [b_0, b_1, \ldots, b_{N-1}]^T$ since the available power P_{budget} is now uniformly distributed over the modulated subcarriers.

The corresponding solution is found iteratively by means of the following bit-removal algorithm with uniform power allocation (BRA-UniPower):

The suboptimum BRA-UniPower algorithm

- *Initialization*:
  1) let $b_n = b_{\max}$;
  2) set $P_n = P_{\text{budget}}/N$ for $n \in \mathcal{N} = \{0, 1, \ldots, N-1\}$ and compute $\overline{p}_e(\boldsymbol{b})$.

- *Bit removal iterations*:
  repeat the following procedure until $\overline{p}_e(\boldsymbol{b}) \leq \overline{p}_{e,\max}$:
  1) $\widetilde{n} = \arg \max_{n \in \mathcal{N}} \{p_{e,n}(b_n, \gamma_n P_n)\}$;
  2) $b_{\widetilde{n}} = b_{\widetilde{n}} - 1$;
  3) if $b_{\widetilde{n}} = 0$, then remove $\widetilde{n}$ from $\mathcal{N}$ and reassign the power so that $P_n = P_{\text{budget}}/\text{card}\{\mathcal{N}\}$ for $n \in \mathcal{N}$;
  4) recompute $\overline{p}_e(\boldsymbol{b})$ for the current bit allocation and power distribution;
  end.

During initialization, the maximum number of bits is tentatively allocated over each subcarrier under a uniform power assignment. At each iteration, the algorithm searches for the subcarrier $\widetilde{n}$ exhibiting the worst BER performance and reduces the corresponding data rate by one single bit. If $b_{\widetilde{n}} = 0$, the index $\widetilde{n}$ is removed from $\mathcal{N}$ so as to exclude the selected subcarrier from transmission and the power P_{budget} is redistributed uniformly over the remaining subcarriers. The average BER $\overline{p}_e(\boldsymbol{b})$ is next computed for the current bit assignment and compared with its maximum admissible value $\overline{p}_{e,\text{max}}$. The process is stopped as soon as $\overline{p}_e(\boldsymbol{b}) \leq \overline{p}_{e,\text{max}}$.

This algorithm allows a certain computational saving with respect to a system in which data rate and transmission power are jointly adjusted according to some specified optimality criterion. However, the need for recomputing the average BER $\overline{p}_e(\boldsymbol{b})$ at each new iteration still represents a serious drawback for practical implementation.

A further reduction of complexity is possible if we adopt a uniform BER constraint $p_{e,n}(b_n, \gamma_n \overline{P}) \leq p_{e,\text{max}}$ instead of specifying the average error rate as in Eq. (6.40). In such a case, b_n is explicitly determined by solving the equation $p_{e,n}(b'_n, \gamma_n \overline{P}) = p_{e,\text{max}}$ with respect to b'_n and taking the integer part of the corresponding solution. This yields

$$b_n = \min \{b_{\text{max}}, \text{int}(b'_n)\}, \tag{6.41}$$

where we have also borne in mind that b_n cannot exceed a prefixed value b_{max}. In this way, bit and power allocation is performed through the following iterative process, which is referred to as uniform-BER and uniform-power loading algorithm (UniBER-UniPower) :

The suboptimum UniBER-UniPower algorithm

- *Initialization*:
  1) let $P_n = P_{\text{budget}}/N$ and
  2) set $b_n = \min \{b_{\text{max}}, \text{int}(b'_n)\}$ for $n \in \mathcal{N} = \{0, 1, \ldots, N-1\}$.

- *subcarrier removal iterations*:
  repeat the following procedure until $b_n > 0$ for all $n \in \mathcal{N}$;
  1) if one or more b_n's are zero, then let $\widetilde{n} = \arg\min_{n \in \mathcal{N}} \{\gamma_n\}$ and remove $\widetilde{n}$ from $\mathcal{N}$;
  2) reassign the power so that $P_n = P_{\text{budget}}/\text{card}\{\mathcal{N}\}$ for $n \in \mathcal{N}$;
  3) recompute $b_n = \min \{b_{\text{max}}, \text{int}(b'_n)\}$ for $n \in \mathcal{N}$ according to the new power distribution;
  end.

As is seen, a preliminary bit distribution is derived from Eq. (6.41) assuming $P_n = P_{budget}/N$ as a tentative power assignment. If some b_n's turn out to be zero, the algorithm iterates by removing the worst quality subcarrier from the set $\mathcal{N}$ and redistributing the overall power P_{budget} across the remaining tones. Bit loading is then recomputed according to the new power distribution. The algorithm is stopped as soon as $b_n > 0$ for all $n \in \mathcal{N}$.

The most demanding task in the described procedure is the need for recomputing the bit allocation each time a subcarrier is excluded from transmission. A simpler yet suboptimal solution is obtained by replacing the subcarrier removal iterations with a single cancellation stage in which all subcarriers presenting an initial zero-bit assignment are simultaneously discarded. This approach results into a significant reduction of complexity since now the final bit assignment is directly derived from Eq. (6.41) after assuming $P_n = P_{budget}/N$ for $n \in \{0, 1, \ldots, N-1\}$, thereby dispensing from any iteration [31]. The final power allocation is eventually obtained by distributing P_{budget} over the modulated subcarriers (*i.e.*, those characterized by a positive bit assignment). In general, this strategy incurs some throughput penalty compared to a system in which the power is redistributed each time a subcarrier is removed from $\mathcal{N}$. The reason is that power redistribution may allow some subcarriers to pass from an initial zero-bit assignment to some positive allocation $b_n > 0$ as a consequence of the increased power level. The suboptimal algorithm excludes these subcarriers from data transmission, even though they could actually be exploited to convey some minimum information with the required reliability.

6.1.7 *Performance comparison*

In this Section we use computer simulations to compare the performance of the discussed bit-loading schemes in terms of achievable data throughput. For this purpose, we assume that a power budget of 10 dBm is available in an uncoded OFDM system with $N = 256$ subcarriers. The signal bandwidth is 10 MHz while the noise power spectral density is -80 dBm. The transmission mode is selected from a set of four possible modulation schemes, namely BPSK, QPSK, 16-QAM and 64-QAM. As a result, the quantities $\{b_n\}$ take values in the set $\{1, 2, 4, 6\}$ for $n = 0, 1, \ldots, 255$. The channel model is the same employed in Sec. 4.4, and comprises four multipath components with fixed path delays and an exponentially decaying power delay profile. A total of 200 snapshots are generated in order to average the simulation results over the channel statistics.

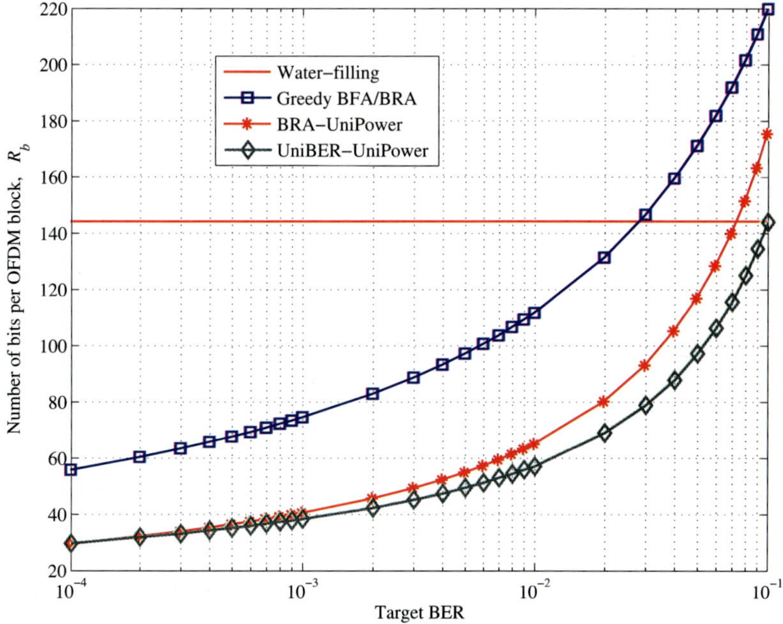

Fig. 6.4   Number of allocated bits as a function of the target BER.

Figure 6.4 illustrates the total bit rate $R_b = \sum_{n=0}^{255} b_n$ achieved by the loading algorithms as a function of the target BER. For comparison, we also show the data throughput provided by the classical water-filling solution. As expected, the greedy bit-filling/bit-removal algorithms (BFA/BRA) outperform their suboptimal BRA-UniPower and UniBER-UniPower versions at the price of a higher computational load. On the other hand, the difference between the two suboptimal schemes with uniform power allocation is quite negligible, particularly at low error probabilities. In the low target-BER region we see that the water-filling policy achieves a significant advantage over the other algorithms due to its implicit assumption of an infinite granularity constellation. As the target-BER grows large, however, this advantage reduces to such a point that greedy BFA/BRA become the leading schemes at BER$> 3 \times 10^{-2}$. This fact can be explained by recalling that the water-filling solution has been derived under the assumption of an arbitrarily small BER, whereas the considered greedy-based techniques can trade data throughput against error-rate probability. This means that a fair comparison between the water-filling policy and other loading schemes can only be made in the low BER region.

6.1.8 *Subband adaptation*

The adaptive techniques illustrated so far operate on a subcarrier basis in that the optimum constellation size and/or power level are individually determined for any subcarrier according to instantaneous channel state information. This approach offers a large amount of flexibility on one hand, but on the other it may entail a prohibitive signaling overhead since the receiver has to be informed as to which modulation parameters are employed over each subcarrier. To alleviate this drawback, system resources can be allocated in a blockwise fashion following a subband adaptation criterion [49,74]. The basic idea behind this approach is to divide the available spectrum into several groups of adjacent subcarriers which are referred to as *subbands*, and use the same set of modulation parameters (constellation size, code rate, power level) over all subcarriers in the same subband. In this way the signaling task is substantially simplified at the price of somewhat reduced flexibility in resource assignment. Roughly speaking, the penalty incurred by subband adaptation in terms of achievable throughput is determined by the extent of channel variations over each subband. If the subband width is smaller than the channel coherence bandwidth, the channel appears as nearly flat across the subband and no significant penalty is incurred with respect to a system that operates at subcarrier level.

In those applications where system complexity is a critical issue, subband loading can be used in conjunction with uniform power distribution over the signal spectrum. In such a case, letting $\mathcal{M} = \{M_1, M_2, \ldots, M_J\}$ be the set of possible transmission modes (each characterized by a given constellation size, code rate and other possible modulation parameters), the problem is to select the best mode over each subband so as to obtain the highest throughput at some specified target BER. Again, the optimization can be performed under either a uniform or average error rate constraint. In the former case the BER over each subcarrier is kept smaller than a given value $p_{e,\max}$, while in the latter an upper limit $\bar{p}_{e,\max}$ is imposed to the average error probability

$$\bar{p}_e(M_j) = \frac{1}{N_s} \sum_n p_{e,n}(M_j), \qquad (6.42)$$

where N_s is the number of subcarriers in the subband, $p_{e,n}(M_j)$ is the

BER over the nth subcarrier for a given transmission mode M_j and the summation is extended to all subcarriers in the considered subband.

In case of a uniform BER constraint, a mode M_j can be activated on a certain subcarrier only if the instantaneous SNR exceeds a given threshold ρ_j which depends on the adopted modulation parameters and target BER. For example, with an uncoded BPSK transmission operating at an error rate of 10^{-3} we have $\rho = 6.8$ dB, while $\rho = 9.8$ dB is requested for an uncoded QPSK. On the other hand, since the channel quality varies across subcarriers and a single mode must be employed in each subband, the transmission parameters in the considered subband are conservatively selected on the basis of the subcarrier which exhibits the lowest SNR. Clearly, this approach results into some performance loss with respect to a system in which the available resources are assigned on a subcarrier basis. The reason is that in each subband the transmission mode and the associated data throughput are exclusively dictated by the most faded subcarrier even though other subcarriers with better channel quality could safely support higher data rates. This problem can be mitigated by a proper design of the subband width, which should be made adequately smaller than the channel coherence bandwidth. In this way all relevant subcarriers undergo similar channel impairments and, in consequence, the selected transmission mode is likely to be optimal over the entire subband.

As anticipated, subband adaptation can also be performed under an average error-rate constraint. In such a case, the average BER $\overline{p}_e(M_j)$ in Eq. (6.42) is computed for all available modes M_j, and in each subband the mode $\widetilde{M}$ exhibiting the highest data rate and satisfying the condition $\overline{p}_e(\widetilde{M}) \leq \overline{p}_{e,\max}$ is selected for transmission. This adaptation strategy is expected to mitigate to some extent the throughput penalty associated with the uniform BER constraint. The reason is that in each subband all subcarriers contribute to the average error rate and, in consequence, the transmission mode is not exclusively selected on the basis of the worst quality subcarrier.

6.1.9 *Open-loop and closed-loop adaptation*

Any link adaptation technique exploits instantaneous channel state information to determine the best set of modulation parameters to be employed in the next transmission. One main assumption behind this approach is that the fading rate is not too rapid since otherwise channel prediction may be obsolete at the time of transmission, thereby resulting into a wrong selection of the modulation parameters.

Roughly speaking, we distinguish between two different classes of adaptation techniques. The former class is suitable for time-division-duplex (TDD) systems, where the same frequency band is used for both uplink and downlink transmissions and the communication channel can reasonably be considered as reciprocal. In this case the receiving station estimates the channel quality during the downlink phase and exploits this estimation to select the best mode for the next uplink transmission. We refer to this operating method as *open-loop adaptation* since the local transmitter adjusts the modulation parameters by only relying on channel measurements acquired during the previous slot and without exploiting any feedback from the remote receiver.

On the other hand, if the communication link is not reciprocal as in frequency-division-duplex (FDD) systems, channel state information derived from the received OFDM blocks cannot be used to determine the modulation parameters for the next transmission stage because of the different propagating conditions encountered in the two communication links. In this case adaptive modulation can be established on condition that the remote receiver performs channel estimation and instructs the transmitter as to which parameters are the best to be used. This policy is known as *closed-loop adaptation* since the transmission mode is activated on the basis of a specific feedback from the remote receiver rather than being autonomously selected by the transmitter. Although closed-loop adaptation is expected to be intrinsically robust against interference and other non-reciprocal effects, it suffers from an inherent feedback delay which might result into outdated information. This makes the Doppler fading rate a rather critical parameter in closed-loop adaptive modulation systems.

6.1.10 *Signaling for modulation parameters*

Signaling plays a major role in the design of an adaptive communication link. In an open-loop system where channel estimation and parameter adaptation are performed by the local transmitter, the remote receiver must be informed as to which transmission mode is currently in use. Vice versa, in a closed loop scenario the modulation parameters are decided by the receiver itself, which therefore has to communicate its choice to the remote transmitter. In any case, it is important that signaling information be exchanged with a high level of reliability since otherwise the receiver might be induced to adopt a wrong detection strategy and would be unable to successfully decode the information data.

One popular signaling scheme is based on the insertion of one or more dedicated subcarriers in each subband to convey information about the set of employed modulation parameters. If N_M is the number of possible transmission modes, a single N_M-PSK symbol would in principle be sufficient for this purpose. However, in order to reduce the probability that signaling information may be corrupted by channel impairments, multiple dedicated symbols can be placed across the subband to take advantage of the channel frequency diversity. A drawback of this signaling method is the throughput penalty that results from the use of dedicated subcarriers.

An alternative approach is based on blind detection algorithms. These schemes try to estimate the currently employed transmission mode from the received signal without requiring any extra overhead. An example of blind algorithm is presented in [73] for systems employing subband adaptation. Let $\{M_j; j = 1, 2, \ldots, J\}$ be the set of possible transmission modes and denote $Y(n) = R(n)/\widehat{H}(n)$ the nth DFT output divided by the corresponding channel estimate $\widehat{H}(n)$. Using (4.6), we can interpret $Y(n)$ as an estimate of the data symbol $c(n)$ transmitted over the nth subcarrier and embedded in additive noise. Then, inside the constellation associated to the transmission mode M_j we select the symbol $\widehat{c}_j(n)$ that is closest to $Y(n)$ and compute the following error signal

$$e_j = \sum_n |Y(n) - \widehat{c}_j(n)|^2, \qquad j = 1, 2, \ldots, J. \qquad (6.43)$$

where the summation is extended to all subcarriers in the considered subband. Clearly, e_j is a measure of the Euclidean distance between the received symbols $\{Y(n)\}$ and the constellation points associated to M_j. To see how the quantities $\{e_1, e_2, \ldots, e_J\}$ can be used to decide which transmission mode is currently in use, we temporarily neglect the noise contribution and assume perfect channel estimation. In this ideal setting we have $Y(n) = c(n)$ and, in consequence, the error signal associated to the actually employed transmission mode turns out to be zero due to a perfect agreement between the received symbols and the corresponding constellation points. Although in the presence of thermal noise and channel estimation inaccuracies this error signal may not be exactly zero, under normal operating conditions it is expected to be relatively small. Hence, it makes sense to argue that the transmission mode employed over the considered subband is the one associated to the minimum error signal. In other words, we decide that $M_{\widehat{j}}$ is currently in use if

$$\widehat{j} = \arg\min_j \{e_j\}. \tag{6.44}$$

Compared to signaling schemes that make use of dedicated subcarriers, this blind method has the advantage of dispensing from any overhead, even though a larger SNR is required to achieve the same level of reliability. In particular, it is found in [73] that the system performance is largely dictated by the number of subcarriers in each subband and by the number of allowable transmission modes, which in practice cannot be greater than four.

6.2 Resource allocation in multiuser OFDM systems

In a typical multiple-access system, users' signals undergo independent fading attenuations because of the different spatial positions occupied by remote terminals. As a consequence, a subcarrier that appears in a deep fade to one terminal may exhibit a much higher channel gain for other users. To take advantage of this *multiuser diversity* effect [78], the available subcarriers should be dynamically assigned to users on the basis of instantaneous channel state information. Compared to conventional OFDMA systems with non-adaptive resource allocation, this approach allows a more efficient use of the system resources. The net result is an increased data throughput since a given subcarrier will be left unused only if it appears in a deep fade to all terminals, a situation that rarely occurs due to the mutual independence of the users' channel responses.

From the above discussion it follows that optimum resource allocation in a multiuser scenario requires the adoption of a dynamic carrier assignment policy in addition to adaptive bit and power loading. This makes the link adaptation task much more challenging than in single-user systems. As users cannot share the same subcarrier, the allocation process results into a combinatorial optimization problem for which no optimal greedy solution exists. This fact has recently stimulated an intense research activity toward the development of suboptimum resource assignment schemes characterized by good performance and affordable complexity. The common idea behind these methods is to consider carrier allocation and bit loading as separate tasks to be performed independently rather than jointly.

The concept of dynamic resource allocation in an OFDMA downlink transmission is illustrated in Fig. 6.5. At the BS, information about the

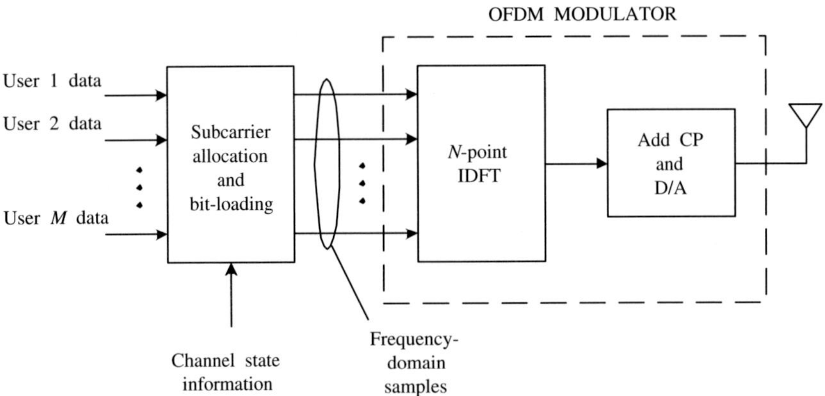

Fig. 6.5   OFDMA downlink transmission with adaptive resource allocation at the BS transmitter.

users' channel responses are passed to the subcarrier allocation and bit loading unit, which maps the users' data over the selected subcarriers using the more appropriate transmission mode (coding and/or modulation scheme). In order to guarantee a specified error rate probability, the power level over each subcarrier is properly adjusted on the basis of the employed transmission mode. The resulting frequency-domain samples are finally fed to an OFDM modulator and transmitted over the channel.

At the mth mobile terminal, the received signal is demodulated and the recovered frequency-domain samples are passed to the subchannel selector, which only retains information from subcarriers assigned to the mth user while discarding all the others. The selected samples are then fed to the decoding unit, which provides final bit decisions using the appropriate detection strategy. Clearly, the BS must inform the users' terminals as to which subcarriers and transmission modes have been assigned to them, otherwise the subchannel selector and data decoding unit cannot properly be configured. This requires the exchange of side information with a corresponding penalty in data throughput due to the transmission overhead. The amount of side information is somewhat reduced by adopting a subband allocation policy where users are given blocks of contiguous subcarriers with similar fading characteristics.

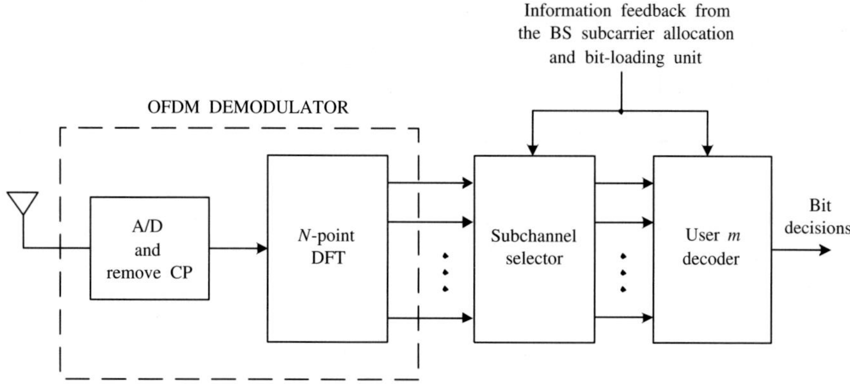

Fig. 6.6   Block diagram of the mth receiving terminal in an OFDMA downlink transmission with adaptive resource allocation.

6.2.1   *Multiaccess water-filling principle*

The extension of the water-filling principle to a multiuser scenario is not straightforward except for the unrealistic case where all users are characterized by the same channel response. The first pioneering results in this area were presented by Cheng and Verdù in their excellent paper [17]. They derived the capacity region and the optimal power allocation for a frequency-selective Gaussian multiaccess channel, where two or more users with independent power constraints transmit data to a common BS receiver. In what follows, the results of [17] are applied to the uplink transmission of a multicarrier system accommodating M simultaneously active users.

Assuming perfect timing and frequency synchronization, the DFT output at the BS receiver takes the form

$$R(n) = \sum_{m=1}^{M} H_m(n)S_m(n) + W(n), \qquad 0 \le n \le N - 1. \tag{6.45}$$

where $H_m(n)$ is the channel frequency response of the mth user over the nth subcarrier, $S_m(n)$ is the corresponding input symbol with power $P_{m,n} = \mathrm{E}\{|S_m(n)|^2\}$ and $W(n)$ is white Gaussian noise with zero-mean and variance σ_w^2. In this uplink scenario, the power constraints are stated as

$$\sum_{n=0}^{N-1} P_{m,n} \le P_{m,\text{budget}}, \tag{6.46}$$

for $m = 1, 2, \ldots, M$, where $P_{m,\text{budget}}$ represents the amount of available power for the mth user and $P_{m,n} \geq 0$ for $n = 0, 1, \ldots, N - 1$.

Unlike the single-user case, the multiaccess channel is characterized by a M-dimensional capacity region $\boldsymbol{C}_R \in \mathbb{R}_+^M$ (we denote $\mathbb{R}_+^M$ the set of M-tuples with non-negative real-valued entries). Each point $\boldsymbol{R} = (R_1, R_2, \ldots, R_M)$ in this region represents a combination of rates at which users can send information with an arbitrarily low error-rate probability. For the sake of simplicity, in the following we limit our attention to a two-user scenario. In this case $\boldsymbol{C}_R$ is a convex set in the positive quadrant of the (R_1, R_2)-plane which can be written as [17]

$$
\boldsymbol{C}_R = \bigcup_{\boldsymbol{P}_1, \boldsymbol{P}_2} \left\{ (R_1, R_2) :
\begin{array}{l}
0 \leq R_1 \leq \displaystyle\sum_{n=0}^{N-1} \log_2 \left(1 + P_{1,n} \gamma_{1,n}\right) \\[2ex]
0 \leq R_2 \leq \displaystyle\sum_{n=0}^{N-1} \log_2 \left(1 + P_{2,n} \gamma_{2,n}\right) \\[2ex]
R_1 + R_2 \leq \displaystyle\sum_{n=0}^{N-1} \log_2 \left(1 + P_{1,n} \gamma_{1,n} + P_{2,n} \gamma_{2,n}\right)
\end{array}
\right\},
$$

$$\tag{6.47}$$

where $\boldsymbol{P}_m = [P_{m,0}, P_{m,1}, \ldots, P_{m,N-1}]^T$ $(m = 1, 2)$ are power vectors satisfying the constraint Eq. (6.46) while $\gamma_{m,n} = |H_m(n)|^2 / \sigma_w^2$ is the channel SNR of the mth user over the nth subcarrier. From the above equation we see that $\boldsymbol{C}_R$ is the union of an infinite number of rate regions, each corresponding to a different pair $(\boldsymbol{P}_1, \boldsymbol{P}_2)$ and representing a pentagon in the (R_1, R_2)-plane.

A possible example of capacity region is depicted in Fig. 6.7. The abscissa of the corner point A indicates the maximum rate at which user 1 can reliably send information over the channel (single-user capacity) when user 2 is not transmitting $(R_2 = 0)$. This point is achieved by optimally allocating the power $P_{1,\text{budget}}$ over the channel $H_1(n)$ according to the classical single-user water-filling principle. The converse is true for the corner point B, which is attained by applying the water-filling policy to $H_2(n)$ assuming that user 1 is turned off. Any other point on the boundary curve connecting A and B is achieved by an appropriate choice of $(\boldsymbol{P}_1, \boldsymbol{P}_2)$ and is optimal in that it maximizes a linear combination of the users' rates, say

$$
R(\alpha) = \alpha R_1 + (1 - \alpha) R_2, \tag{6.48}
$$

with $\alpha \in [0, 1]$. This can readily be seen by considering the family of parallel straight lines in the (R_1, R_2)-plane over which $R(\alpha)$ keeps constant. These lines have a common slope $\alpha/(\alpha - 1)$ and, due to the convexity of $\boldsymbol{C}_R$,

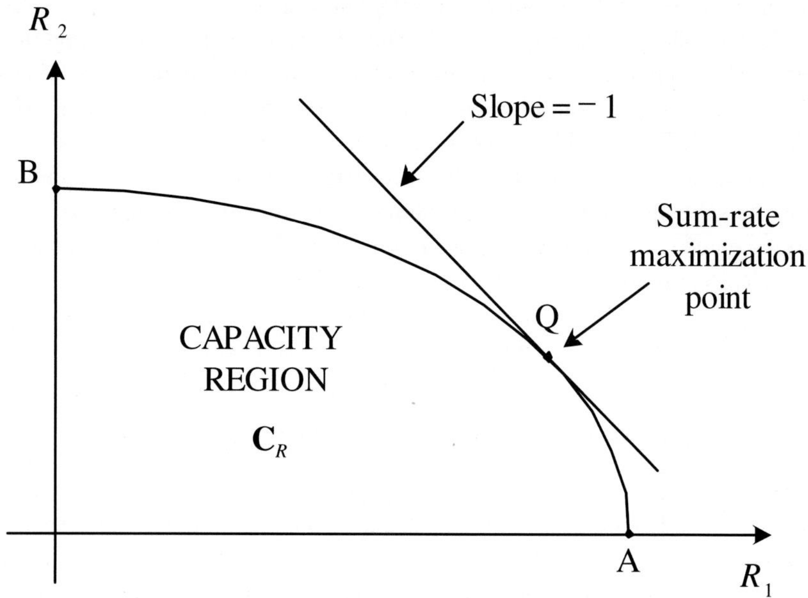

Fig. 6.7   Example of capacity region in a two-user scenario.

only one of them is tangent to the boundary curve in some point $Q(\alpha)$. The coordinates of $Q(\alpha)$ provide the values R_1 and R_2 that maximize $R(\alpha)$ over the capacity region.

Inspection of Eq. (6.48) provides a useful interpretation of α as a parameter that determines the relative users' priorities. Specifically, as α approaches unity the priority given to user 1 increases and the point $Q(\alpha)$ moves on the boundary curve toward A. When $\alpha = 1/2$ both users are given the same priorities. In this case the corresponding boundary point results in the maximization of the sum-rate $R_1 + R_2$ and is graphically determined by considering the tangent line with slope -1 as illustrated in Fig. 6.7.

From the above discussion it appears that in a two-user scenario different users' priorities result into different optimum operating points, each located on the boundary of the capacity region. Hence, the task is to find, for any given value of α, the optimum pair (P_1, P_2) that allows one to achieve the boundary point where $R(\alpha)$ is maximum. A geometrical solution to this problem has been presented in [17] and consists of two fundamental steps. In the first step, an equivalent transfer function $H^{(eq)}(n)$ is computed from $H_1(n)$ and $H_2(n)$, and the classical water-filling principle is then applied to $H^{(eq)}(n)$. This provides the optimum allocation of the *total* available power $P_{1,\text{budget}} + P_{2,\text{budget}}$ in the frequency-domain. The second step determines how the total power $P_n = P_{1,n} + P_{2,n}$ allocated over each subcarrier should be optimally split among the active users. The result is that in general each subcarrier has to be shared by both users, who therefore interfere with each other. In this case, the successive decoding idea suggests that the user with the lowest priority (say user 1) should be decoded first while treating the other user's signal as noise. The receiver then regenerates the signal of user 1 and subtracts it from the received waveform. This results into an expurgated signal which is eventually employed to detect the information sent by user 2.

An interesting situation occurs when both users are given the same priority. As mentioned earlier, in this case the optimum power assignment maximizes the sum-rate $R_1 + R_2$ over the capacity region and achieves the boundary point Q depicted in Fig. 6.7. A prominent result of [17] is that the optimum power split among equal-priority users corresponds to the classical OFDMA concept in which subcarriers are grouped into disjoint clusters that are exclusively assigned to users. This means that OFDMA is capable of achieving the sum-rate capacity promised by the Gaussian multiaccess channel.

In case of only two users with equal priorities, the optimum power assignment $(\boldsymbol{P}_1, \boldsymbol{P}_2)$ is found through a geometrical procedure which is reminiscent of the water-filling argument. The basic idea behind this method is to properly scale the water-filling diagrams associated with the channel responses $H_1(n)$ and $H_2(n)$ such that they present the same water level and can thus be combined into a single diagram. More specifically, letting ρ_1 and ρ_2 be the scaling coefficients, we arbitrarily fix the water level to unity and plot the curves $\rho_1/\gamma_{1,n}$ and $\rho_2/\gamma_{2,n}$ as a function of n on the same diagram. As indicated in Fig. 6.8, we treat the minimum of the two curves as the bottom of the vessel where water is poured, and adjust ρ_1 and ρ_2 such that: 1) the total amount of water is $\rho_1 P_{1,\text{budget}} + \rho_2 P_{2,\text{budget}}$; 2) the amount of water in the region where $\rho_1/\gamma_{1,n} \leq \rho_2/\gamma_{2,n}$ is equal to $\rho_1 P_{1,\text{budget}}$.

In general, the coefficients ρ_1 and ρ_2 can only be obtained graphically or numerically as they depend on the channel transfer functions and power constraints in a rather complicated fashion which makes their analytical

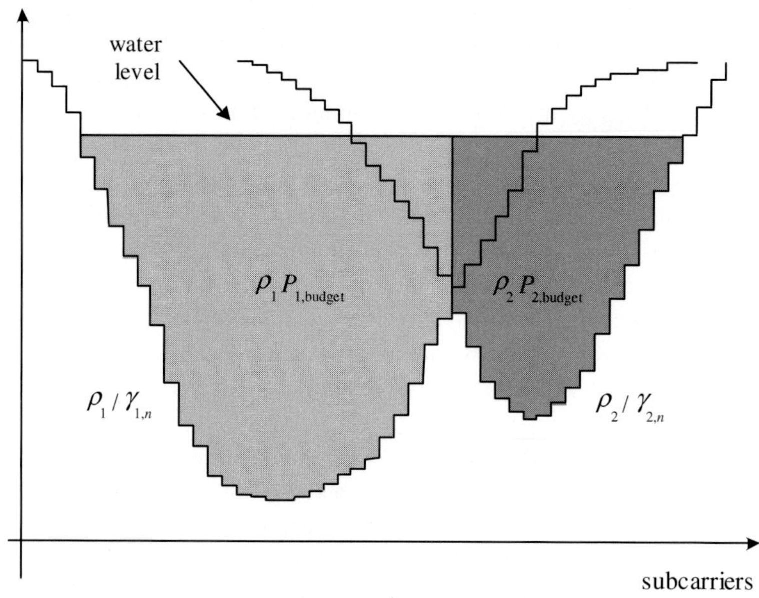

Fig. 6.8   The water-filling principle in a two-user scenario.

derivation a rather difficult task. Anyway, assuming that these parameters have been derived in some manner, the optimum power assignment for the two users is eventually found after scaling the shaded regions in Fig. 6.8 by ρ_1 and ρ_2. As anticipated, different users are given different subcarriers according to the OFDMA principle. In particular, the frequency band where $\rho_1/\gamma_{1,n} \leq \rho_2/\gamma_{2,n}$ is assigned to user 1 while the remaining part is available for user 2. Clearly, if $\min\{\rho_1/\gamma_{1,n}, \rho_2/\gamma_{2,n}\}$ exceeds the water surface for some n, the corresponding subcarriers are left unused as they cannot support reliable data transmission.

6.2.2   *Multiuser rate maximization*

Although relevant from an information theoretic perspective, the multiuser water-filling policy turns out to be too complex for practical purposes due to lack of efficient methods for determining the scaling coefficient of each individual channel response. As in the single-user case, a more convenient approach for dynamic resource allocation is based on the rate-maximization concept (RMC). This strategy aims at maximizing the *aggregate* data rate of all active users under fixed constraints in terms of total transmission power and error-rate performance.

To see how the RMC can be extended to a typical OFDMA downlink scenario with M active users, we denote $b_{m,n}$ the number of bits of the mth user that are allocated over the nth subcarrier. We also assume that $b_{m,n} \in \{0, 1, \ldots, b_{\max}\}$, where $b_{\max}$ is determined by the maximum allowable constellation size. Since each subcarrier cannot be shared by more than one user, for any index n only one single $m \in \{1, 2, \ldots, M\}$ may exist for which $b_{m,n} \neq 0$. The performance requirement of the mth user is specified by the maximum tolerable BER $p_{m,\max}$. In order to maintain the desired quality of service, the power allocated to the mth user over the nth subcarrier must equal $P_{m,n} = f(b_{m,n}, p_{m,\max})/\gamma_{m,n}$, where $f(b, p)$ is the rate-power function indicating the minimum SNR that is required to detect b information bits at a target BER p. Note that in this way the same error probability $p_{m,\max}$ is maintained over all subcarriers assigned to the mth user (uniform BER constraint).

Under the above assumptions and statements, the multiuser RMC problem is mathematically formulated as

$$\text{maximize} \quad R_b = \sum_{m=1}^{M} \sum_{n=0}^{N-1} b_{m,n} \tag{6.49}$$

with respect to the bit assignments $\{b_{m,n}\}$, where maximization is subject to

$$P_T = \sum_{m=1}^{M} \sum_{n=0}^{N-1} \frac{f(b_{m,n}, p_{m,\max})}{\gamma_{m,n}} \leq P_{\text{budget}}, \tag{6.50}$$

and

$$\text{if } b_{m',n} \neq 0 , \text{ then } b_{m,n} = 0 \quad \text{for all } m \neq m'. \tag{6.51}$$

The constraint Eq. (6.50) specifies an upper limit P_{budget} to the total transmission power while Eq. (6.51) ensures that each subcarrier is exclusively assigned to only one user, as demanded by the OFDMA concept.

From Eqs. (6.49)-(6.51) we see that extending the RMC criterion to a multiuser scenario results into a combinatorial maximization problem for which no practical solution is available. Things become easier if all users are characterized by a common BER constraint $p_{m,\max} = p_{\max}$ for $m = 1, 2, \ldots, M$. This particular situation is considered in [78], where the optimum solution to the RMC problem is found in two successive steps. In the first step each subcarrier is exclusively assigned to the user exhibiting the highest channel SNR over it. More precisely, the m'th user is given the nth subcarrier on condition that

$$m' = \arg \max_{1 \leq m \leq M} \{\gamma_{m,n}\} . \tag{6.52}$$

In the second step, the number of bits allocated over any assigned subcarrier is determined so as to maximize the objective function R_b in Eq. (6.49) under the power constraint Eq. (6.50). This task is accomplished in the same way as in single-user OFDM transmissions. Indeed, after all subcarriers have been assigned, the OFDMA downlink can be viewed as an equivalent single-user system with channel SNRs given by $\gamma_n^{(eq)} = \max_{1 \leq m \leq M} \{\gamma_{m,n}\}$ for $n = 0, 1, \ldots, N-1$ and with a data rate that equals the aggregate data rate of the original multiuser scenario. Optimum bit assignment is thus achieved by means of RMC-based greedy techniques as those discussed in Sec. 6.1.5.

Numerical results illustrated in [78] indicate that for a given power consumption P_T the achievable sum-rate R_b increases with the number of users due to multiuser diversity effects [65]. However, a fundamental drawback of the RMC criterion as stated in Eqs. (6.49)-(6.51) is that it does not provide any guarantees on the minimum achievable data rate of each individual user. Actually, in some extreme situations maximizing the aggregate data rate may result into the assignment of all available subcarriers to only a subset of users exhibiting good channel quality, thereby excluding all other users from transmission.

6.2.3 *Max-min multiuser rate maximization*

One possible approach to overcome the inherent limitations associated with the sum-rate maximization criterion is described in [130]. The idea is to distribute system resources so as to maximize the *minimum* data rate amongst all users for a fixed transmission power and assigned error probabilities. The resulting strategy is called the max-min rate-maximization concept and is mathematically formulated as

$$\text{maximize} \quad R_b^{(\min)} = \min_{1 \leq m \leq M} \left\{ \sum_{n=0}^{N-1} b_{m,n} \right\} \tag{6.53}$$

with respect to the bit assignments $\{b_{m,n}\}$ and subject to the constraints Eqs. (6.50), (6.51). The rationale behind the "max-min" operation in

Eq. (6.53) is to assign more power to users exhibiting poor channel conditions so that they can achieve a data rate comparable to that of other users with better channel quality.

Unfortunately, the problem stated in Eq. (6.53) is not convex and can only be solved through a numerical search over all admissible bit assignments satisfying Eqs. (6.50) and (6.51). In practical applications this search turns out to be prohibitively complex due to the large number of possible candidate assignments. A way out is offered by the use of Lagrangian relaxation (LR) techniques, where the Lagrange method of optimization is applied to an integer parameter which is relaxed to take on noninteger values. The LR approach is adopted in [130] to transform Eq. (6.53) into a similar but more tractable optimization problem. In particular, the requirement $b_{m,n} \in \{0, 1, \ldots, b_{\max}\}$ is relaxed by allowing $b_{m,n}$ to take on any noninteger value within the interval $[0, b_{\max}]$. In addition, a new set of variables $\{\alpha_{m,n}\}$ is introduced to indicate the percentage of times each subcarrier is shared by a given user. This amounts to considering a very large number of OFDM blocks (say J_B) where users are allowed to *time*-share the available subcarriers. In this respect, $\alpha_{m,n}$ represents the ratio between the number of blocks where the nth subcarrier is assigned to the mth user and the total number of blocks J_B. Clearly, the assumption behind this approach is that the users' channel responses do not change significantly over a timing interval spanning J_B blocks.

After scaling both the transmit power and data rate by the corresponding time-sharing factor $\alpha_{m,n}$, the new optimization problem is stated as

$$\text{maximize} \quad \min_{1 \leq m \leq M} \left\{ \sum_{n=0}^{N-1} \alpha_{m,n} b_{m,n} \right\} \tag{6.54}$$

with respect to $\{b_{m,n}\}$ and $\{\alpha_{m,n}\}$, where maximization is subject to

$$\sum_{m=1}^{M} \sum_{n=0}^{N-1} \alpha_{m,n} \frac{f(b_{m,n}, p_{m,\max})}{\gamma_{m,n}} \leq P_{\text{budget}}, \tag{6.55}$$

and

$$\sum_{m=1}^{M} \alpha_{m,n} = 1, \tag{6.56}$$

for $n = 0, 1, \ldots, N-1$, with $b_{m,n} \in [0, b_{\max}]$ and $\alpha_{m,n} \in [0, 1]$. As indicated in [130], the solution to the above problem is found iteratively by means of standard optimization software as long as the rate-power function $f(b, p)$ is convex with respect to b. However, this solution cannot directly be used for a couple of reasons. A first difficulty is that in general the number $b_{m,n}$ of allocated bits is noninteger and may not correspond to any practical modulation/coding scheme. In addition, some of the quantities $\alpha_{m,n}$ may be within $(0,1)$, thereby indicating a time-sharing allocation policy. This represents a potential problem in most wireless communication systems since the channel responses are typically time-varying and do not keep unchanged long enough to make time-sharing a feasible solution.

6.2.4 *Multiuser margin maximization*

In real-time multimedia communications, the users' bit rates are generally dictated by the employed data compression algorithms. In such a case the system resources cannot be assigned according to the RMC criterion as there is no guarantee that each user can meet its individual rate requirement. When a specified throughput must be retained for each user, the margin maximization concept (MMC) turns out to be the most appropriate approach for adaptive resource allocation. This strategy aims at minimizing the total power consumption under fixed constraints in terms of individual bit rates and error probabilities. This feature makes it particularly suited for applications where different classes of services must simultaneously be supported.

To fix the ideas, we denote R_m the number of information bits of the mth user that must be conveyed by each OFDM block and call $p_{m,\max}$ the maximum admissible BER. Then, recalling that the power allocated to the mth user over the nth subcarrier is given by $P_{m,n} = f(b_{m,n}, p_{m,\max})/\gamma_{m,n}$, we state the multiuser MMC optimization problem as

$$\text{minimize} \quad P_T = \sum_{m=1}^{M} \sum_{n=0}^{N-1} \frac{f(b_{m,n}, p_{m,\max})}{\gamma_{m,n}}, \tag{6.57}$$

with respect to the bit assignments $\{b_{m,n}\}$, where $b_{m,n} \in \{0, 1, \ldots, b_{\max}\}$ and subject to

$$\sum_{n=0}^{N-1} b_{m,n} \geq R_m, \tag{6.58}$$

for $m = 1, 2, \ldots, M$, and

$$b_{m,n} = 0, \tag{6.59}$$

if $b_{m',n} \neq 0$ for all $m' \neq m$.

The constraints Eq. (6.58) specify the users' rate requirements while Eq. (6.59) avoids that a given subcarrier is shared by more than one user.

It is worth noting that in some works related to DSL applications the individual rate requirements in Eq. (6.58) are replaced by a single sum-rate constraint [83–85]. Although this approach has the advantage of increasing the number of degrees of freedom for the minimization of P_T, it has the fundamental drawback of not considering fairness among users.

Similarly to the RMC policy, the multiuser MMC criterion results into a combinatorial optimization problem whose solution requires an exhaustive search over all possible bit assignments. The complexity associated with the exhaustive search turns out to be prohibitive for practical implementation. Again, the use of Lagrangian relaxation techniques proves to be useful as it provides a computationally manageable (yet suboptimum) solution. Following this approach, we still allow users to time-share each subcarrier over a number J_B of OFDM blocks and assume that $b_{m,n}$ can take any noninteger value within the interval $[0, b_{\max}]$. Then, calling $\alpha_{m,n}$ $(m = 1, 2, \ldots, M)$ the time-sharing factors for the nth subcarrier, we formulate a modified MMC-based optimization problem as

$$\text{minimize} \quad P_T = \sum_{m=1}^{M} \sum_{n=0}^{N-1} \alpha_{m,n} \frac{f(b_{m,n}, p_{m,\max})}{\gamma_{m,n}} \tag{6.60}$$

with respect to $\{b_{m,n}\}$ and $\{\alpha_{m,n}\}$, subject to

$$\sum_{n=0}^{N-1} \alpha_{m,n} b_{m,n} = R_m, \quad \text{for} \quad m = 1, 2, \ldots, M. \tag{6.61}$$

and

$$\sum_{m=1}^{M} \alpha_{m,n} = 1, \quad \text{for} \quad n = 0, 1, \ldots, N - 1. \tag{6.62}$$

where $\alpha_{m,n} \in [0, 1]$ and $b_{m,n} \in [0, b_{\max}]$. A numerical solution to the above problem is found in [172] using convex optimization techniques. The only requirements are that $f(b, p)$ is convex with respect to b and the aggregate data rate is less than $N b_{\max}$ (which is the maximum number of bits that one OFDM block can convey). As mentioned previously, however, a time-sharing allocation policy is hardly usable in a wireless scenario as a consequence of the time-varying nature of the channel responses. Furthermore, the fact that $b_{m,n}$ can take any value within $[0, b_{\max}]$ poses some difficulties in the selection of a practical modulation scheme that may attain the required bit rate. Note that simply quantizing $b_{m,n}$ and $\alpha_{m,n}$ does not provide a feasible solution since the resulting bit allocation is not guaranteed to satisfy the individual rate requirements specified in Eq. (6.58).

One possible approach to overcome these problems is based on a two-step suboptimal procedure in which subcarrier assignment and bit loading are performed *separately* instead of *jointly*. This strategy has been suggested in many works, including [78] and [172]. In particular, in [172] the available subcarriers are *exclusively* allocated to users on the basis of the optimum time-sharing factors $\alpha_{m,n}$ satisfying Eqs. (6.60)-(6.62). The allocation criterion is that any subcarrier must be assigned to the user who exhibits the largest time-sharing factor over it. After subcarrier allocation, bit loading is independently performed for each user over the assigned subcarriers. Any conventional greedy algorithm based on the MMC criterion may be used for this purpose.

6.2.5 *Subcarrier assignment through average channel signal-to-noise ratio*

As mentioned previously, a suboptimum yet practical approach for adaptive resource allocation in OFDMA systems is based on a strict separation between the subcarrier assignment and bit loading tasks. Even in this case, however, allocating the available subcarriers to the active users on the basis of some optimality criterion remains a difficult problem. The relaxation-based solution described in [172] requires knowledge of the optimum time-sharing factors $\alpha_{m,n}$, which can only be determined iteratively by means of convex optimization methods. A potential drawback of this approach is the large number of iterations that may be required to achieve convergence.

A simpler scheme suggested in [77] divides the subcarrier assignment task in two successive steps. The first step, known as *bandwidth allocation*, determines the number of subcarriers that each user will get on the basis of the individual rate requirements and *average* channel SNRs. In the second stage, full channel state information is exploited to properly allocate the subcarriers to each user. By solving these subproblems separately, a good assignment of system resources is possible with affordable complexity.

The bandwidth allocation step operates in accordance to the MMC principle of minimizing the total power consumption under individual constraints in terms of data rate and error probability. From a mathematical viewpoint, the problem is that of determining the number N_m of subcarriers that must be reserved to the mth user ($m = 1, 2, \ldots, M$) for reliable transmission of R_m bits per OFDMA block. To simplify the derivation, we temporarily assume that each user signal undergoes flat-fading distortion and experiences the same channel SNR over each subcarrier. The latter is set equal to the average SNR across the signal bandwidth and reads

$$\overline{\gamma}_m = \frac{1}{N\sigma_w^2} \sum_{n=0}^{N-1} |H_m(n)|^2, \qquad \text{for} \quad m = 1, 2, \ldots, M. \tag{6.63}$$

In the above hypothesis, the optimal loading strategy results into a uniform bit distribution, which amounts to transmitting $\overline{b}_m(N_m) = R_m/N_m$ bits over each allocated subcarrier. The total transmission power associated to the mth user is thus given by

$$P_m(N_m) = \frac{N_m}{\overline{\gamma}_m} \, f(R_m/N_m, p_{m,\max}), \tag{6.64}$$

where $f(b, p)$ is the rate-power function and $p_{m,\max}$ denotes the maximum tolerable BER.

Note that $P_m(N_m)$ decreases with N_m if $f(b, p)$ is strictly convex and uniformly increasing as illustrated in Fig. 6.3. Under the above assumptions, the objective of the bandwidth allocation process is to find the set of integers $\{N_1, N_2, \ldots, N_M\}$ that solves the following optimization problem:

$$\text{minimize} \quad \overline{P}_T = \sum_{m=1}^{M} \frac{N_m}{\overline{\gamma}_m} \, f(R_m/N_m, p_{m,\max}) \tag{6.65}$$

subject to

$$\sum_{m=1}^{M} N_m = N, \tag{6.66}$$

and

$$N_m \in \left\{ \left\lceil \frac{R_m}{b_{\max}} \right\rceil, \ldots, N \right\}, \tag{6.67}$$

where $b_{\max}$ is the maximum number of bits that can be allocated over any subcarrier and the notation $\lceil x \rceil$ indicates the smallest integer greater than or equal to x. The constraint Eq. (6.66) indicates that no more than N subcarriers are available for all active users, while Eq. (6.67) specifies that a minimum of $\lceil R_m/b_{\max} \rceil$ subcarriers is needed for the mth user to satisfy a rate requirement of R_m bits per OFDMA block.

The solution to the above problem is found through the following iterative procedure:

Bandwidth allocation based on average SNR (BABS) algorithm

- *Initialization:*
  1) let $N_m = \lceil R_m / b_{\max} \rceil$ and
  2) let $\Delta P_m = P_m(N_m) - P_m(N_m + 1)$ for each $m \in \mathcal{M} = \{1, 2, \ldots, M\}$.

- *Resource allocation iterations*: repeat the following procedure:
  1) if $\sum_{m=1}^{M} N_m = N$ then stop the algorithm;
  2) $\widetilde{m} = \arg \max_{m \in \mathcal{M}} \{\Delta P_m\}$;
  3) $N_{\widetilde{m}} = N_{\widetilde{m}} + 1$;
  4) $\Delta P_{\widetilde{m}} = P_{\widetilde{m}}(N_{\widetilde{m}}) - P_{\widetilde{m}}(N_{\widetilde{m}} + 1)$;
  end.

As is seen, in the initialization stage each user is given the minimum number of subcarriers that is needed to satisfy its rate requirement. The power saving ΔP_m resulting from the assignment of one additional subcarrier is also computed for all users. Assuming that there is enough bandwidth to satisfy all individual rate requirements, after initialization a total of

$$N - \sum_{m=1}^{M} \left\lceil \frac{R_m}{b_{\max}} \right\rceil \tag{6.68}$$

subcarriers are still available for further assignment. Then, at each iteration one additional subcarrier is given to the user $\widetilde{m}$ that allows the maximum power saving and the new saving $\Delta P_{\widetilde{m}}$ is evaluated for the selected user. The procedure terminates as soon as the number of allocated subcarriers is equal to N.

It is worth noting that the BABS algorithm only determines the *number* of subcarriers that must be reserved to each user. After its application, the next step is to specify *which* subcarriers are actually to be assigned. This task is accomplished by exploiting knowledge of the users' channel responses across the transmission bandwidth. One feasible solution based on heuristic arguments is presented in [77]. This scheme is known as the amplitude craving greedy (ACG) algorithm as each subcarrier is assigned to the user exhibiting the highest channel gain over it. Clearly, once a user has obtained the number of subcarriers specified by the BABS algorithm, it cannot bid for any more.

Let $\mathcal{I}_m$ be the set of subcarrier indices assigned to the mth user and denote card$\{\cdot\}$ the cardinality of the enclosed set. Then, the ACG proceeds as follows:

Amplitude craving greedy (ACG) algorithm

- *Initialization*:
  1) let $\mathcal{I}_m = \varnothing$ for each $m \in \mathcal{M} = \{1, 2, \ldots, M\}$.

- *Subcarrier assignment iterations*: repeat the following procedure for each subcarrier $n \in \{0, 1, \ldots, N - 1\}$:
  1) $\widetilde{m} = \arg \max_{m \in \mathcal{M}} \left\{ |H_m(n)|^2 \right\}$;
  2) $\mathcal{I}_{\widetilde{m}} = \mathcal{I}_{\widetilde{m}} \cup \{n\}$;
  3) if card$\{\mathcal{I}_{\widetilde{m}}\} = N_{\widetilde{m}}$, then remove $\widetilde{m}$ from $\mathcal{M}$;
  end.

After initializing all sets $\mathcal{I}_m$ to $\varnothing$, at each iteration a subcarrier is assigned to that user $\widetilde{m}$ exhibiting the maximum channel gain in the set $\mathcal{M}$. If the selected user has obtained the desired number $N_{\widetilde{m}}$ of subcarriers, its index is removed from $\mathcal{M}$ so as to exclude the user from any further assignment. To counteract the effect of channel correlation between adjacent subcarriers, it is recommended that the latter be processed in some random order rather than in the natural order $n = 0, 1, \ldots, N - 1$. In addition, the users' channel responses should be normalized to a common average energy before starting the assignment process so that weak users may have a fair chance when bidding against more powerful users.

Simulations indicate that BABS and ACG algorithms perform well under realistic channel and data traffic scenarios, thereby providing a computationally efficient method for subcarrier allocation in OFDMA systems. After this operation has been completed, bit and power loading is independently performed for each user over the corresponding set of assigned subcarriers. Again, greedy techniques based on the MMC criterion can be resorted to if the objective is to guarantee a target throughput under a specified BER constraint.

6.3 Dynamic resource allocation for MIMO-OFDMA

In recent years, the multiple-input multiple-output (MIMO) technology with multiple antennas deployed at both the transmit and receive ends has been shown capable of achieving much higher spectral efficiency than conventional single-input single-output (SISO) transmission schemes [152]. This fact has inspired considerable research interest on dynamic resource allocation for MIMO-OFDMA. In these applications users are still separated on a subcarrier basis, but each subcarrier is now characterized by a channel matrix of dimensions $N_R \times N_T$, with N_T and N_R denoting the number of transmit and receive antennas, respectively. After diagonalizing this channel matrix by means of singular-value-decomposition (SVD), each subcarrier is converted into a set of parallel flat-fading SISO subchannels which are commonly referred to as *eigenchannels* or *eigenmodes*. This means that we can view a MIMO channel as a source of spatial diversity. The latter can be exploited to improve reliability and coverage by means of space-time coding techniques [151] and/or to increase the data rate through spatial multiplexing [46]. In particular, the presence of several eigenmodes for each subcarrier offers the opportunity of simultaneously transmitting parallel data streams over the same frequency band, thereby increasing the achievable data throughput to a large extent.

As mentioned previously, in MIMO-OFDMA each subcarrier is exclusively assigned to only one user, who can therefore access all the associated eigenchannels. One possible drawback of this approach is that if some of these eigenchannels are deeply faded, they are definitively wasted as no other user is allowed to exploit them. An alternative strategy relies on the possibility of separating users in the spatial domain so that all of them can access the same set of subcarriers. This technique is commonly known as space division multiple-access (SDMA), and is characterized by increased spectral efficiency due to the opportunity of frequency reuse. In practice, SDMA is implemented by adopting a beamforming approach where multiple antennas deployed at the BS are used to transmit information over orthogonal spatial channels. The combination of SDMA and OFDMA results in a new technology called SDMA-OFDMA [157]. The research on dynamic resource allocation for SDMA-OFDMA was first pioneered by Koutsopoluos [80] and later investigated in [181] under the constraint of instantaneous QoS provisioning.

Recent advances on resource allocation for MIMO multicarrier transmissions have motivated further investigations on the performance penalty induced by imperfect channel state information (CSI). In TDD systems, the BS can exploit the reciprocity between alternative downlink and uplink transmissions to get information about the downlink channel, whereas in a FDD network CSI must be fed back by the mobile terminals on a dedicated control channel. In MIMO multicarrier systems, the number of spatial eigenchannels increases linearly with the number of antennas so that the amount of CSI that must be returned to the BS is much greater than in SISO transmissions and may far exceed the capacity of the control channel. As a result, in most cases only imperfect or partial channel information is available at the BS, which may greatly degrade the performance of existing resource allocation schemes. A few methods have recently been proposed to cope with imperfect channel knowledge in adaptive MIMO multiuser systems. Several sources of degradation have been considered, including outdated information due to feedback delay, channel estimation inaccuracies induced by Gaussian noise [174] and quantized CSI for bandwidth-constrained control channels [175].

6.4 Cross-layer design

The research on dynamic resource allocation is closely related to some recent developments in the field of cross-layer design for wireless networks. In a conventional communication system the network protocol is divided into several layers, each of which is designed independently of the others to accomplish some specific tasks. While such an approach reduces the complexity involved in the design of a complicated network, it ignores any possible interaction among different layers. For instance, in a conventional network protocol the channel estimation process is performed in the physical (PHY) layer while subcarrier assignment is handled by the multiple-access control (MAC) layer without exploiting the interdependence of these two tasks. As discussed throughout this chapter, however, the system performance is greatly improved if subcarriers are dynamically allocated to the active users on the basis of instantaneous channel state information. Some pioneering works in the field of cross-layer design have recently appeared in the literature [81, 147].

The need for a cross-layer design approach has been further driven by the success of wireless networks. In contrast to wired systems, wireless networks are characterized by time-varying channel transfer functions. As a result, a close collaboration between the PHY layer and upper layers is highly required in order to more efficiently distribute the available system resources among users. Some novel approaches have been proposed for general communication systems in which channel information is exploited to improve either the carrier-sense multiple-access (CSMA) scheme used in the MAC layer [14] or the transmission scheduling of multiple users in the network layer [165]. However, most of these techniques have only been devised for single-carrier modulations. Their extension to multicarrier systems is still largely unexplored and needs further investigations.

Chapter 7

Peak-to-Average Power Ratio (PAPR) Reduction

One of the major obstacles to the practical implementation of a multicarrier system is represented by the relatively high peak-to-average power ratio (PAPR) of the transmitted waveform. Recalling that the OFDM signal is a superposition of N sinusoids modulated by possibly coded data symbols, the peak power can theoretically be up to N times larger than the average power level. This fact poses two different problems. The first one is related to the A/D and D/A converters, which must be equipped with a sufficient number of bits to cover a potentially broad dynamic range. The second difficulty is that the transmitted signal may suffer significant spectral spreading and in-band distortion as a consequence of intermodulation effects induced by a non-linear power amplifier (PA). One possible method to circumvent this problem is the use of a large power backoff which allows the amplifier to operate in its linear region. However, this results into considerable power efficiency penalty, which translates into expensive transmitter equipments and reduced battery lifetime at the user's terminal. It is thus of interest to look for some efficient schemes that can reduce the occurrence of large signal peaks at the input of the PA so as to minimize the detrimental effects of non-linear distortions without sacrificing the power efficiency.

In this chapter we present basic material related to the PAPR mitigation problem in OFDM transmissions. After defining the PAPR and analyzing its statistical properties, some of the most representative PAPR-reduction techniques available in the literature are reviewed in detail. We also show how large amplitude fluctuations of the received signal may affect the design of the automatic gain control (AGC) unit at the receive side.

201

7.1 PAPR definitions

The continuous-time baseband representation of an OFDM signal with N subcarriers is given by

$$s(t) = \frac{1}{\sqrt{N}} \sum_{n=0}^{N-1} c(n) \, e^{j2\pi n f_{cs} t}, \quad 0 \leq t < T, \tag{7.1}$$

where $c(n)$ is the data symbol transmitted onto the nth subchannel, f_{cs} denotes the subcarrier spacing and $T = 1/f_{cs}$ is the data block duration (excluded the cyclic prefix). As indicated in Eq. (7.1), $s(t)$ is the superposition of N modulated complex sinusoidal waveforms, each corresponding to a given subcarrier. In the extreme situation where all sinusoids interfere constructively, their sum will result into a large signal peak that greatly exceeds the average power level. Furthermore, assuming that N is adequately large, we can reasonably approximate $s(t)$ as a Gaussian random process by virtue of the central limit theorem (CLT). As shown later, this assumption plays an important role in the statistical characterization of the signal amplitude.

After baseband processing, $s(t)$ is up-converted to a higher carrier frequency f_c. The resulting RF waveform is expressed by

$$s_{RF}(t) = \Re\left\{ s(t) e^{j2\pi f_c t} \right\}, \tag{7.2}$$

which represents the actual input to the PA. Thus, strictly speaking the PAPR should be defined over $s_{RF}(t)$ rather than over $s(t)$. However, since this approach would lead to some mathematical complications, it is a common practice to measure the PAPR at baseband. This procedure provides accurate results as long as $f_c \gg 1/T$, a condition that is always met in all practical systems.

With the above assumption, the *continuous-time* PAPR is defined as

$$\gamma_c \stackrel{\text{def}}{=} \frac{\max\limits_{0 \leq t < T} |s(t)|^2}{\mathrm{E}\{|s(t)|^2\}}, \tag{7.3}$$

and is sometimes referred to as the peak-to-mean envelope power (PMEPR) [144, 150]. Without loss of generality, one can normalize $s(t)$ such that $\mathrm{E}\{|s(t)|^2\} = 1$. In this case γ_c reduces to

$$\gamma_c = \max_{0 \leq t < T} |s(t)|^2. \tag{7.4}$$

In principle, the maximum of $|s(t)|^2$ can be computed by setting its derivative to zero. Unfortunately, this operation is not trivial since the derivative is a sum of sinusoidal functions and its roots cannot easily be found. To overcome this difficulty, it is expedient to replace the continuous-time waveform $s(t)$ by its samples $\{s_k^{(L)}\}$ taken at some rate L/T_s, where $T_s = T/N$ while L is a suitable integer which is commonly referred to as *oversampling factor*. This leads to the definition of the following *discrete-time* PAPR

$$\gamma_d \overset{\text{def}}{=} \max_{0 \le k \le LN-1} \left| s_k^{(L)} \right|^2, \tag{7.5}$$

where $s_k^{(L)}$ is obtained after setting $t = kT_s/L$ into Eq. (7.1), *i.e.*,

$$s_k^{(L)} = \frac{1}{\sqrt{N}} \sum_{n=0}^{N-1} c(n)\, e^{j2\pi nk/LN}, \quad 0 \le k \le LN - 1. \tag{7.6}$$

Inspection of Eq. (7.5) reveals that the discrete-time PAPR is computed through a numerical search over the set $\left\{ \left| s_k^{(L)} \right|^2 ;\ k = 0, 1, \ldots, NL - 1 \right\}$, thereby avoiding the need for solving highly non-linear equations. Moreover, comparing Eqs. (7.4) and (7.5) it is easily seen that γ_d approaches γ_c as L grows large. For this reason, γ_d is normally employed as a practical metric for evaluating the performance of PAPR-reduction techniques. An interesting question is how large the oversampling factor must be chosen to make γ_d a sufficiently accurate approximation of the continuous-time PAPR. This issue was first assessed by Tellambura in [155], and represents the subject of the next section.

7.2 Continuous-time and discrete-time PAPR

For notational simplicity, in the ensuing discussion we denote $P_a(t) = |s(t)|^2$ the instantaneous envelope power of the transmitted signal. Then, from Eq. (7.1) we have

$$P_a(t) = \frac{1}{N} \sum_{n=0}^{N-1} \sum_{\ell=0}^{N-1} c(n)c^*(\ell)\, e^{j2\pi(n-\ell)t/T}, \quad 0 \le t < T \tag{7.7}$$

which can also be rewritten as

$$P_a(t) = \frac{1}{N} \sum_{n=0}^{N-1} |c(n)|^2 + \frac{2}{N} \sum_{m=1}^{N-1} \sum_{\ell=0}^{N-1-m} \Re\left\{ c(\ell)c^*(\ell+m)\, e^{-j2\pi mt/T} \right\}. \tag{7.8}$$

A necessary condition for $P_a(t)$ to achieve its maximum at $t = t^*$ is that

$$\left.\frac{\partial P_a(t)}{\partial t}\right|_{t=t^*} = 0. \tag{7.9}$$

Thus, the global maximum of $P_a(t)$ is found by evaluating the roots of $\partial P_a(t)/\partial t = 0$ and comparing the values taken by $P_a(t)$ over these roots. As mentioned previously, solving the equation $\partial P_a(t)/\partial t = 0$ is in general a formidable task due to the non-linear nature of the trigonometric functions in Eq. (7.8). One possible method to circumvent this obstacle is proposed in [155] by transforming $P_a(t)$ into a linear sum of Chebyshev polynomials. Unfortunately, this procedure can only be used in conjunction with real-valued data symbols and does not apply to general PSK or QAM constellations. For this reason, in the sequel we limit our attention to a BPSK modulation where $c(n) \in \{\pm 1\}$.

We begin by rewriting Eq. (7.8) as

$$P_a(t) = \sum_{m=0}^{N-1} \beta_m \cos(2\pi m t/T), \tag{7.10}$$

with

$$\beta_m = \begin{cases} 1 & m = 0 \\ \dfrac{2}{N} \displaystyle\sum_{\ell=0}^{N-1-m} c(\ell)c(\ell + m) , & m = 1, 2, \ldots, N - 1. \end{cases} \tag{7.11}$$

Then, we recall that the mth order Chebyshev polynomial is defined as $T_m(t) = \cos(m \cos^{-1} t)$ and can be computed through the following recursion [122]

$$T_m(t) = 2t \cdot T_{m-1}(t) - T_{m-2}(t), \qquad m = 2, 3, \ldots \tag{7.12}$$

with $T_0(t) = 1$ and $T_1(t) = t$. Hence, using the identity $T_m(\cos\theta) = \cos(m\theta)$, we rewrite Eq. (7.10) as

$$P_a(t) = \sum_{m=0}^{N-1} \beta_m T_m\left[\cos(2\pi t/T)\right], \tag{7.13}$$

which provides an expression of $P_a(t)$ as a combination of Chebyshev polynomials in the variable $\xi = \cos(2\pi t/T)$. Computing the derivative of Eq. (7.13) with respect to t yields

$$\frac{\partial P_a(t)}{\partial t} = -\frac{2\pi}{T} \sin(2\pi t/T) \cdot Q\left[\cos(2\pi t/T)\right], \tag{7.14}$$

where $Q(\xi)$ is the following polynomial of order $N - 2$

$$Q(\xi) = \sum_{m=0}^{N-1} \beta_m \frac{\partial T_m(\xi)}{\partial \xi}. \tag{7.15}$$

Inspection of Eq. (7.14) reveals that the stationary points of $P_a(t)$ satisfy either $\sin(2\pi t/T) = 0$ or $Q\left[\cos(2\pi t/T)\right] = 0$. Recalling that $0 \le t < T$, the former equation is solved by $t/T = 0$ or $1/2$. As for the latter equation, its solutions are given by

$$\frac{t_i}{T} = \frac{\cos^{-1}(\xi_i)}{2\pi}, \qquad i = 1, 2, \ldots, I. \tag{7.16}$$

where $\{\xi_i; \quad i = 1, 2, \ldots, I\}$ are all real-valued roots of $Q(\xi)$ lying in the interval $[-1, +1]$. Clearly, $I \le N - 2$ since $Q(\xi)$ is a polynomial of degree $N - 2$ and has a total of $N - 2$ roots. The latter can be computed numerically using one of the many software packages that provide the roots of a polynomial equation.

Once the stationary points of $P_a(t)$ have been found, they are collected into a set

$$\Lambda = \left\{ 0, 1/2, \frac{\cos^{-1}(\xi_1)}{2\pi}, \frac{\cos^{-1}(\xi_2)}{2\pi}, \cdots, \frac{\cos^{-1}(\xi_I)}{2\pi} \right\}. \tag{7.17}$$

The continuous-time PAPR γ_c is eventually computed by evaluating $P_a(t)$ over the entries of Λ and picking up the maximum, *i.e.*,

$$\gamma_c = \max_{\lambda \in \Lambda} \left\{ P_a(\lambda) \right\}. \tag{7.18}$$

Figure 7.1 illustrates the complementary cumulative distribution function (CCDF) of γ_c and γ_d for a BPSK-OFDM signal with $N = 32$ subcarriers. The quantity γ_c is computed from Eq. (7.18) using the discussed procedure based on Chebyshev polynomials, while γ_d is obtained by looking for the maximum in the set $\left\{ \left| s_k^{(L)} \right|^2 ; \ k = 0, 1, \ldots, NL - 1 \right\}$ with either $L = 1, 2$ or 4. In any case, the curves represent the probability that the measured PAPR exceeds the threshold γ indicated on the horizontal axis. We see that the discrete-time PAPR obtained without oversampling ($L = 1$) is only a rough estimate of γ_c, meaning that some caution must be taken when γ_d is used as a measure of the PAPR. As expected, the difference between γ_d and γ_c reduces as L increases and becomes negligible when $L = 4$. These results validate the rule-of-thumb idea that the discrete-time PAPR with four-time oversampling provides an accurate approximation of the continuous-time PAPR [52].

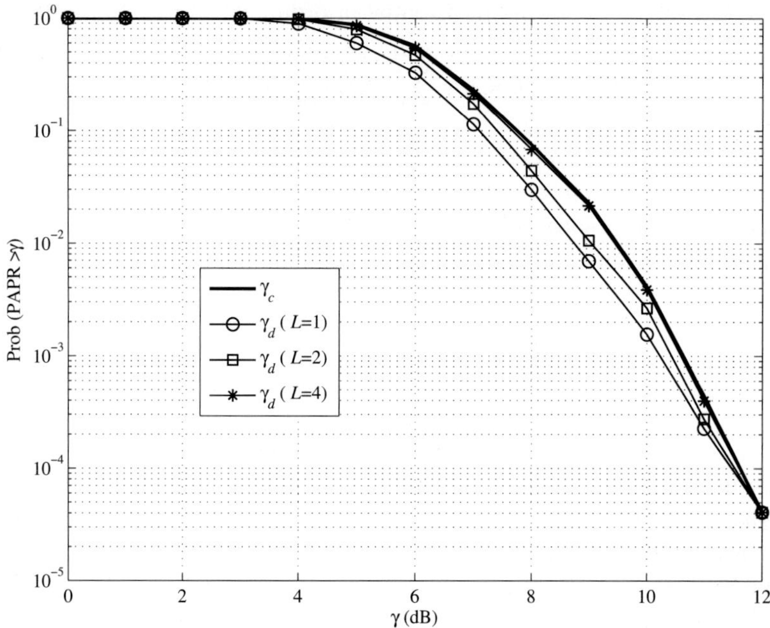

Fig. 7.1   CCDF of γ_c and γ_d for a BPSK-OFDM signal with different oversampling factors.

It is worth pointing out that the results reported in [155] has only been verified on BPSK systems. Whether the same conclusions are valid with higher-order modulations is still an open question which is worth for further investigations. Nevertheless, the pioneering work of [155] has laid down the fundamental guidelines for the identification of a practical PAPR metric. In the sequel, the quantity γ_d obtained from the oversampled time-domain sequence $\{s_k^{(L)}\}$ is used as a measure of the true PAPR and exploited to assess the performance of the PAPR-reduction techniques considered throughout this chapter.

7.3   Statistical properties of PAPR

The statistical properties of the PAPR are normally given in terms of the corresponding CCDF. From the central limit theorem we know that the real and imaginary parts of the time-domain samples $\{s_k^{(L)}\}$ can reasonably be approximated as statistically independent Gaussian random variables with zero-mean and variance $\sigma^2 = 1/2$ (recall that the signal has been normalized such that $E\{|s(t)|^2\} = 1$). This means that $\left|s_k^{(L)}\right|^2$ follows a central chi-square distribution with two degrees of freedom [115], and its cumulative distribution function (CDF) is thus given by

$$\Pr\left\{\left|s_k^{(L)}\right|^2 \leq \gamma\right\} = 1 - e^{-\gamma}, \quad \text{for } \gamma \geq 0. \tag{7.19}$$

The CDF of γ_d is easily computed when $L = 1$ [109] since in this case the N samples $\{s_k^{(L)}\}$ are mutually independent and we can write

$$\Pr\left\{\gamma_d \leq \gamma\right\} = \prod_{k=0}^{N-1} \Pr\left\{\left|s_k^{(L)}\right|^2 \leq \gamma\right\}. \tag{7.20}$$

Substituting Eq. (7.19) into Eq. (7.20) provides the CCDF of γ_d in the form

$$\begin{aligned} F(\gamma) &= 1 - \Pr\left\{\gamma_d \leq \gamma\right\} \\ &= 1 - \left(1 - e^{-\gamma}\right)^N. \end{aligned} \tag{7.21}$$

Unfortunately, results obtained with $L = 1$ have a rather scarce practical relevance since in this case the quantity γ_d is only a rough approximation of the true PAPR. The use of an oversampling factor $L \geq 2$ introduces a statistical correlation among neighboring samples, which makes Eq. (7.21) a poor approximation of the true CCDF. Many attempts have been made to derive more accurate expressions of the CCDF in situations where the signal samples are statistically dependent. For instance, in [104] the oversampling effect is taken into account by introducing an ad-hoc parameter α in the exponent of Eq. (7.21), yielding

$$F(\gamma) = 1 - \left(1 - e^{-\gamma}\right)^{\alpha N}. \tag{7.22}$$

Numerical results indicate that $\alpha = 2.8$ is a good choice when N is adequately large and $L \geq 4$.

Figure 7.2 compares the analytical result Eq. (7.22) with the simulated CCDF of γ_d for a QPSK-OFDM signal. The oversampling factor is $L = 8$ and the number of subcarriers is $N = 64, 256$ or 1024. As expected, the probability that the signal power exceeds a given threshold increases with N. Furthermore, we see that Eq. (7.22) represents a reasonable approximation of the true CCDF, especially for large values of N.

Since the occurrence of large peaks in the envelope of OFDM signals is an event with non-negligible probability, effective PAPR-mitigation techniques are essential to enable the use of efficient non-linear power amplifiers without incurring severe spectral spreading and/or in-band distortion. Some popular methods for PAPR reduction are described in the subsequent sections.

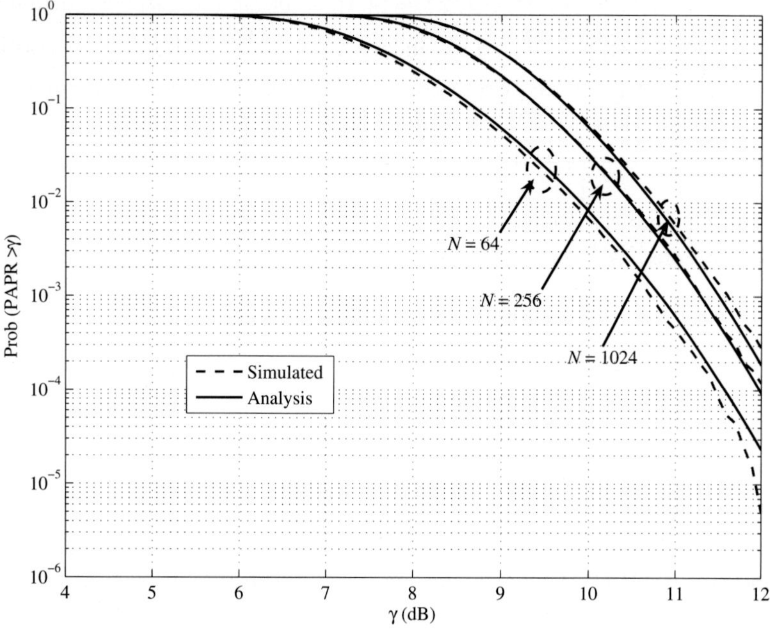

Fig. 7.2 CCDF of γ_d for a QPSK-OFDM signal with oversampling factor $L = 8$ and different number of subcarriers.

7.4 Amplitude clipping

The simplest approach to limit the amplitude peaks in a multicarrier waveform is to deliberately clip the signal before amplification [111]. This operation is normally accomplished at baseband using a soft envelope limiter. If clipping is directly applied to the analog signal $s(t)$ (*i.e.*, after D/A conversion), the output $y(t)$ of the limiter appears as indicated in Fig. 7.3, where A is the maximum permissible amplitude over which the signal is clipped.

The distortion caused by the clipping process is mathematically expressed as

$$d(t) = y(t) - s(t), \tag{7.23}$$

and is viewed as an additional source of noise. Since the derivative of $d(t)$ exhibits discontinuities at the clipping instants, its bandwidth is theoretically infinite. This means that in general amplitude clipping gives rise to in-band distortion as well as out-of-band emission. The former degrades the bit-error-rate (BER) performance while the latter reduces the spectral

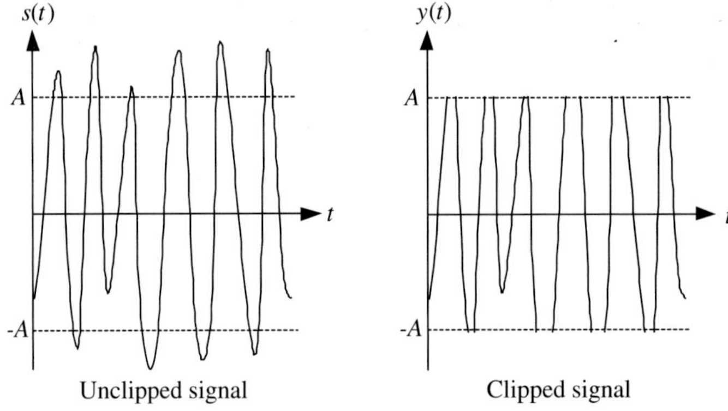

Fig. 7.3   The effect of clipping on the transmitted signal.

efficiency of the communication system. Filtering after clipping can reduce out-of-band radiation to a large extent, but may also produce some peak regrowth in the filtered signal [88].

7.4.1   *Clipping and filtering of oversampled signals*

In practical applications clipping and filtering is performed digitally (*i.e.*, before D/A conversion) on an oversampled version of the OFDM signal [52]. Letting $J \geq 1$ be the employed oversampling factor, we denote

$$s(k) = \frac{1}{\sqrt{N}} \sum_{n=0}^{N-1} c(n) \, e^{j2\pi nk/JN}, \qquad 0 \leq k \leq JN - 1. \qquad (7.24)$$

the samples of $s(t)$ corresponding to a given block of data $\boldsymbol{c} = [c(0), c(1), \ldots, c(N-1)]^T$. Note that J should not be confused with parameter L defined in Sec. 7.1. Indeed, the former is the oversampling factor that is actually used in the OFDM transmitter to execute clipping and filtering operations, while the latter is just a parameter employed in computer simulations for PAPR measurements.

As illustrated in Fig. 7.4, the oversampled data sequence, $\boldsymbol{s} = [s(0), s(1), \ldots, s(JN - 1)]^T$, can be efficiently generated as the IDFT of the zero-padded data block $\boldsymbol{c}^{(ZP)}$, which is obtained by extending $\boldsymbol{c}$ with

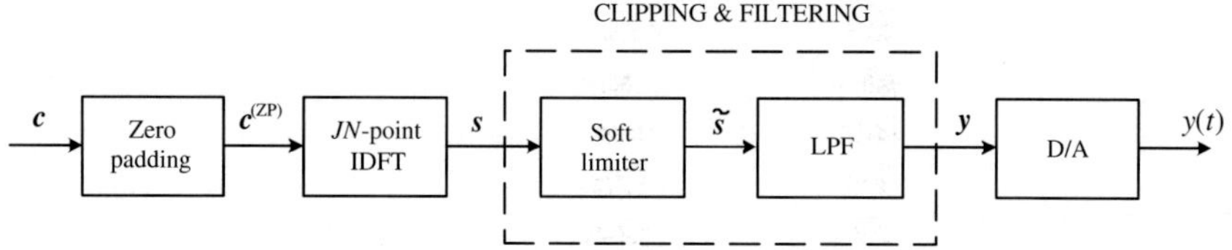

Fig. 7.4   Clipping and filtering operations on the oversampled OFDM signal.

$(J-1)N$ zeros

$$c^{(ZP)} = [c(0), c(1), \ldots, c(N-1), \underbrace{0, 0, \ldots, 0}_{(J-1)N \ \text{zeros}}]^{T}. \qquad (7.25)$$

Each sample $s(k)$ is then clipped by a soft envelope limiter. Letting $\rho_k e^{j\varphi_k}$ be the representation of $s(k)$ in polar coordinates, the output from the limiter is given by

$$\tilde{s}(k) = \begin{cases} s(k), & \text{if } \rho_k \leq A \\ Ae^{j\varphi_k}, & \text{if } \rho_k > A. \end{cases} \qquad (7.26)$$

It is a common practice to normalize the clipping level A to the root-mean-square (rms) value of the input signal. This results into the following clipping ratio (CR)

$$\mu = \frac{A}{\sqrt{P_{in}}}, \qquad (7.27)$$

where $P_{in} = \text{E}\{|s(k)|^2\}$ is the average power of the unclipped samples.

As is intuitively clear, the clipping process leads to a certain reduction of the output power. If the OFDM signal can be modeled as a zero-mean circularly symmetric complex Gaussian process, the amplitude ρ_k is Rayleigh distributed and the average power of the clipped samples turns out to be

$$P_{out} = (1 - e^{-\mu^2})P_{in}. \qquad (7.28)$$

Note that the difference between P_{out} and P_{in} reduces as μ grows large and becomes zero when $\mu = \infty$, which corresponds to an ideal system without clipping.

As mentioned earlier, in general the power spectral density (PSD) of the non-linear distortion introduced by the amplitude limiter has a theoretically infinite bandwidth. Hence, aliasing will occur if clipping is carried out on the samples $\{s(k)\}$ rather than on the continuous-time signal $s(t)$. In particular, when clipping is done at the Nyquist rate ($J = 1$), the spectrum of the resulting distortion is folded back into the signal bandwidth. This gives rise to considerable in-band distortion, with ensuing limitations of the error-rate performance. Furthermore, extensive simulations indicate that the PAPR reduction capability of Nyquist-rate clipping is *not* so significant due to considerable peak regrowth after D/A conversion [108, 110]. As a result, clipping is normally performed on an oversampled version of the OFDM signal ($J > 1$).

The oversampled approach has the advantage of reducing in-band distortion and peak regrowth to some extent, but inevitably generates out-of-band radiation that must be removed in some way. The conventional solution to this problem is to pass the clipped samples $\widetilde{\boldsymbol{s}} = [\widetilde{s}(0), \widetilde{s}(1), \ldots, \widetilde{s}(JN - 1)]^T$ through a low-pass filter (LPF) as indicated in Fig. 7.4. This produces a vector $\boldsymbol{y} = [y(0), y(1), \ldots, y(N - 1)]^T$ of time-domain samples, which are extended by the cyclic prefix (CP) and fed to the D/A converter. The resulting baseband waveform is then upconverted and passed to the power amplifier before being launched over the channel.

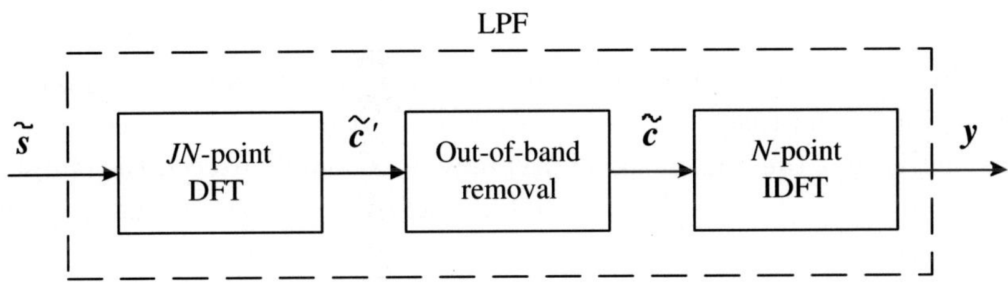

Fig. 7.5   Filtering process to remove out-of-band radiation.

The filtering process is outlined in Fig. 7.5. The sequence $\widetilde{\boldsymbol{s}}$ is transformed in the frequency domain through a DFT operation which produces the following vector of length JN

$$\widetilde{\boldsymbol{c}}' = [\underbrace{\widetilde{c}(0), \widetilde{c}(1), \ldots, \widetilde{c}(N - 1)}_{\text{in-band components}}, \underbrace{\widetilde{c}(N), \widetilde{c}(N + 1), \ldots, \widetilde{c}(JN - 1)}_{\text{out-of-band components}}]^T, \quad (7.29)$$

with entries

$$\widetilde{c}(n) = \frac{1}{J\sqrt{N}} \sum_{k=0}^{JN-1} \widetilde{s}(k) e^{-j2\pi nk/JN}. \qquad 0 \le n \le JN - 1. \qquad (7.30)$$

Next, out-of-band radiation is suppressed by discarding the last $(J-1)N$ elements of $\tilde{c}'$ (out-of-band components) while leaving the first N elements unaltered (in-band components). This yields a vector of N *modified frequency-domain samples* $\tilde{c} = [\tilde{c}(0), \tilde{c}(1), \ldots, \tilde{c}(N-1)]^T$, which is in fact a distorted version of the original data block c. Vector $\tilde{c}$ is then transformed back in the time domain through an N-point IDFT, which yields the sequence y of N *modified time-domain samples*. After D/A conversion, the analog signal $y(t)$ can be expressed in terms of the modified symbols $\{\tilde{c}(n)\}$ as

$$y(t) = \frac{1}{\sqrt{N}} \sum_{n=0}^{N-1} \tilde{c}(n)e^{j2\pi n f_{cs}t}, \quad 0 \leq t < T. \qquad (7.31)$$

It is worth noting that the filtering procedure sketched in Fig. 7.5 is equivalent to an ideal brick-wall low-pass filter which *totally* eliminates out-of-band radiation regardless of the oversampling factor J. Clearly, the entire filtering process becomes useless when clipping is performed at the Nyquist rate. The reason is that in this case there are no out-of-band components to be suppressed in $\tilde{c}'$ and, in consequence, the architecture in Fig. 7.5 reduces to a pair of N-points DFT/IDFT units, which simply provides $y = \tilde{s}$.

Albeit necessary for suppressing out-of-band emission, the filtering-after-clipping approach results into some peak regrowth. A consequence of this fact is that the analog signal $y(t)$ may occasionally exceed the clipping level A at some instants. As reported in many works, however, filtering the oversampled and clipped version of the OFDM signal produces much less peak regrowth than clipping at Nyquist rate. This conclusion is also supported by the simulation results shown in Fig. 7.6, illustrating the CCDF of the PAPR for a clipped QPSK-OFDM signal with $N = 256$ subcarriers. The clipping-ratio is set to $\mu = 1$ while the oversampling factor is $J = 1, 2$ or 4. The curve pertaining to the unclipped signal ($\mu = \infty$) is also shown for comparison. The PAPR of the analog signal $y(t)$ is measured as

$$\gamma_d = \frac{\max\limits_{0 \leq k \leq LN-1} \left|y_k^{(L)}\right|^2}{\widehat{P}}, \qquad (7.32)$$

where $\{y_k^{(L)}\}$ are samples of $y(t)$ taken at rate L/T_s while $\widehat{P}$ is the power of the current OFDM block *after* clipping and filtering, which is given by

$$\widehat{P} = \frac{1}{N} \sum_{n=0}^{N-1} |\tilde{c}(n)|^2. \qquad (7.33)$$

As discussed in Sec. 7.2, the quantity γ_d provides an accurate measure of the PAPR as long as parameter L is properly designed. The value $L = 8$ is adopted throughout simulations.

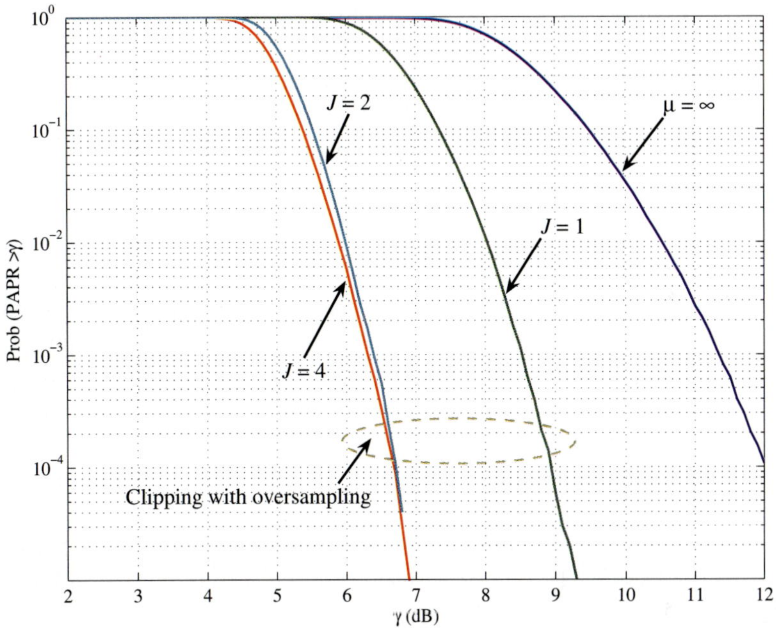

Fig. 7.6   PAPR CCDF for a clipped and filtered OFDM signal with oversampling.

Inspection of Fig. 7.6 reveals that clipping at Nyquist rate considerably reduces the PAPR of the transmitted signal as compared to a system without clipping. However, much better results are obtained if clipping is executed on the oversampled waveform. In particular, a PAPR reduction of approximately 2 dB is achieved when J is increased from 1 to 4. Clearly, this advantage comes at the expense of a higher computational complexity due to the larger dimension of the IDFT unit in Fig. 7.4 and the need for filtering the signal after clipping. Theoretical analysis [145] and computer simulations [88] indicate that in many cases a good trade-off between performance and complexity is obtained with an oversampling factor of 4. Repeated clipping and filtering operations can also be used to further reduce the overall peak regrowth after D/A conversion [2].

7.4.2   *Signal-to-clipping noise ratio*

The in-band distortion affecting the clipped signal is normally measured in terms of signal-to-clipping noise ratio (SCNR) [2]. This quantity is defined as the ratio of the average received signal power to the average power of the clipping distortion, and can be computed by resorting to the Bussgangs' theorem [134]. To see how this comes about, we consider a conventional OFDM receiver in which the incoming waveform is low-pass filtered and sampled at Nyquist rate. After discarding the CP, the remaining samples are passed to an N-point DFT unit to retrieve the information symbols. In case of ideal timing and frequency synchronization, the DFT output takes the form

$$R(n) = H(n)\widetilde{c}(n) + W(n), \qquad 0 \leq n \leq N - 1. \tag{7.34}$$

where $H(n)$ is the channel response over the nth subcarrier, $W(n)$ accounts for thermal noise and $\widetilde{c}(n)$ is a distorted version of the original symbol $c(n)$. The relationship between $\widetilde{c}(n)$ and the clipped sequence $\{\widetilde{s}(k)\}$ is provided by Eq. (7.30). A more useful expression for $\widetilde{c}(n)$ is found by observing that $\widetilde{s}(k)$ is the output of a memoryless non-linearity driven by the unclipped signal $s(k)$, as indicated in Eq. (7.26). If the number of subcarriers is adequately large, from the central limit theorem we know that the sequence $\{s(k)\}$ is approximately Gaussian distributed with zero-mean. Hence, by applying the Bussgangs' theory, we can write the output of the non-linearity as [134]

$$\widetilde{s}(k) = \eta s(k) + d(k), \qquad 0 \leq k \leq JN - 1. \tag{7.35}$$

where $d(k)$ is a zero-mean distortion term uncorrelated with $s(k)$, while η is an attenuation factor given by

$$\eta = \frac{\mathrm{E}\left\{\widetilde{s}(k)s^*(k)\right\}}{\mathrm{E}\{|s(k)|^2\}}. \tag{7.36}$$

For a soft envelope limiter characterized by a clipping ratio μ, it can be shown that [108]

$$\eta = 1 - e^{-\mu^2} + \frac{\sqrt{\pi}\mu}{2}\mathrm{erfc}(\mu), \tag{7.37}$$

with

$$\mathrm{erfc}(x) = \frac{2}{\sqrt{\pi}} \int_x^\infty e^{-t^2}\, dt. \tag{7.38}$$

Substituting Eq. (7.35) into Eq. (7.30) and bearing in mind Eq. (7.24), yields

$$\widetilde{c}(n) = \eta c(n) + D(n), \qquad 0 \leq n \leq N - 1. \tag{7.39}$$

where

$$D(n) = \frac{1}{J\sqrt{N}} \sum_{k=0}^{JN-1} d(k)\, e^{-j2\pi nk/JN}, \tag{7.40}$$

represents the in-band distortion over the nth subcarrier. It is worth noting that, although the probability density function of $d(k)$ is in general very non-Gaussian due to the presence of a large peak at zero corresponding to the unclipped samples, the distribution of $D(n)$ is approximately Gaussian as long as the number of clips occurring in each OFDM block is adequately large. The reason is that in the latter case $D(n)$ is the sum of several non-zero random variables $d(k)$ as indicated in Eq. (7.40), and the central limit theorem can thus be applied.

Inspection of Eq. (7.39) reveals that in general the clipping process results into a shrinking of the signal constellation plus an added noise-like effect. Calling $C_2 = \mathrm{E}\{|c(n)|^2\}$ the power of the original data symbols, the SCNR over the nth subcarrier is found to be

$$\mathrm{SCNR}_n = \frac{\eta^2 C_2}{P_{D,n}}, \tag{7.41}$$

where $P_{D,n} = \mathrm{E}\{|D(n)|^2\}$ is the PSD of $d(k)$. Obviously, SCNR_n is independent of the channel frequency response since clipping noise is introduced at the transmitter side and, in consequence, it fades along with the signal.

Substituting Eq. (7.39) into Eq. (7.34) yields

$$R(n) = H(n)\left[\eta c(n) + D(n)\right] + W(n), \qquad 0 \le n \le N-1. \tag{7.42}$$

which represents the equivalent model of a clipped OFDM transmission channel as depicted in Fig. 7.7.

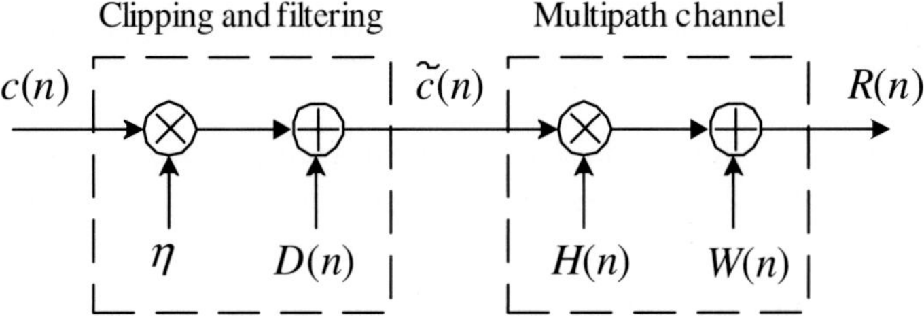

Fig. 7.7   Equivalent model of a clipped OFDM multipath channel.

Although some attempts have been made in the literature to derive theoretical expressions of $P_{D,n}$, computer simulations are normally employed for SCNR measurements. In Fig. 7.8 the SCNR is shown as a function of the clipping ratio μ for an OFDM signal with 256 subcarriers. The oversampling factor is $J = 4$ and data symbols are taken from a QPSK constellation with unit power, *i.e.*, $C_2 = 1$. The results are numerically obtained by averaging the right-hand-side of Eq. (7.41) over the available subcarriers. As is seen, SCNR increases with μ and remains quite large even in the presence of severe clipping.

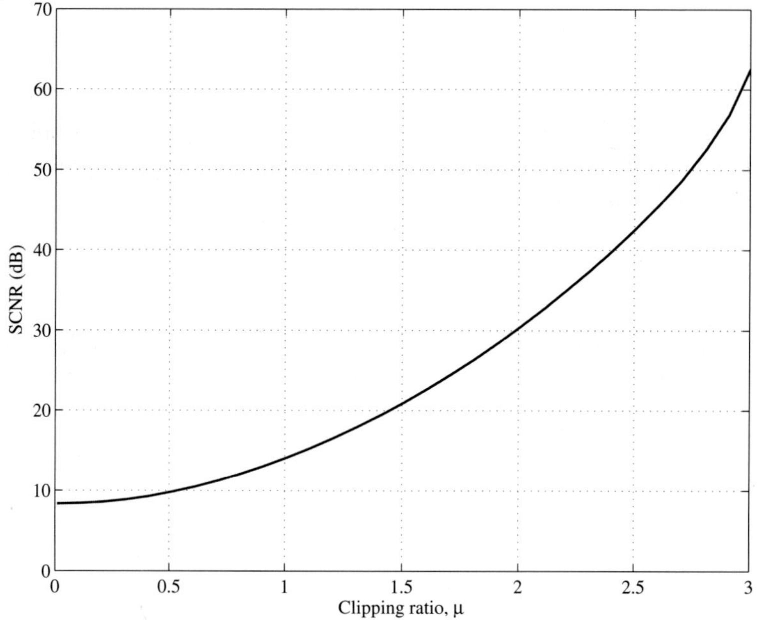

Fig. 7.8   SCNR as a function of μ for a QPSK-OFDM signal with 256 subcarriers and oversampling factor $J = 4$.

The impact of clipping noise on the error-rate performance is shown in Fig. 7.9. Here, the BER obtained with several values of μ over an AWGN channel is illustrated as a function of E_b/N_0, where E_b is the average energy per bit after clipping and filtering while $N_0/2$ is the two-sided noise PSD. Compared to the unclipped signal ($\mu = \infty$), the SNR penalty incurred with $\mu = 1.0$ is 3.5 dB at a target BER of 10^{-4}. The degradation reduces to approximately 0.5 dB when $\mu \geq 1.5$, while an irreducible error floor is observed with $\mu \leq 0.5$. As shown in [110], the distortion caused by the clipping process can be alleviated by means of suitable coding techniques.

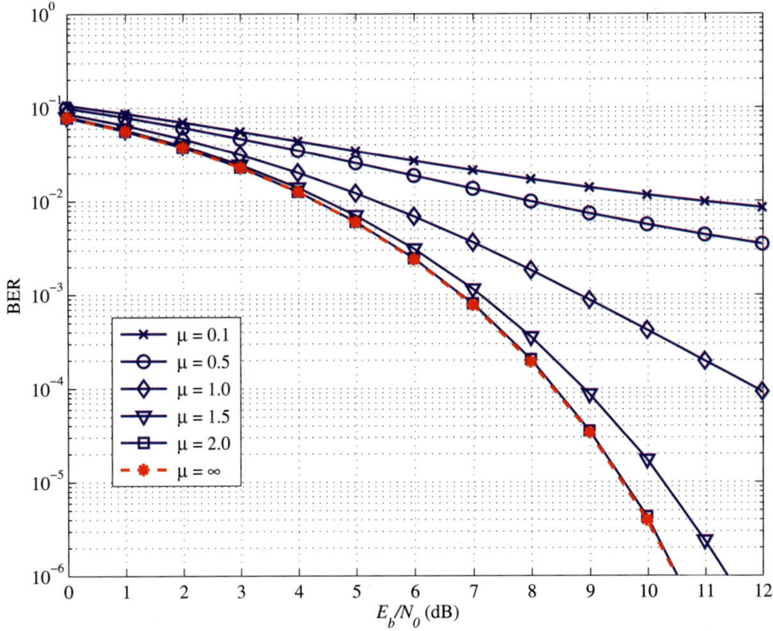

Fig. 7.9   The impact of clipping noise on the error-rate performance of a QPSK-OFDM transmission over an AWGN channel.

7.4.3   *Clipping noise mitigation*

Several methods have been proposed in the literature to mitigate the harmful effects of clipping noise in OFDM systems. Some of them attempt to retrieve the original amplitude of clipped samples by interpolating the received oversampled signal [137]. However, correct interpolation requires some out-of-band emission at the transmit side, thereby leading to a reduction of the spectral efficiency. An alternative scheme that does not require any bandwidth expansion is derived in [16] making use of iterative interference cancellation techniques. This method operates in the frequency domain and employs detected data to regenerate the clipping noise distortion. The latter is then subtracted from the DFT output at each new iteration.

To better illustrate this approach, we assume that the receiver has perfect knowledge of the channel frequency response and collect data decisions taken at the jth iteration into a vector $\widehat{\boldsymbol{c}}^{(j)} = [\widehat{c}^{(j)}(0), \widehat{c}^{(j)}(1), \ldots, \widehat{c}^{(j)}(N-1)]^T$. Then, the clipping-noise canceler proceeds as follows:

(1) The detected symbols $\widehat{c}^{(j)}(n)$ undergo the same clipping and filtering operations as those performed at the transmitter (see Fig. 7.10). This produces the sequence of N samples $\widetilde{\boldsymbol{c}}^{(j)} = [\widetilde{c}^{(j)}(0), \widetilde{c}^{(j)}(1), \ldots, \widetilde{c}^{(j)}(N-1)]^T$ which, similarly to Eq. (7.39), can be represented as

$$\widetilde{c}^{(j)}(n) = \eta\, \widehat{c}^{(j)}(n) + \widehat{D}^{(j)}(n), \qquad 0 \leq n \leq N-1. \tag{7.43}$$

where η is given in Eq. (7.37).

(2) The clipping noise terms $\widehat{D}^{(j)}(n)$ are derived from Eq. (7.43) in the form

$$\widehat{D}^{(j)}(n) = \widetilde{c}^{(j)}(n) - \eta\, \widehat{c}^{(j)}(n), \tag{7.44}$$

and are subtracted from the DFT output so as to obtain a refined observation sequence

$$\widehat{R}^{(j)}(n) = R(n) - H(n)\widehat{D}^{(j)}(n). \tag{7.45}$$

Substituting Eqs. (7.42) and (7.44) into Eq. (7.45) yields

$$\widehat{R}^{(j)}(n) = \eta H(n)c(n) + H(n)\left[D(n) - \widehat{D}^{(j)}(n)\right] + W(n), \tag{7.46}$$

where $D(n) - \widehat{D}^{(j)}(n)$ is the residual clipping noise over the nth subcarrier.

(3) The refined DFT output $\widehat{\boldsymbol{R}}^{(j)} = [\widehat{R}^{(j)}(0), \widehat{R}^{(j)}(1), \ldots, \widehat{R}^{(j)}(N-1)]^T$ is fed to the channel equalization and data detection unit, which delivers new data decisions $\widehat{c}^{(j+1)}(n)$ $(n = 0, 1, \ldots, N-1)$ to be employed in the next iteration.

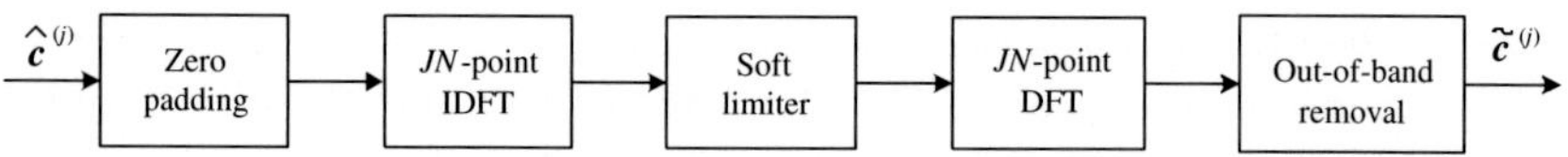

Fig. 7.10   Regeneration of the clipped and filtered signal at the receiver.

Simulation results reported in [16] indicate that the accuracy of the estimated clipping noise component $\widehat{D}^{(j)}(n)$ increases with the number of iterations, thereby improving the error-rate performance. From Fig. 7.10 it turns out that the crux in the computation is represented by the JN-point IDFT and DFT pair, which must be performed at each iteration. However, in many cases the required complexity is moderate since incremental gains diminish after the first iteration and a couple of iterations are often sufficient to restore the system performance.

7.5   Selected mapping (SLM) technique

One possible approach for PAPR control in multicarrier systems is based on the idea of mapping the data block $\boldsymbol{c} = [c(0), c(1), \ldots, c(N-1)]^T$ into a set of adequately different signals and then choosing the most favorable one for transmission. This technique is called selected mapping (SLM) and its main concept is shown in Fig. 7.11.

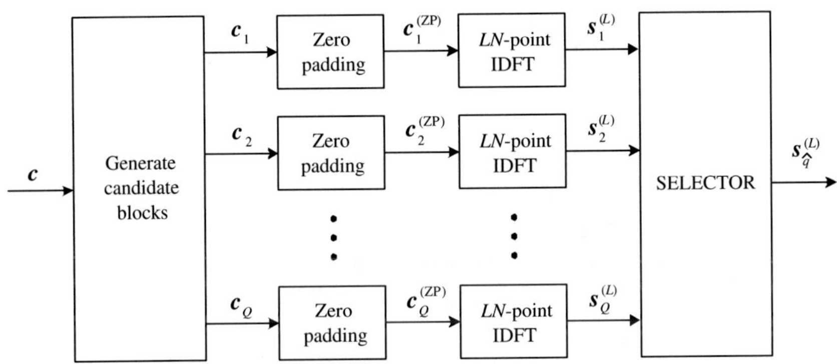

Fig. 7.11   Block diagram of the SLM technique.

As is seen, the transmitter generates a number Q of candidate data blocks $\boldsymbol{c}_q = [c_q(0), c_q(1), \ldots, c_q(N-1)]^T$ $(q = 1, 2, \ldots, Q)$ using some suitable algorithm. Each block has length N and conveys the same information as the original data sequence $\boldsymbol{c}$. The latter is normally included into the set of candidate blocks by letting $\boldsymbol{c}_1 = \boldsymbol{c}$. After transforming all blocks $\boldsymbol{c}_q$ in the time-domain, the one exhibiting the lowest PAPR is selected for transmission.

Since the PAPR of the continuous-time waveform cannot precisely be computed from its Nyquist-rate samples, each candidate block is padded with $(L-1)N$ zeros and fed to a LN-point IDFT unit. This provides Q

oversampled sequences $\boldsymbol{s}_q^{(L)}$ $(q = 1, 2, \ldots, Q)$ with entries

$$s_{q,k}^{(L)} = \frac{1}{\sqrt{N}} \sum_{n=0}^{N-1} c_q(n)\, e^{j2\pi nk/LN}, \qquad\qquad 0 \leq k \leq NL - 1. \qquad (7.47)$$

and characterized by the following discrete-time PAPRs

$$\gamma_q = \frac{\displaystyle\max_{0 \leq k \leq LN-1} \left|s_{q,k}^{(L)}\right|^2}{\widehat{P}_q}, \qquad q = 1, 2, \ldots, Q. \qquad (7.48)$$

with

$$\widehat{P}_q = \frac{1}{N} \sum_{n=0}^{N-1} |c_q(n)|^2. \qquad (7.49)$$

As mentioned in Sec. 7.2, setting $L = 4$ is sufficient to capture the peaks of the continuous-time waveform.

The selector in Fig. 7.11 computes the quantities γ_q and chooses the sequence $\boldsymbol{s}_{\widehat{q}}^{(L)}$ such that

$$\widehat{q} = \arg \min_{1 \leq q \leq Q} \{\gamma_q\}. \qquad (7.50)$$

The selected sequence is then passed to the D/A converter and the corresponding waveform is finally launched over the channel after up-conversion and power amplification.

To better illustrate the PAPR-reduction capability of the SLM technique, we denote $F_q(\gamma) = \Pr\{\gamma_q \geq \gamma\}$ the CCDF of γ_q and observe that

$$F_{\widehat{q}}(\gamma) = \Pr\left\{\bigcap_{q=1}^{Q} (\gamma_q \geq \gamma)\right\}, \qquad (7.51)$$

since $\gamma_{\widehat{q}}$ is the minimum of the set $\{\gamma_q\}$. If the candidate sequences $\boldsymbol{s}_q^{(L)}$ are sufficiently "different", the random variables γ_q may be considered as nearly independent and Eq. (7.51) reduces to

$$F_{\widehat{q}}(\gamma) = \prod_{q=1}^{Q} F_q(\gamma). \qquad (7.52)$$

Figure 7.12 illustrates function $F_{\widehat{q}}(\gamma)$ for $N = 256$ and some values of Q. The results are derived analytically under the simplifying assumption that each factor $F_q(\gamma)$ in Eq. (7.52) can be expressed as indicated in Eq. (7.22). In this case we have

$$F_{\widehat{q}}(\gamma) = \left[1 - \left(1 - e^{-\gamma}\right)^{\alpha N}\right]^Q, \qquad (7.53)$$

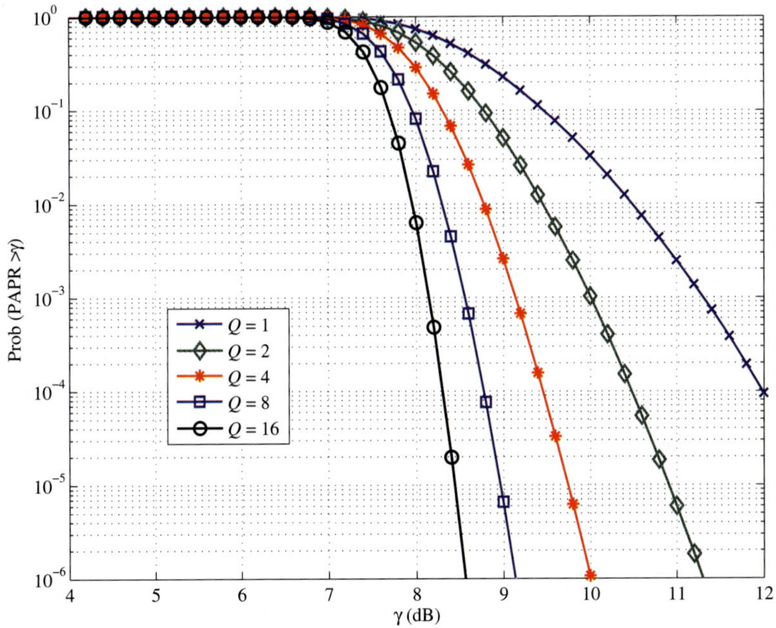

Fig. 7.12   Function $F_{\hat{q}}(\gamma)$ for different values of Q.

with $\alpha = 2.8$. As expected, the amount of PAPR reduction depends on the number Q of candidate sequences. We see that significant gains are achieved in passing from $Q = 1$ to $Q = 4$, while only marginal improvements are observed with higher values of Q.

Unfortunately, the result Eq. (7.53) is only an approximation of the CCDF of $\gamma_{\hat{q}}$. The reason is that in practice the quantities γ_q are not truly independent as they are derived from sequences $\boldsymbol{s}_q^{(L)}$ that convey the same information $\boldsymbol{c}$. Thus, the question arises as to how candidate blocks $\boldsymbol{c}_q$ that result into adequately different sequences $\boldsymbol{s}_q^{(L)}$ can be generated.

The solution suggested in [66] employs a set of Q pseudo-random interleavers to get permuted versions of the original data block $\boldsymbol{c}$. In such a case, the entries of $\boldsymbol{c}_q$ are given by

$$c_q(n) = c(\pi_q(n)), \qquad n = 0, 1, \ldots, N - 1. \tag{7.54}$$

where $n \rightarrow \pi_q(n)$ is a one-to-one mapping, with $\pi_q(n) \in \{0, 1, \ldots, N - 1\}$ for all n.

An alternative approach is sketched in Fig. 7.13, where the candidate

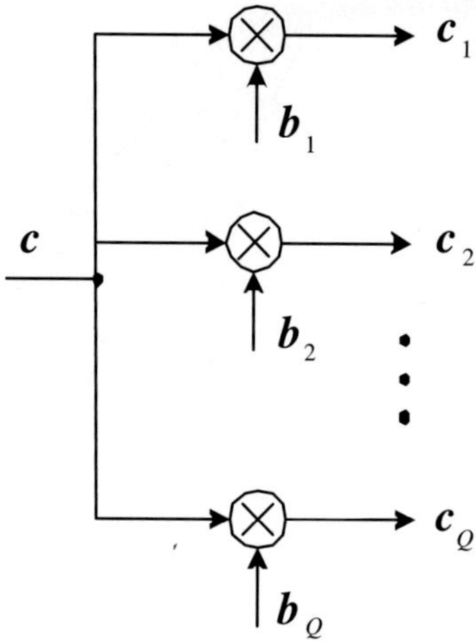

Fig. 7.13   Generation of candidate sequences through pseudo-random phase shifts.

blocks are obtained through an element-wise multiplication of c by Q different pseudo-random phase sequences $b_q = [e^{j\varphi_q(0)}, e^{j\varphi_q(1)}, \ldots, e^{j\varphi_q(N-1)}]^T$ [7]. This produces the following modified symbols

$$c_q(n) = c(n)e^{j\varphi_q(n)}, \qquad n = 0, 1, \ldots, N-1. \qquad (7.55)$$

To reduce the system complexity, the phase shifts $\varphi_q(n)$ are normally chosen as multiples of $\pi/2$. In this way $c_q(n)$ is obtained from $c(n)$ by means of simple sign inversions, thereby dispensing from any multiplication.

The computational requirement of the SLM technique is mainly related to the generation of the sequences $s_q^{(L)}$. Since this operation involves Q oversampled IDFTs for each OFDM block, in practice parameter Q must be carefully designed so as to guarantee a reasonable trade-off in terms of system complexity and PAPR-reduction capability. Compared with amplitude clipping, SLM has the considerable advantage of being distortionless as it does not produce any inter-modulation among subcarriers nor undesired out-of-band emission. Clearly, in order to recover the original data symbols, the receiver must be informed as to which interleaver or phase sequence has been employed at the transmitter to generate the selected block $c_{\hat{q}}$. Since both the transmitter and the receiver can store the permutation indices $\{\pi_q(n)\}$ or phase vectors $\{b_q\}$ in memory, the integer $\hat{q}$ represents the minimum side information that must be sent to the receiver for each OFDM block. This operation requires $\log_2 Q$ dedicated bits that must carefully be protected against channel impairments since an error in the reception of $\widehat{q}$ would entail the loss of the entire data block. An SLM technique that eliminates the need for any exchange of side information is discussed in [9].

7.6 Partial transmit sequence (PTS) technique

In the SLM technique, the data block is mapped into different sequences of frequency-domain samples. As indicated in Fig. 7.11, in such a case a dedicated IDFT operation is required to measure the PAPR associated with each candidate sequence. In applications where system complexity is a critical issue, this approach limits the number of possible candidate waveforms to only a few units, with a corresponding decrease of the PAPR reduction capability. To circumvent this problem, the partial transmit sequence (PTS) technique generates candidate sequences in the time-domain rather than in the frequency-domain. In this way, a large set of candidates is obtained with only a few IDFT operations as it is now explained.

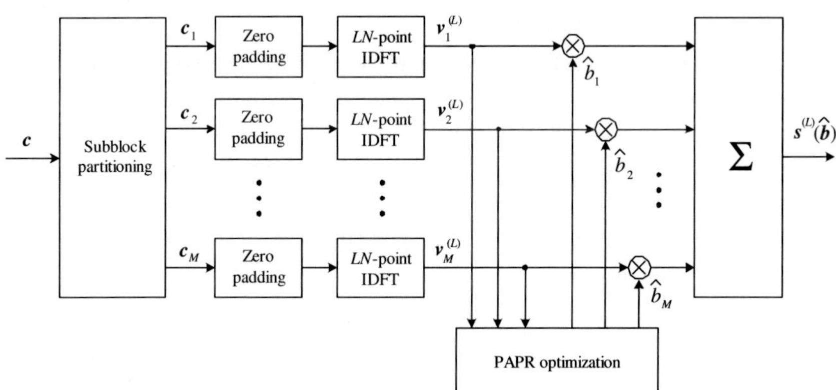

Fig. 7.14　Block diagram of the PTS technique.

Figure 7.14 illustrates the basic idea behind the PTS approach. The input data vector $\boldsymbol{c} = [c(0), c(1), \ldots, c(N-1)]^T$ is partitioned into M disjoint subblocks $\boldsymbol{c}_m = [c_m(0), c_m(1), \ldots, c_m(N-1)]^T$ $(m = 1, 2, \ldots, M)$

with entries

$$c_m(n) = \begin{cases} c(n), & \text{if } n \in J_m \\ 0, & \text{otherwise.} \end{cases} \qquad (7.56)$$

The sets $\{J_m\}$ collect the indices of subcarriers assigned to the various subblocks and satisfy the identities

$$\bigcup_{m=1}^{M} J_m = \{0, 1, \ldots, N-1\}, \qquad (7.57)$$

and

$$J_m \cap J_\ell = \varnothing, \quad \text{for} \quad m \neq \ell. \qquad (7.58)$$

Hence, from Eq. (7.56) we have

$$c = \sum_{m=1}^{M} c_m. \qquad (7.59)$$

Three different strategies can be adopted for generating the M subblocks. In the *subband* design the subcarriers of any subblock occupy adjacent positions in the signal spectrum, while in the *interleaved* design they are uniformly spaced over the signal bandwidth. A more versatile approach is based on a *pseudo-random* design, where subcarriers are randomly partitioned into M clusters. In any case, subblocks of equal size are normally employed even though in principle an arbitrary number of subcarriers might be included in each subblock.

Returning to Fig. 7.14, we see that vectors c_m are concatenated with $(L-1)N$ zeros and transformed in the time-domain through a bank of M separate and parallel IDFT units. This operation provides a set of oversampled vectors $\{v_m^{(L)}; m = 1, 2, \ldots, M\}$ which are referred to as *partial transmit sequences* (PTSs). The latter are next combined using M complex rotating factors $\widehat{\boldsymbol{b}} = [\widehat{b}_1, \widehat{b}_2, \ldots, \widehat{b}_M]^T$, with $\widehat{b}_m = e^{j\widehat{\varphi}_m}$. After combining, the time-domain samples

$$s^{(L)}(\widehat{\boldsymbol{b}}) = \sum_{m=1}^{M} \widehat{b}_m v_m^{(L)} \qquad (7.60)$$

are fed to the D/A converter and transmitted over the channel. The objective of the PAPR optimization block is to find the set of phase shifts $\{\widehat{\varphi}_m\}$ that minimize the PAPR of the transmitted sequence $s^{(L)}(\widehat{\boldsymbol{b}})$.

To reduce the complexity associated with the optimization problem, the phase shifts are normally constrained to vary in a finite set of W elements. In this case the optimum weighting vector $\widehat{\boldsymbol{b}}$ is computed as

$$\widehat{\boldsymbol{b}} = \arg\min_{\boldsymbol{b}} \left\{ \text{PAPR}[s^{(L)}(\boldsymbol{b})] \right\}, \qquad (7.61)$$

where $b_m = e^{j\varphi_m}$ and $\varphi_m \in \{2\pi\ell/W;\ \ell = 0, 1, \ldots, W - 1\}$. It is worth noting that in practice the number of phase shifts that must be optimized is $M-1$ since we can arbitrarily set $\widehat{b}_1 = 1$ without incurring any performance penalty. Hence, a total of W^{M-1} permissible vectors $\boldsymbol{b}$ is to be tested in Eq. (7.61), with a complexity that increases exponentially with the number M of PTSs.

Various techniques have been suggested to reduce the complexity of the optimization problem stated in Eq. (7.61) [22, 51, 156]. In the iterative flipping algorithm [22], the weighting factors $\{\widehat{b}_m\}$ are determined one by one in $M - 1$ steps following the natural order $m = 2, 3, \ldots, M$. For illustration purposes, we assume $W = 2$ so that $b_m \in \{\pm 1\}$ and recall that b_1 is arbitrarily set to unity without any loss of performance. Then, after initializing $b_m = 1$ for $m = 1, 2, \ldots, M$, during the first step the algorithm flips the sign of b_2 and evaluates the PAPRs of the two signals generated with weighting factors $[1, 1, 1, \ldots, 1]^T$ and $[1, -1, 1, \ldots, 1]^T$. The value $\widehat{b}_2$ that yields the lowest PAPR is then retained and used in the next step, where signals obtained with $[1, \widehat{b}_2, 1, \ldots, 1]^T$ and $[1, \widehat{b}_2, -1, \ldots, 1]^T$ are tested to find $\widehat{b}_3$. The iterative process continues in this fashion until all factors $\{\widehat{b}_2, \widehat{b}_3, \ldots, \widehat{b}_M\}$ have been determined.

The flipping algorithm can easily be generalized to any value of W. In this case the rotating factors are taken from the set $P = \{e^{j2\pi\ell/W};\ \ell = 0, 1, \ldots, W - 1\}$ and W different alternatives are explored at each step. The search complexity associated with the flipping procedure is thus proportional to $W(M - 1)$, which translates into considerable computational saving with respect to the ordinary PTS technique. The price for this advantage is a certain degradation of the system performance in terms of PAPR reduction. Better results are obtained by allowing $r > 1$ weighting factors to be simultaneously flipped at each new iteration [51]. In general, a suitable design of r allows one to achieve a reasonable trade-off between performance and complexity.

As is intuitively clear, the PAPR reduction capability of the PTS technique improves with M and W due to the increased number of candidate sequences $\boldsymbol{s}^{(L)}(\boldsymbol{b})$. In order to keep the system complexity to a tolerable level, in practice the number of PTSs cannot exceed a few units, while W is normally set to 4 since in this case $\widehat{\varphi}_m$ is a multiple of $\pi/2$ and no multiplication is required when rotating and combining the PTSs in Eq. (7.60). Another factor that may considerably affect the system performance is the particular strategy adopted for generating the M subblocks. Although numerical simulations indicate that the pseudo-random criterion represents the best choice in terms of PAPR minimization, the subband design is normally preferred for its simplicity. It is worth noting that when M is a power of two and an interleaved design is adopted for subblock partitioning, a computationally efficient implementation of the IDFT algorithms is possible by taking into account that the majority of elements in each subblock is zero.

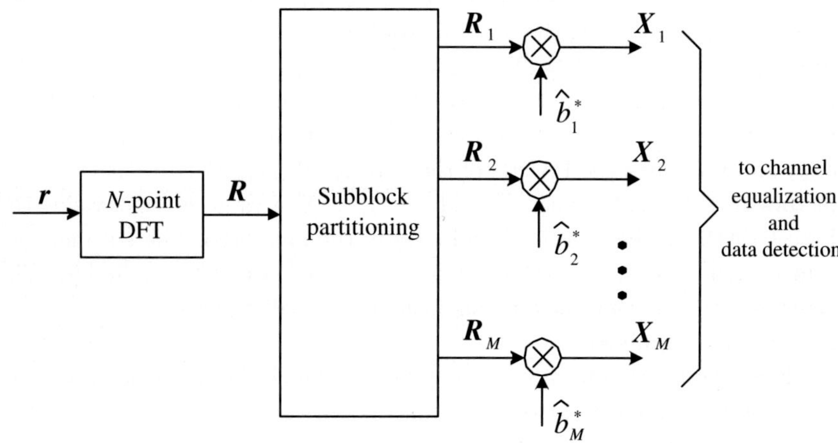

Fig. 7.15   Coherent receiver for an OFDM system employing the PTS technique.

Figure 7.15 illustrates the block diagram of an OFDM receiver for a system employing the PTS technique. The received samples r are transformed in the frequency-domain through an N-point DFT operation and the resulting vector R is partitioned into M subblocks $\{R_1, R_2, \ldots, R_M\}$ using the same partitioning policy employed at the transmitter. The entries of R_m are given by

$$R_m(n) = \begin{cases} H(n)c(n)\widehat{b}_m + W(n), & \text{if } n \in J_m, \\ 0, & \text{otherwise.} \end{cases} \qquad (7.62)$$

where $H(n)$ is the channel frequency response over the nth subcarrier while $W(n)$ represents thermal noise.

The subblocks are then rotated back so as to generate M vectors $\{X_1, X_2, \ldots, X_M\}$, with $X_m = \widehat{b}_m^* R_m$. Recalling that $\widehat{b}_m = e^{j\widehat{\varphi}_m}$, from

Eq. (7.62) it follows that $\boldsymbol{X}_m$ has entries

$$X_m(n) = \begin{cases} H(n)c(n) + W'(n), & \text{if } n \in J_m, \\ 0, & \text{otherwise.} \end{cases} \qquad (7.63)$$

with $W'(n) = \widehat{b}_m^* W(n)$. The non-zero elements of $\boldsymbol{X}_m$ are then passed to the channel equalization and data detection unit, which provides final decisions on the information symbols conveyed by the mth subblock. From the above discussion it turns out that, similarly to SLM, the PTS is a distortionless technique in which the receiver must be informed about the specific set of rotation factors that have been employed at the transmitter to generate the time-domain samples. An unambiguous representation of $\widehat{\boldsymbol{b}}$ has thus to be sent to the receiver as side information. Since $\widehat{\boldsymbol{b}}$ is taken from a set of W^{M-1} admissible vectors, a total of $(M-1)\log_2 W$ bits is required to represent this side information.

An interesting alternative to the coherent receiver architecture of Fig. 7.15 is represented by a differential decoding approach which, however, can only be used on condition that a subband strategy is adopted for generating the M subblocks. Since the entries of any given subblock are rotated by the same angle, the phase relations among subcarriers remain unchanged in each subblock. Hence, if the transmitted information is mapped as phase differences between adjacent subcarriers, differential decoding can be applied on a subblock-by-subblock basis without requiring knowledge of the rotation vector $\widehat{\boldsymbol{b}}$. Clearly, in this case one additional carrier must be inserted in each subblock to provide the necessary phase reference. This calls for a total of M redundant subcarriers, with a corresponding overhead that is independent of W.

Figure 7.16 illustrates the performance of the SLM and PTS techniques in terms of CCDF of the corresponding PAPR levels. The OFDM system has $N = 256$ QPSK modulated subcarriers and the candidate transmit signals in the SLM algorithm are obtained as depicted in Fig. 7.13 using $Q = 8$ different phase vectors. To make comparisons with the same number of IDFT units, $M = 8$ subblocks are generated in the PTS scheme. A subband design criterion has been adopted with clusters of 32 adjacent subcarriers assigned to any subblock. For simplicity, only binary phase shifts are employed in both SLM and PTS, meaning that $b_m \in \{\pm 1\}$. The oversampling factor is $L = 4$, which results into 1024-points IDFT operations. In addition to the ordinary PTS scheme, the possibility of reducing the search complexity by means of the flipping algorithm is also investigated. We see that the ordinary PTS performs remarkably better than the SLM

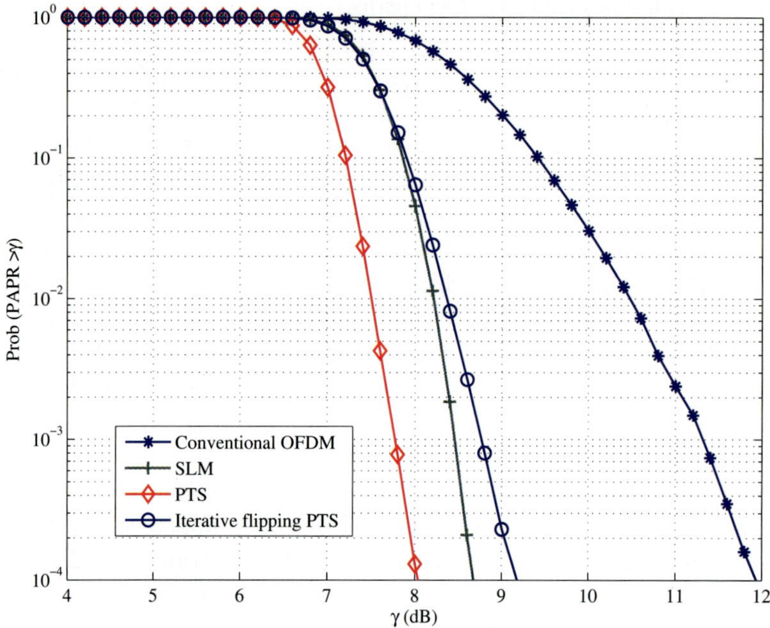

Fig. 7.16   Comparison between SLM and PTS in terms of PAPR reduction.

technique. The reason is that the former minimizes the PAPR by exploring among $W^{M-1} = 2^7$ candidate signals while in the latter the number of alternative waveforms is limited to $Q = 2^3$. However, both schemes ensure considerable PAPR reduction as compared to a conventional system where nothing is done to control amplitude fluctuations. Although the use of the flipping algorithm can significantly reduce the system complexity, a penalty of approximately 1 dB is incurred with respect to the ordinary PTS.

7.7   Coding

It is a well recognized fact that the frequency diversity offered by the multipath channel cannot fully be exploited in OFDM systems without employing some form of channel coding. A natural question is whether the redundancy introduced by channel coding can be exploited not only for error correction purposes, but also as a means for minimizing the PAPR of the transmitted waveform. The possibility of using block coding for

Table 7.1 PAPR γ_d of BPSK-modulated codewords with $N = 4$.

Code words				BPSK symbols				PAPR (dB)
$b(0)$	$b(1)$	$b(2)$	$b(3)$	$c(0)$	$c(1)$	$c(2)$	$c(3)$	γ_d
0	0	0	0	1	1	1	1	**6.02**
1	0	0	0	-1	1	1	1	2.32
0	1	0	0	1	-1	1	1	2.32
1	1	0	0	-1	-1	1	1	**3.72**
0	0	1	0	1	1	-1	1	2.32
1	0	1	0	-1	1	-1	1	**6.02**
0	1	1	0	1	-1	-1	1	**3.72**
1	1	1	0	-1	-1	-1	1	2.32
0	0	0	1	1	1	1	-1	2.32
1	0	0	1	-1	1	1	-1	**3.72**
0	1	0	1	1	-1	1	-1	**6.02**
1	1	0	1	-1	-1	1	-1	2.32
0	0	1	1	1	1	-1	-1	**3.72**
1	0	1	1	-1	1	-1	-1	2.32
0	1	1	1	1	-1	-1	-1	2.32
1	1	1	1	-1	-1	-1	-1	**6.02**

PAPR reduction was originally proposed in the seminal work [69], where only codewords exhibiting the lowest PAPR are selected for transmission while discarding all the others. Table 7.1 illustrates the highly-cited example given in [69], where the discrete-time PAPR is listed for all possible data blocks in a BPSK-OFDM system with $N = 4$ subcarriers and oversampling factor $L = 4$.

We see that four data blocks are characterized by a maximum PAPR of 6.02 dB and another set of four blocks results into a PAPR of 3.72 dB. Clearly, using a suitable coding scheme that avoids transmitting these sequences helps to reduce the PAPR of the transmitted signal. In the particular example shown in Table 7.1, this goal is achieved with an odd parity check code of rate 3/4 where the first three elements $b(0), b(1), b(2)$ in each codeword represent the information bits while the fourth element is computed as $b(3) = b(0) \oplus b(1) \oplus b(2) \oplus 1$, with $\oplus$ denoting the arithmetic addition in the binary Galois field. In this way the PAPR becomes 2.32 dB for all codewords, thereby leading to a reduction of 3.70 dB with respect to the uncoded system. It is shown in [69] that higher gains of 4.58 and 6.02 dB are possible in case of $N = 8$ subcarriers using coding schemes with rates 7/8 and 3/4, respectively. Clearly, these benefits are achieved at the price of some penalty in terms of spectral efficiency due to the inherent redundancy introduced in the transmitted signal. Note that the latter is only exploited for PAPR reduction purposes rather than to protect information against channel impairments. In addition, the method in [69] becomes impractical for large values of N since the best codes can only be found through an exhaustive search and prohibitively large look-up tables are required for the encoding and decoding operations.

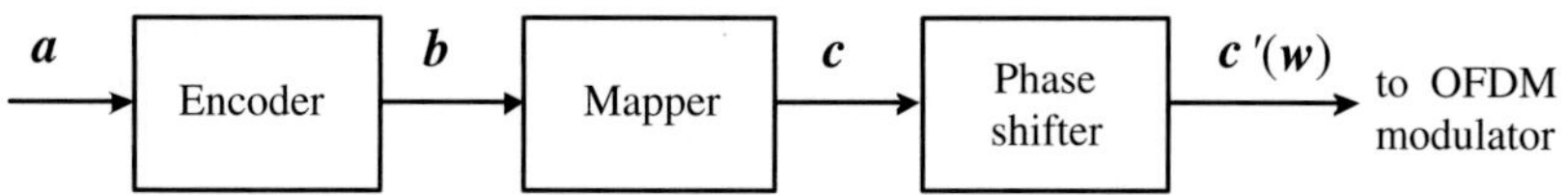

Fig. 7.17   Coding and phase rotation for simultaneous error control and PAPR reduction.

A more sophisticated approach proposed by Jones and Wilkinson in [68] relies on the design of combined coding schemes for *simultaneous* error control and PAPR reduction. This solution employs conventional linear block codes to achieve the desired level of error protection and the code redundancy is subsequently exploited to minimize the PAPR. The basic idea behind this method is sketched in Fig. 7.17. Let ϑ be the number of points in the employed constellation and assume that a $(N\vartheta, k)$ binary block code has been chosen for its correction property. As is seen, a block $\boldsymbol{a}$ of k information bits is first transformed into a vector $\boldsymbol{b}$ of $N\vartheta$ coded bits. The latter is next divided into N adjacent segments of length ϑ, where each segment is independently mapped onto a modulation symbol $c(n)$. This produces a codeword $\boldsymbol{c} = \{c(0), c(1), \ldots, c(N-1)\}$ of length N for each block of k information bits. We denote $\mathcal{C} = \{\boldsymbol{c}_m; m = 1, 2, 3, \ldots, 2^k\}$ the set of all possible codewords. Then, in an attempt of reducing the PAPR, the codewords are element-wise multiplied by a same rotating vector $\boldsymbol{w} = \left[e^{j\psi(0)}, e^{j\psi(1)}, \ldots, e^{j\psi(N-1)}\right]^T$, where the phase shifts $\{\psi(n)\}$ vary in the compact set $[0, 2\pi] \times [0, 2\pi] \times \cdots \times [0, 2\pi]$. The rotated version of $\boldsymbol{c}_m$ is denoted $\boldsymbol{c}'_m(\boldsymbol{w})$ and reads

$$\boldsymbol{c}'_m(\boldsymbol{w}) = \{c_m(0)e^{j\psi(0)}, c_m(1)e^{j\psi(1)}, \ldots, c_m(N-1)e^{j\psi(N-1)}\}. \qquad (7.64)$$

Since distances among codewords remain unchanged after rotation, the new code $\mathcal{C}'(\boldsymbol{w}) = \{\boldsymbol{c}'_m(\boldsymbol{w}); m = 1, 2, 3, \ldots, 2^k\}$ has the same error correction capability as the original code $\mathcal{C}$. However, it may exhibit a lower PAPR if the phase shifts are suitably chosen. Hence, for a given code $\mathcal{C}$, the problem is to find an optimal vector $\widehat{\boldsymbol{w}} = [e^{j\hat{\psi}(0)}, e^{j\hat{\psi}(1)}, \ldots, e^{j\hat{\psi}(N-1)}]^T$ such that

$$\widehat{\boldsymbol{w}} = \arg\min_{\boldsymbol{w}} \left\{ \text{PAPR}\left[\mathcal{C}'(\boldsymbol{w})\right] \right\}, \tag{7.65}$$

where $\text{PAPR}[\mathcal{C}'(\boldsymbol{w})]$ is defined as

$$\text{PAPR}\left[\mathcal{C}'(\boldsymbol{w})\right] = \max_{\boldsymbol{c}'_m(\boldsymbol{w}) \in \mathcal{C}'(\boldsymbol{w})} \left\{ \text{PAPR}\left[\boldsymbol{c}'_m(\boldsymbol{w})\right] \right\}, \tag{7.66}$$

with $\text{PAPR}[\boldsymbol{c}'_m(\boldsymbol{w})]$ denoting the PAPR of the waveform associated to the mth rotated codeword $\boldsymbol{c}'_m(\boldsymbol{w})$.

It is worth noting that in this way PAPR reduction comes for free since, as mentioned previously, both $\mathcal{C}$ and $\mathcal{C}'(\boldsymbol{w})$ are perfectly equivalent in terms of error rate performance and decoding complexity. At the receive side, the phase shifts introduced by $\widehat{\boldsymbol{w}}$ can easily be compensated for by appropriate counter-rotation of the DFT output. For this purpose, $\widehat{\boldsymbol{w}}$ must be known to the receiver.

The main drawback of the described approach is the heavy computational load that is required to solve the optimization problem Eq. (7.65). An algorithm for finding the optimum rotation vector is discussed in [68] under the assumption that the phase shifts belong to a finite set $\Psi = \{2\pi\ell/W; \ \ell = 0, 1, \ldots, W - 1\}$. Unfortunately, this method is only applicable to relatively short codes because of the huge complexity involved in computing the PAPR of all phase-shifted codewords. A computationally efficient solution to this problem is outlined in [150], where a simplified method is proposed to identify codewords characterized by the highest PAPR and a gradient-based iterative minimization technique is next used to search for the optimum rotation vector. A possible shortcoming is that in general the objective function in Eq. (7.65) presents various local minima which may attract the gradient algorithm toward spurious locks.

A third approach for the design of low-PAPR coding schemes was motivated by the observation that the PAPR of an OFDM signal is at most 3 dB if the modulation sequence is constrained to be a member of a Golay complementary pair [48, 119]. For a long time these sequences were not recognized to possess sufficient structure to form a practical coding scheme until a theoretical connection has been established between them and the first- and second-order Reed-Muller codes [32]. This connection offers the opportunity to combine the error correcting capability of classical Reed-Muller codes with the attractive PAPR control property of Golay complementary sequences. Further improvements to this approach are found in [33], where a range of flexible coding schemes using binary, quaternary and higher order modulations has been designed to achieve desired tradeoffs in terms of PAPR control, spectral efficiency and error-correcting capability. Computationally efficient decoding algorithms have also been developed based on the fast Hadamard transform (FHT). A unified theory linking Golay complementary sets of polyphase sequences and Reed-Muller codes has been presented by Paterson in [117] and exploited to design a broad range of coding options employing high-order modulations. Unfortunately, the usefulness of all these techniques is somewhat limited by the fact that they can only be applied to multicarrier systems with a small number of subcarriers in order to keep the computational complexity to a tolerable level. One possible advantage is that no side information is required at the receiver to recover the transmitted data symbols.

7.8   Tone reservation and injection techniques

An efficient family of PAPR reduction methods is based on the idea of adding a *data-dependent* vector $\boldsymbol{e} = [e(0), e(1), \ldots, e(N-1)]^T$ in the *frequency domain* to the original data block $\boldsymbol{c} = [c(0), c(1), \ldots, c(N-1)]^T$ so as to reduce the peaks of the resulting OFDM signal. The most representative examples in this family are the tone reservation (TR) and tone injection (TI) techniques [153] which are discussed below. Both schemes have the remarkable advantage of being distortionless, as the added vector can easily be canceled out at the receiver without incurring any performance loss.

7.8.1   *Tone reservation (TR)*

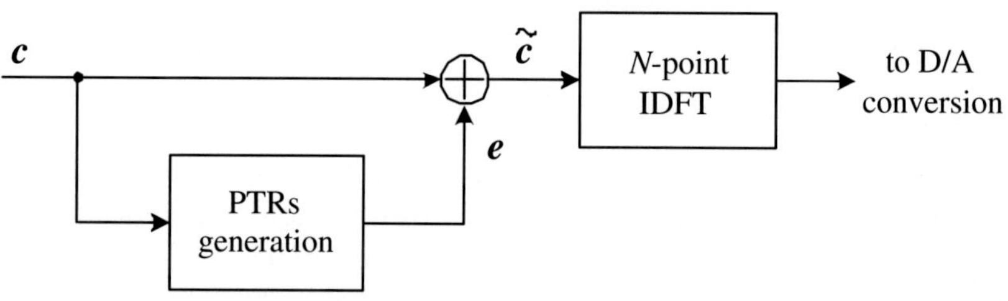

Fig. 7.18   Block diagram of the TR technique.

In the TR approach, the transmitter does not send information over a small set of Q subcarriers which are reserved for PAPR control. These subcarriers are referred to as *peak reduction tones* (PRTs) and are normally distributed in a pseudo-random fashion across the signal bandwidth. For illustration purposes, we denote $J^c = \{i_1, i_2, \ldots, i_Q\}$ the set collecting the indices of the PRTs while data-bearing subchannels have indices in the set $J = \{0, 1, \ldots, N-1\} - J^c$. As shown in Fig. 7.18, at the transmitter vectors $\boldsymbol{e}$ and $\boldsymbol{c}$ are summed up to form a block $\widetilde{\boldsymbol{c}} = [\widetilde{c}(0), \widetilde{c}(1), \ldots, \widetilde{c}(N-1)]^T$ of frequency-domain samples with entries

$$\widetilde{c}(n) = \begin{cases} c(n), & n \in J, \\ e(n), & n \in J^c. \end{cases} \tag{7.67}$$

The sequence $\{\widetilde{c}(n)\}$ is then transformed in the time-domain through an N-point IDFT unit and passed to the D/A converter, which provides the continuous-time signal

$$s(t) = \frac{1}{\sqrt{N}} \sum_{n=0}^{N-1} \widetilde{c}(n) \, e^{j2\pi n f_{cs} t}, \quad 0 \leq t < T. \tag{7.68}$$

Since $\boldsymbol{e}$ and $\boldsymbol{c}$ are constrained to lie into disjoint frequency subspaces, at the receiver the information symbols are simply recovered by selecting the outputs of the DFT with indices in the set J. Clearly, this requires that the receiver be informed as to which subcarriers are reserved to the PRTs.

Collecting the non-zero entries of $\boldsymbol{e}$ into a Q-dimensional vector $\widetilde{\boldsymbol{e}} = [e(i_1), e(i_2), \ldots, e(i_Q)]^T$, the goal of the TR scheme is to find the optimal $\widetilde{\boldsymbol{e}}$ that minimizes the PAPR of $s(t)$. As we know, a practical approach to accomplish this task is to replace $s(t)$ by its samples $s_k^{(L)}$ ($k = 0, 1, \ldots, NL-1$) taken with oversampling factor $L \geq 4$. The TR optimization problem can thus be cast as a constrained quadratic program

$$\text{minimize } \{\gamma\} \tag{7.69}$$

with respect to $\widetilde{\boldsymbol{e}} \in \mathcal{E}$ and subject to

$$\left| s_k^{(L)} \right|^2 \leq \gamma, \qquad \text{for all } k = 0, 1, \ldots, NL-1. \tag{7.70}$$

where $\mathcal{E}$ is the multidimensional space of admissible vectors $\widetilde{\boldsymbol{e}}$ and $s_k^{(L)}$ is given by

$$s_k^{(L)} = \frac{1}{\sqrt{N}} \sum_{n \in J} c(n) e^{j2\pi nk/LN} + \frac{1}{\sqrt{N}} \sum_{n \in J^c} e(n) e^{j2\pi nk/LN}, \quad 0 \leq k \leq NL-1. \tag{7.71}$$

Finding the exact solution to the above problem is in general a computationally expensive task. However, since we are minimizing a linear function under quadratic constraints, the problem is also convex. This property may be exploited to obtain a good, yet suboptimal, solution. For instance, an efficient method to iteratively approach the optimum $\widetilde{e}$ has been suggested in [38] using the sub-gradient algorithm.

Increasing the number Q of PRTs provides the optimization process with more degrees of freedom. In this way, the PAPR reduction capability of the TR technique is improved at the price of a throughput penalty due to the reduced number of data-bearing subcarriers. In general, a tradeoff between these conflicting requirements is sought through a careful design of parameter Q. Computer simulations indicate that gains of approximately 3 dB and 6 dB in terms of PAPR reduction can be achieved with a loss in data rate of less than 0.2% and 5%, respectively.

Another factor that remarkably affects the system performance is the set J^c of PRT positions. Finding the optimal J^c that minimizes the PAPR results into a combinatorial optimization problem which cannot be solved with affordable complexity. However, experimental results indicate that a good selection is obtained by generating a sufficiently large number of pseudo-random sets and choosing the best one.

In wireline DSL applications, the throughput penalty associated with the TR technique is partly alleviated by placing the PRTs over frequency subchannels that would go otherwise unused because of their relatively poor SNRs. Unfortunately, a similar approach cannot be pursued in wireless systems since in these applications no fast channel state feedback is available to adaptively decide which subcarriers should be used to send information and which others should be reserved to PRTs.

7.8.2 *Tone injection (TI)*

Tone injection can be viewed as an improvement of the TR technique in that it aims at reducing the PAPR without sacrificing the spectral efficiency. The basic idea is to send information over *all* subcarriers using an expanded non-bijective constellation set, where each data symbol is mapped into a subset of equivalent points. The signal peaks are then reduced through appropriate selection of the constellation point within the subset.

To explain the TI principle, we consider a conventional M-ary QAM constellation $\mathcal{A}$ where $2d$ is the minimum distance between neighboring points. In this case, the real and imaginary parts of each symbol take values in the set $\{\pm d, \pm 3d, \ldots, \pm(\sqrt{M} - 1)d\}$, with $\sqrt{M}$ denoting the number of levels per dimension. In the expanded constellation set $\widetilde{\mathcal{A}}$, any symbol c

of the original constellation is mapped into one of several equivalent points $\widetilde{c} = c + pD + jqD$, where p and q are suitable integers while D is a positive real number known at the receiver. In the ensuing discussion, we refer to $S(c) = \{c + pD + jqD;\ p, q \in I \subset \mathbb{Z}\}$ as the *subset* associated to c. The integers p and q provide extra degrees of freedom that are exploited to reduce the PAPR of the transmitted signal. Clearly, to ensure that the information symbol c can be recovered from $\widetilde{c}$ without any ambiguity, it is necessary that *different* points of $\mathcal{A}$ are mapped onto *disjoint* subsets of $\widetilde{\mathcal{A}}$. As explained in [154], this condition requires a careful design of parameter D. In particular, setting $D = 2\rho d\sqrt{M}$ with $\rho \geq 1$ yields disjoint subsets and results into approximately the same error-rate probability as a conventional OFDM system without TI.

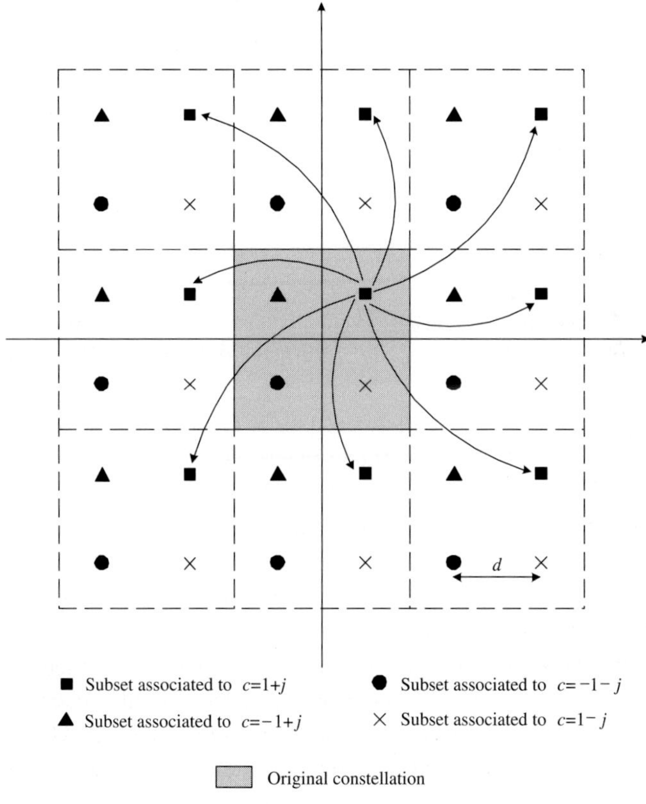

Fig. 7.19   The expanded constellation set in the TI technique.

Figure 7.19 depicts the expanded constellation $\widetilde{\mathcal{A}}$ in case of $\rho = 1$ and QPSK symbols ($M = 4$). For illustration purposes, the integers p and q are constrained to the set $I = \{-1, 0, 1\}$. As is seen, $\widetilde{\mathcal{A}}$ is obtained by replicating the original constellation $\mathcal{A}$ through known translation vectors. Four subsets are present, each containing nine symbols and corresponding to a different symbol of $\mathcal{A}$. Note that the original symbol c can perfectly be recovered from *anyone* of the points $\widetilde{c} \in S(c)$ by simply using a modulo operator that acts independently over the real and imaginary parts of its input according to the following rule

$$MOD(x) = x - 2\rho d\sqrt{M} \left\lfloor \frac{x + \rho d\sqrt{M}}{2\rho d\sqrt{M}} \right\rfloor, \qquad (7.72)$$

where the notation $\lfloor z \rfloor$ represents the smallest integer not exceeding z. In practice, $MOD(x)$ performs a periodic mapping of the complex plane into the square region $\{x_R + jx_I \, | \, x_R, x_I \in (-\rho d\sqrt{M}, \rho d\sqrt{M}]\}$ with side length $2\rho d\sqrt{M}$.

Denoting $\widetilde{c}(n) = c(n) + p_n D + jq_n D$ the selected point in the subset $S(c(n))$ associated with $c(n)$, the oversampled sequence of time-domain samples can be written as

$$s_k^{(L)} = \frac{1}{\sqrt{N}} \sum_{n=0}^{N-1} \widetilde{c}(n) \, e^{j2\pi nk/LN}, \qquad 0 \leq k \leq NL - 1. \qquad (7.73)$$

Note that the vector $\widetilde{c} = [\widetilde{c}(0), \widetilde{c}(1), \ldots, \widetilde{c}(N-1)]^T$ of frequency-domain samples is obtained as shown in Fig. 7.18 after defining $e(n) = p_n D + jq_n D$ for $n = 0, 1, \ldots, N - 1$.

Since it is desirable to reduce the peaks of the transmitted signal as much as possible, we look for the integers $p = \{p_0, p_1, \ldots, p_{N-1}\}$ and $q = \{q_0, q_1, \ldots, q_{N-1}\}$ that minimize the PAPR of the sequence $s_k^{(L)}$. This results into an integer programming problem whose complexity grows exponentially with the number N of available subcarriers. Fortunately, good approximations to the optimum solution can be obtained through efficient iterative methods that dispense one from exploring all candidate vectors p and q. A further reduction of complexity is possible if the expanded constellation set is only employed over a small fraction of the available subcarriers. Clearly, this approach reduces the number of candidate vectors to be explored at the price of some performance loss in terms of PAPR reduction.

Inspection of Fig. 7.19 reveals that the modified symbol $\widetilde{c}(n)$ has more energy than $c(n)$ whenever $p_n \neq 0$ or $q_n \neq 0$. This means that the TI technique reduces the PAPR at the expense of a certain increase of the total transmission power. However, no loss in data rate is incurred since, contrarily to the TR scheme, all subcarriers are employed to transmit data. At the receiver side, the original symbols $\{c(n)\}$ are recovered by passing the decoded sequence $\{\widetilde{c}(n)\}$ through the modulo operator Eq. (7.72), thereby avoiding the need for any exchange of side information between the transmitter and receiver.

7.9 PAPR reduction for OFDMA

In OFDMA systems, the available subcarriers are divided into mutually exclusive subchannels that are assigned to distinct users for simultaneous transmission. As illustrated in Fig. 3.14, three different strategies can be adopted to accomplish the subcarrier assignment task. In the subband CAS each subchannel is composed by a set of adjacent subcarriers while in the interleaved CAS the subcarriers of each user are uniformly spaced over the signal bandwidth to take advantage of the channel frequency diversity. The more flexible strategy is represented by the generalized CAS, where users are provided with the best quality subcarriers that are currently available.

From a physical layer perspective, the OFDMA downlink is essentially equivalent to an OFDM system. The only difference is that in OFDMA each block conveys simultaneous information for multiple subscribers while in OFDM the transmitted data are intended for a single specific user. This suggests that statistical PAPR characterization as well as PAPR reduction methods devised for single-user OFDM systems can be extended to the OFDMA downlink in a straightforward fashion. A rather different situation occurs in the OFDMA uplink. Here, each signal employs only a fraction of the available subcarriers and the underlying subcarrier assignment scheme is expected to play a major role in determining the PAPR of the transmitted waveform. A theoretical analysis presented in [166] indicates that, on average, the generalized CAS results into higher signal peaks than the subband or interleaved CAS. In any case, the PAPR problem in the OFDMA uplink is not as serious as in the downlink because of the relatively small number of modulated subcarriers. This explains why the topic of PAPR reduction in uplink transmissions has remained largely unexplored up to now.

In what follows we revisit some of the PAPR control methods described throughout this chapter and show how they can be extended to an OFDMA downlink.

7.9.1   *SLM for OFDMA*

The SLM technique applies to the OFDMA downlink without any substantial modification with respect to the single-user case. The only difference is that in OFDMA the candidate signals are exclusively generated by shifting the phase of the original data symbols while in OFDM they can also be obtained through pseudo-random permutations. The latter approach is not suited for OFDMA as it would result into a modification of the subcarrier allocation scheme, which is clearly unfeasible in systems employing rigid subband or interleaved CAS. Information about the employed phase sequence is broadcasted to all active terminals using some dedicated subcarriers. This information is exploited by each user to retrieve its own data.

7.9.2   *PTS for OFDMA*

The PTS technique employed in OFDM systems can easily be modified for OFDMA downlink transmissions. In such a case the subcarriers of each user are grouped into one or more subblocks, and PTSs are next obtained by transforming these subblocks in the time-domain. One subcarrier per subblock is reserved to provide information about the phase factor employed over that subblock. At the receiving terminal this subcarrier is extracted and used as a phase reference for data detection over the corresponding subblock.

7.9.3   *TR for OFDMA*

The TR approach is applied to the OFDMA downlink exactly in the same way as in single-user OFDM systems. As suggested in [166], however, a certain reduction of complexity is possible if a set of PRTs is exclusively assigned to each user and optimized for the data sequence of that user only. This results into a suboptimal optimization process in which data of different users are processed independently at the transmit side for PAPR mitigation. To further reduce the computational load, the amplitude of the PRTs may be optimized over a finite set of values and stored in a look-up table for every possible information sequence [182]. In this way, there is no need to recompute the optimum PRT values at each new transmitted block since the latter are simply obtained from the look-up table with the information sequence serving as a memory address.

7.10   Design of AGC unit

The presence of large amplitude fluctuations in the OFDM signal requires a careful design of the automatic gain control (AGC) unit and A/D converter at the receiver side. Figure 7.20 illustrates the front-end of a conventional two-branch receiver for digital transmissions.

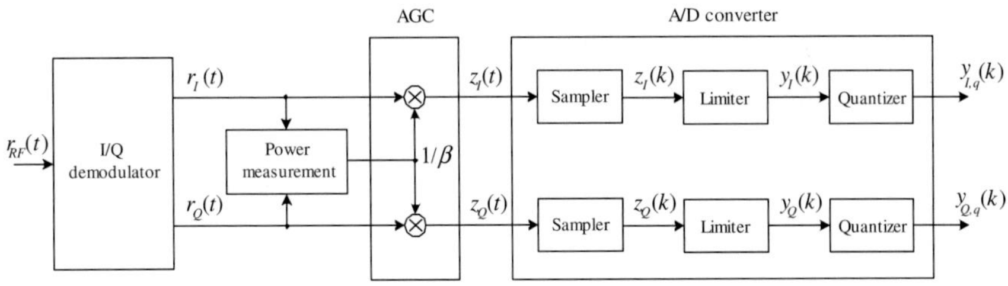

Fig. 7.20   Front-end of a typical two-branch receiver.

After I/Q demodulation, the baseband signals $r_I(t)$ and $r_Q(t)$ are passed to the AGC unit, where they are scaled by a factor $1/\beta$. The resulting signals $z_I(t)$ and $z_Q(t)$ are next fed to the A/D converter, which consists of a sampling device plus a quantizer operating over a finite dynamic range $[-A, A]$. As is intuitively clear, the scaling factor β must properly be designed so as to minimize the distortions introduced by the quantization process. For this purpose, it is convenient to model the overall quantization unit as the cascade of a limiter with cutting level A followed by a quantizer with infinite dynamic range. This approach offers the opportunity of separately assessing the impact of clipping distortions and quantization errors on the system performance.

Without loss of generality, in the ensuing discussion we let $A = 1$ and denote N_b the number of bits reserved to the A/D conversion. The design of N_b depends on many parameters, including the computational requirement as well as the accuracy needed for a given constellation size. In practice, $N_b = 10$ is commonly adopted for a 64-QAM constellation, while smaller values of N_b are used with lower order modulations. The AGC gain is adaptively adjusted on the basis of appropriate power measurements in an attempt of achieving a balanced trade-off between two conflicting requirements. On one hand, small values of β enlarge the dynamic range of the signal at the input of the A/D converter, thereby reducing the effects of quantization errors. On the other hand, a too large signal dynamic is undesirable as it increases the occurrence of clipping events. In what follows, we look for the optimum AGC gain that maximizes the SNR at the quantizer output. In doing so we limit our attention to the I branch in Fig. 7.20 and neglect the index I for notational simplicity.

If the number of modulated subcarriers is adequately large, we know that $r(t)$ can be approximated as a zero-mean Gaussian random process with some power σ_r^2. In this case, samples $\{z(k)\}$ at the input of the limiter are Gaussian distributed with probability density function

$$p_Z(z) = \frac{1}{\sigma_z\sqrt{2\pi}}e^{-z^2/(2\sigma_z^2)}, \tag{7.74}$$

where $\sigma_z = \sigma_r/\beta$ is the rms value of $z(k)$. The output of the limiter is mathematically described as

$$y(k) = \begin{cases} 1, & \text{if } z(k) \geq 1, \\ z(k), & \text{if } -1 < z(k) < 1, \\ -1, & \text{if } z(k) \leq -1. \end{cases} \tag{7.75}$$

In practice, we can view $y(k)$ as the sum of the useful signal $z(k)$ plus a clipping noise term $w_c(k)$, *i.e.*,

$$y(k) = z(k) + w_c(k), \tag{7.76}$$

where $w_c(k)$ is obtained after substituting Eq. (7.75) into Eq. (7.76), and reads

$$w_c(k) = \begin{cases} 1 - z(k), & \text{if } z(k) \geq 1, \\ 0, & \text{if } -1 < z(k) < 1, \\ -1 - z(k), & \text{if } z(k) \leq -1. \end{cases} \tag{7.77}$$

To proceed further, we define the clipping noise power as $P_c = \text{E}\{|w_c(k)|^2\}$. Then, from Eq. (7.77) it follows that

$$P_c = \int_{-\infty}^{-1} (1+z)^2 p_Z(z)dz + \int_{1}^{\infty} (1-z)^2 p_Z(z)dz, \tag{7.78}$$

or, equivalently,

$$P_c = 2\int_{1}^{\infty} (1-z)^2 p_Z(z)dz. \tag{7.79}$$

Substituting Eq. (7.74) into Eq. (7.79) and performing standard computations, yields

$$P_c = \left(\frac{1}{\mu^2} + 1 \right) \text{erfc} \left(\frac{\mu}{\sqrt{2}} \right) - \frac{1}{\mu} \sqrt{\frac{2}{\pi}} e^{-\mu^2/2}, \tag{7.80}$$

where $\text{erfc}(x)$ is the complementary error function given in Eq. (7.38) while $\mu = 1/\sigma_z$ is the clipping crest factor, which is defined as the ratio between the maximum allowable amplitude $A = 1$ and the rms of $z(k)$. Recalling that $\sigma_z = \sigma_r/\beta$, we also have $\mu = \beta/\sigma_r$.

Next, we consider the quantization error $e_q(k) = y(k) - y_q(k)$ introduced by the N_b-bit quantizer. Letting $\Delta = 2/2^{N_b}$ be the *quantization step-size*, we can approximate $e_q(k)$ as a random variable with uniform distribution in the interval $[-\Delta/2, \Delta/2)$ [123]. The power of the quantization noise is thus given by

$$P_q = \frac{\Delta^2}{12} = \frac{1}{3 \cdot 2^{2N_b}}. \tag{7.81}$$

Neglecting for simplicity the effect of thermal noise, the SNR at the output of the A/D converter is found to be

$$\gamma_{A/D} = \frac{P_Z}{P_c + P_q}, \tag{7.82}$$

where $P_Z = \sigma_z^2$ is the power of $z(k)$.

Bearing in mind that $\sigma_z^2 = 1/\mu^2$, after substituting Eqs. (7.80) and (7.81) into Eq. (7.82) we obtain

$$\gamma_{A/D} = \left[(\mu^2 + 1) \, \text{erfc} \left(\frac{\mu}{\sqrt{2}} \right) - \mu \sqrt{\frac{2}{\pi}} e^{-\mu^2/2} + \frac{\mu^2}{3 \cdot 2^{2N_b}} \right]^{-1}. \tag{7.83}$$

Figure 7.21 illustrates $\gamma_{A/D}$ as a function of μ for $N_b = 8$ and 10. As expected, at low values of μ clipping noise dominates the system performance and $\gamma_{A/D}$ increases with μ. However, when the crest factor goes beyond its optimal value μ_{opt}, the SNR starts to decrease since in this case the quantization error becomes the most critical impairment to the system performance. Inspection of Fig. 7.21 reveals that μ_{opt} is close to 4 with either $N_b = 8$ or 10. These results indicate that optimum performance is achieved when the I/Q components of the received waveform are scaled such that their rms is approximately four times smaller than the clipping level A. This is a consequence of the large amplitude fluctuations characterizing the OFDM signal.

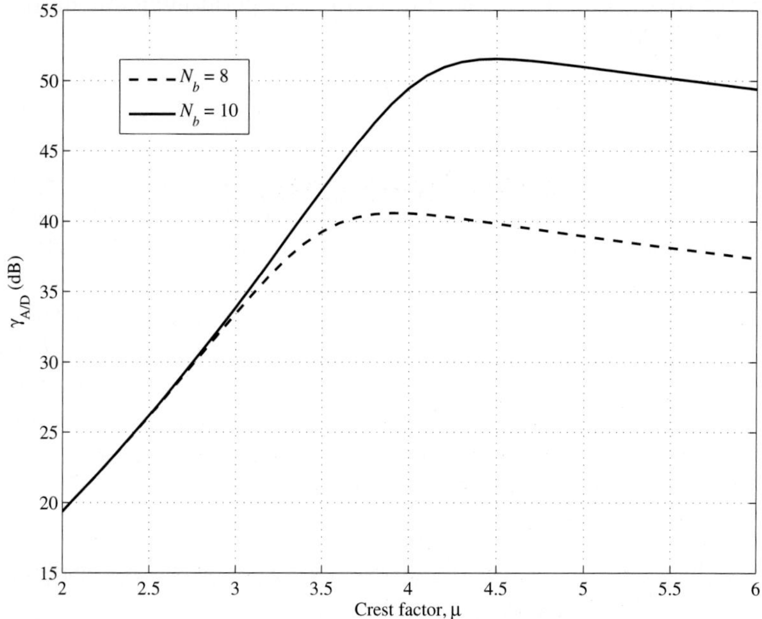

Fig. 7.21　Output SNR vs. μ for an A/D converter with N_b bits.

Recalling that $\mu = \beta/\sigma_r$, we can use the quantities μ_{opt} and σ_r to determine the optimum AGC coefficient in the form

$$\beta_{opt} = \mu_{opt}\sigma_r. \tag{7.84}$$

While μ_{opt} can be inferred from the theoretical curves of Fig. 7.21, an estimate of σ_r is normally obtained by measuring the average power of the received signal as indicated in Fig. 7.20.

Bibliography

[1] ANSI (1995). Asymmetric digital subscriber line (ADSL) metallic interface. draft american national standard for telecommunications, .

[2] Armstrong, J. (2002). Peak-to-average power reduction for OFDM by repeated clipping and frequency-domain filtering, *Elect. Letters* **38**, pp. 246–247.

[3] Baccarelli, E. and Biagi, M. (2004). Optimal integer bit-loading for multicarrier ADSL systems subject to spectral-compatibility limits, *Signal Processing, Elsevier* **84**, pp. 729–741.

[4] Bagheri, R., Mirzaei, A., Heidari, M. E., Chehrazi, S., Lee, M., Mikhemar, M., Tang, W. and Abidi, A. A. (2006). Software-defined radio receiver: dream to reality, *IEEE Commun. Magazine* **44**, pp. 111–118.

[5] Bahl, L., Cocke, J., Jelinek, F. and Raviv, J. (1974). Optimal decoding of linear codes for minimizing symbol error rate, *IEEE Trans. Inform. Theory* **IT-20**, pp. 284–287.

[6] Barbarossa, S., Pompili, M. and Giannakis, G. (2002). Channel-independent synchronization of orthogonal frequency division multiple access systems, *IEEE Journal Select. Areas Commun.* **20**, 2, pp. 474–486.

[7] Bauml, R., Fischer, R. and Huber, J. (1996). Reducing the peak-to-average power ratio of multicarrier modulation by selective mapping, *Electronics Letters* **32**, pp. 2056–2057.

[8] Bingham, J. (1990). Multicarrier modulation for data transmission: an idea whose time has come, *IEEE Communications Magazine* **28**, pp. 5–14.

[9] Breiling, M., Muller-Weinfurtner, S. and Huber, J. (2001). SLM peak-power reduction without explicit side information, *IEEE Commun. Letters* **5**, pp. 239–241.

[10] Campello, J. (1999). Practical bit loading for DMT, *In Proc. Int. Conf. Commun. (ICC99), Vancouver, Canada* , pp. 801–805.

[11] Cao, Z., Tureli, U. and Yao, Y. D. (2004a). Deterministic multiuser carrier-frequency offset estimation for interleaved OFDMA uplink, *IEEE Trans. Commun.* **52**, 9, pp. 1585–1594.

[12] Cao, Z., Tureli, U., Yao, Y. D. and Honan, P. (2004b). Frequency synchronization for generalized OFDMA uplink, *in Proc. Globecom 2004, Dallas,*

Texas , pp. 1071–1075.

[13] Catreux, S., Erceg, V., Gesbert, D. and R.W. Heath, J. (2002). Adaptive modulation and MIMO coding for broadband wireless data networks, *IEEE Commun. Magazine* **40**, pp. 108–115.

[14] Chan, D. and Berger, T. (2004). Performance and cross-layer design of CSMA for wireless networks with multipacket reception, *In Proc. Asilomar 2004, Pacific Grove, California* **2**, pp. 1917–1921.

[15] Chang, R. W. (1966). Synthesis of band-limited orthogonal signals for multipath channel data transmission, *Bell Syst. Tech. Journal* **46**, pp. 1775–1796.

[16] Chen, H. and Haimovich, A. (2003). Iterative estimation and cancellation of clipping noise for OFDM signals, *IEEE Commun. Letters* **7**, pp. 305–307.

[17] Cheng, R. and Verdu, S. (1993). Gaussian multiaccess channels with ISI: capacity region and multiuser water-filling, *IEEE Trans. Info. Theory* **39**, pp. 773–785.

[18] Choi, J., Lee, C., Jung, H. and Lee, Y. (2000). Carrier frequency offset compensation for uplink of OFDM-FDMA systems, *IEEE Commun. Letters* **4**, 12, pp. 414–416.

[19] Chow, J., Tu, J. and Cioffi, J. (1991). A discrete multitone transceiver system for HDSL applications, *IEEE Journal Select. Areas Commun.* **9**, pp. 895–908,.

[20] Chow, P. S., Cioffi, J. M. and Bingham, J. A. C. (1995). A practical discrete multitone transceiver loading algorithm for data transmission over spectrally shaped channels, *IEEE Trans. Commun.* **43**, pp. 773–775.

[21] Cimini, L. (1985). Analysis and simulation of a digital mobile channel using orthogonal frequency division multiplexing, *IEEE Trans. Commun.* **COM-33**, pp. 665–675.

[22] Cimini, L. J. and Sollenberger, N. (2000). Peak-to-average power ratio reduction of an OFDM signal using partial transmit sequences, *Electronics Letters* **4**, pp. 86–88.

[23] Cioffi, J. (1997). Lecture notes for advanced digital communications, *Stanford* .

[24] Classen, F. and Meyr, H. (1994). Frequency synchronization algorithms for OFDM systems suitable for communication over frequency selective fading channels, *In Proc. IEEE Vehicular Technology Conference (VTC) Fall 1994, Los Angeles, California* **3**, pp. 1655–1659.

[25] Cooper, M. (Granted on September 1975). *US Patent number 3 906 16.*

[26] Costello, D. and Lin, S. (1983). *Error control coding* (Prentice-Hall, N.J.).

[27] Cover, T. and Thomas, J. (1991). *Elements of information theory* (John Wiley & Sons, Inc).

[28] Crochiere, R. and Rabiner, L. (1983). *Multirate Digital Signal Processing* (Prentice Hall, Englewood Cliffs, New Jersey).

[29] Daffara, F. and Adami, O. (1996). A novel carrier recovery technique for orthogonal multicarrier systems, *European Trans. on Telecommunications* **7**, pp. 323–334.

[30] Daffara, F. and Chouly, A. (1993). Maximum likelihood frequency detectors

for orthogonal multicarrier systems, *In Proc. ICC93, Geneva, Switzerland* , pp. 766–771.

[31] Daly, D., Heneghan, C. and Fagan, A. (2003). Power and bit-loading algorithms for multitone systems, *In Proc. 3rd Int. Symposium on Image and Signal Processing and Analysis* , pp. 639–644.

[32] Davis, J. and Jedwab, J. (1997). Peak-to-mean power control and error correction for OFDM transmission using Golay sequences and Reed–Muller codes, *Elect. Letters* **33**, 4, pp. 267–268.

[33] Davis, J. and Jedwab, J. (1999). Peak-to-mean power control in OFDM, Golay complementary sequences and Reed–Muller codes, *IEEE Trans. Info. Theory* **45**, 7, pp. 2397–2417.

[34] Dempster, A., Laird, N. and Rubin, D. (1977). Maximum likelihood from incomplete data via the EM algorithm, *J. Royal Stat. Soc.* **39**, pp. 1–38.

[35] Deng, J. and Lee, T. (2003). An iterative maximum SINR receiver for multicarrier CDMA systems over a multipath fading channel with frequency offset, *IEEE Trans. Wireless Commun.* **2**, pp. 560–569.

[36] Dong, M. and Tong, L. (2002). Optimal design and placement of pilot symbols for channel estimation, *IEEE Trans. Signal Proc.* **50**, pp. 3055–3069.

[37] Edfors, O., Sandell, M., van de Beek, J., Wilson, S. and Borjesson, P. (1998). OFDM channel estimation by singular value decomposition, *IEEE Trans. Commun.* **46**, pp. 931–939.

[38] Erdogan, A. (2004). A subgradient algorithm for low complexity DMT PAR minimization, *In Proc. ICASSP 2004, Montreal, Canada* **4**, pp. 1077–1080.

[39] ETSI (1995). *Radio Broadcasting Systems: Digital Audio Broadcasting to Mobile, Portable and Fixed Receivers* (European Telecommunication Standard, ETS 300 401).

[40] ETSI (1997). *Digital Video Broadcasting (DVB-T); Frame structure, Channel Coding, and Modulation for Digital Terrestrial Television* (European Telecommunication Standard, ETS 300 744).

[41] ETSI (1999). *Broadband Radio Access Network (BRAN): HIPERLAN type 2 functional specification Part I: Physical layer* (ETSI Std. ETS/BRAN 030 003-1).

[42] Fazel, K. and Kaiser, S. (2003). *Multi-carrier and spread spectrum systems* (John Wiley & Sons, Inc).

[43] Feder, M. and Weinstein, E. (1988). Parameter estimation of superimposed signals using the EM algorithm, *IEEE Trans. Acoustics, Speech and Signal Processing* **36**, 4.

[44] Ferro, E. and Potorti, F. (2005). Bluetooth and Wi-Fi wireless protocols: a survey and a comparison, *IEEE Communications Magazine* **12**, pp. 12–26.

[45] Fessler, J. and Hero, A. (1994). Space-alternating generalized expectation-maximization algorithm, *IEEE Trans. Signal Proc.* **42**, 10, pp. 2664–2677.

[46] Foschini, G. and Gans, M. (1998). On the limits of wireless communication in a fading environment when using multiple antennas, *Wireless Personal Communications* **6**, pp. 311–335.

[47] Gallo, A., Vitetta, G. and Chiavaccini, E. (2004). A BEM-based algorithm

for soft-in soft-output detection of co-channel signals, *IEEE Trans. Wireless Commun.* **3**, 5, pp. 1533–1542.

[48] Golay, M. (1961). Complementary series, *IEEE Trans. Info. Theory* **7**, pp. 82–87.

[49] Grunheid, R., Bolinth, E. and Rohling, H. (2001). A blockwise loading algorithm for the adaptive modulation technique in OFDM systems, *In Proc. Vehicular Technol. Conf. (VTC) Fall 2001, Atlantic City, New Jersey* **2**, pp. 948–951.

[50] Haindl, B., Sajatovic, M., Rihacek, C., Prinz, J. and Schnell, M. (2005). B-VHF: A multi-carrier based broadband VHF communications concept for air traffic management, *in Proc. IEEE Aerospace Conference 2005, Vienna, Austria* , pp. 1894–1904.

[51] Han, S. and Lee, J. (2004). PAPR reduction of OFDM signals using a reduced complexity PTS technique, *IEEE Sig. Proc. Letters* **11**, pp. 887–890.

[52] Han, S. and Lee, J. (2005). An overview of peak-to-average power ratio reduction techniques for multicarrier transmission, *IEEE Wireless Commun.* **12**, pp. 56–65.

[53] Hara, S. and Prasad, R. (1997). Overview of multicarrier CDMA, *IEEE Communications Magazine* **35**, 12, pp. 126–133.

[54] Hoeher, P., Kaiser, S. and Robertson, P. (1997). Two-dimensional pilot-symbol-aided channel estimation by Wiener filtering, *In Proc. IEEE ICASSP 1997, Munich, Germany* **3**, pp. 21–24.

[55] Huang, D. and Letaief, K. (2005). An interference-cancellation scheme for carrier frequency offsets correction in OFDMA systems, *IEEE Trans. Commun.* **53**, 7, pp. 1155–1165.

[56] Hughes-Hartogs, D. (1987-1989). Ensemble modem structure for imperfect transmission media, *U.S. Patents Nos. 4,679,227 (July 1987), 4,731,816 (march 1998) and 4,833,706 (May 1989)* .

[57] Hui, S. and Yeung;, K. (2003). Challenges in the migration to 4G mobile systems, *IEEE Commun. Magazine* **41**, pp. 54–59.

[58] IEEE802.11 (1997). *IEEE Standard for Wireless LAN Medium Access Control (MAC) and Physical Layer (PHY) Specifications.*

[59] IEEE802.11a (1999). *Part 11: Wireless LAN Medium Access Control (MAC) and Physical Layer (PHY) Specifications, Higher-Speed Physical Layer Extension in the 5 GHz Band.*

[60] IEEE802.11b (1999). *Part 11: Wireless LAN Medium Access Control (MAC) and Physical Layer (PHY) Specifications, Higher-Speed Physical Layer Extension in the 2.4 GHz Band.*

[61] IEEE802.11g (2003). *Part 11: Wireless LAN Medium Access Control (MAC) and Physical Layer (PHY) Specifications, Further Higher-Speed Physical Layer Extension in the 2.4 GHz Band.*

[62] IEEE802.15 (2002). *Information Technology – Telecommunications and Information Exchange between Systems – Local and Metropolitan Area Networks – Specific Requirements Part 15.1: Wireless Medium Access Control (MAC) and Physical Layer (PHY) Specifications for Wireless Personal*

Area Networks (WPANs).

[63] IEEE802.16-2001 (2002). *IEEE Standard for Local and Metropolitan Area Networks – Part 16: Air Interface for Fixed Broadband Wireless Access Systems*.

[64] Jakes, W. (1974). *Microwave Mobile Communications* (Wiley, New York, NY).

[65] Jang, J. and Lee, K. (2003). Transmit power adaptation for multiuser OFDM systems, *IEEE Journal Select. Areas Commun.* **17**, pp. 171–178.

[66] Jayalath, A. and Tellambura, C. (2000). Reducing the peak-to-average power ratio of orthogonal frequency division multiplexing signals through bit or symbol interleaving, *Elect. Letters* **36**, 13, pp. 1161–1163.

[67] Joint Technical Committee (JTC) on Wireless Access (1993). *Final Report on RF Channel Characterization JTC(AIR)/93.09.23-238R2*.

[68] Jones, A. and Wilkinson, T. (1996). Combined coding error control and increased robustness to system nonlinearities in OFDM, *In Proc. IEEE 46th VTC 1996, Atlanta, Georgia* , pp. 904–908.

[69] Jones, A., Wilkinson, T. and Barton, S. (1994). Block coding scheme for reduction of peak to mean envelope power ratio of multicarrier transmission schemes, *Electronics Letters* **30**, pp. 2098–2099.

[70] Jr., R. W. H. and Giannakis, G. (2001). Exploiting input cyclostationarity for blind channel identification in OFDM systems, *IEEE Trans. Signal Proc.* **47**, pp. 848–856.

[71] Kalet, I. (1989). The multitone channel, *IEEE Trans. Commun.* **37**, pp. 119–124.

[72] Kay, S. (1993). *Fundamentals of Statistical Signal Processing : Estimation Theory* (Prentice Hall).

[73] Keller, T. and Hanzo, L. (1998). Adaptive orthogonal frequency division multiplexing schemes, *In Proc. ACST Summit, Rhodos, Greece* , pp. 794–799.

[74] Keller, T. and Hanzo, L. (1999). Blind-detection assisted sub-band adaptive turbo-coded OFDM schemes, *In Proc. Vehicular Technol. Conf. (VTC) 1999, Houston, Texas* , pp. 489–493.

[75] Keller, T. and Hanzo, L. (2000). Adaptive multicarrier modulation: A convenient framework for time-frequency processing in wireless communications, *Proceedings of the IEEE* **88**, pp. 611–640.

[76] Keller, T., Piazzo, L., Mandarini, P. and Hanzo, L. (2001). Orthogonal frequency division multiplex synchronization techniques for frequency-selective fading channels, *IEEE Journal Select. Areas Commun.* **19**, 6, pp. 999–1008.

[77] Kivanc, D., Li, G. and Liu, H. (2003). Computationally efficient bandwidth allocation and power control for OFDMA, *IEEE Trans. Wireless Commun.* **2**, pp. 1150–1158.

[78] Knopp, R. and Humblet, P. A. (1995). Information capacity and power control in single-cell multiuser communications, *In Proc. IEEE ICC'95, Seattle, WA* , pp. 331–335.

[79] Koetter, R., Singer, A. and Tuchler, M. (2004). Turbo equalization, *IEEE*

Signal Processing Magazine **21**, pp. 67–80.

[80] Koutsopoulos, I. and Tassiulas, L. (2002). Adaptive resource allocation in SDMA-based wireless broadband networks with OFDM signaling, *In Proc. IEEE INFOCOM 2002, New York City, New York* **3**, pp. 1376–1385.

[81] Koutsopoulos, I. and Tassiulas, L. (2006). Cross-layer adaptive techniques for throughput enhancement in wireless OFDM-based networks, *IEEE/ACM Trans. Networking* **14**, pp. 1056–1066.

[82] Krongold, B., Ramchandran, K. and Jones, D. (2000). Computationally efficient optimal power allocation algorithms for multicarrier communication systems, *IEEE Trans. Commun.* **48**, pp. 23–27.

[83] Lee, J., Sonalkar, R. and Cioffi, J. (2002a). Multi-user discrete bit-loading for DMT-based DSL systems, *In Proc. IEEE GLOBECOM 2002, Taipei, Taiwan* **2**, pp. 1259–1263.

[84] Lee, J., Sonalkar, R. and Cioffi, J. (2002b). A multi-user rate and power control algorithm for VDSL, *In Proc. IEEE GLOBECOM 2002, Taipei, Taiwan* **2**, pp. 1264–1268.

[85] Lee, J., Sonalkar, R. and Cioffi, J. (2006). Multiuser bit loading for multicarrier-systems, *IEEE Trans. Commun.* **54**, pp. 1170–1174.

[86] Li, C. and Roy, S. (2003). Subspace-based blind channel estimation for OFDM by exploiting virtual carriers filters, *IEEE Trans. Wireless Commun.* **2**, pp. 141–150.

[87] Li, J., Kim, H., Lee, Y. and Kim, Y. (2003). A novel broadband wireless OFDMA scheme for downlink in cellular communications, *In Proc. IEEE WCNC 2003, New Orleans, Louisiana* , pp. 1907–1911.

[88] Li, X. and Cimini, L. J. (1998). Effects of clipping and filtering on the performance of OFDM, *IEEE Commun. Letters* **2**, pp. 131–133.

[89] Li, Y., Chuang, J. and Sollenberger, N. (1999). Transmitter diversity for OFDM systems and its impact on high-rate data wireless networks, *IEEE Journal Select. Areas Commun.* **17**, pp. 1233–1243.

[90] Li, Y., Cimini, L. J. and Sollenberger, N. (1998). Robust channel estimation for OFDM systems with rapid dispersive fading channels, *IEEE Trans. Commun.* **46**, pp. 902–915.

[91] Luise, M., Reggiannini, R. and Vitetta, G. (1998). Blind equalization/detection for OFDM signals over frequency-selective channels, *IEEE Journal Select. Areas Commun.* **16**, pp. 1568–1578.

[92] Lutkepohl, H. (1996). *Handbook of Matrices* (John Wiley & Sons Ltd, West Sussex).

[93] Manton, J. (2001). Optimal training sequences and pilot tones for OFDM systems, *IEEE Commun. Letters* **5**, pp. 151–153.

[94] McLachlan, G. J. and Krishnan, T. (1997). *The EM Algorithm and Extensions* (John Wiley & Sons, Inc).

[95] Minn, H., Bhargava, V. and Letaief, K. (2003). A robust timing and frequency synchronization for OFDM systems, *IEEE Trans. on Wireless Commun.* **2**, 4, pp. 822–839.

[96] Moose, P. (1994). A technique for orthogonal frequency division multiplexing frequency offset correction, *IEEE Trans. Commun.* **42**, 10, pp. 2908–

2914.

[97] Morelli, M. (2004). Timing and frequency synchronization for the uplink of an OFDMA system, *IEEE Trans. Commun.* **52**, 2, pp. 296–306.

[98] Morelli, M., D'Andrea, A. and Mengali, U. (2001). Feedback frequency synchronization for OFDM applications, *IEEE Commun. Letters* **5**, 1, pp. 28–30.

[99] Morelli, M. and Mengali, U. (1999). An improved frequency offset estimator for OFDM applications, *IEEE Commun. Letters* **3**, 3, pp. 75–77.

[100] Morelli, M. and Mengali, U. (2000). Carrier-frequency estimation for transmissions over selective channels, *IEEE Trans. Commun.* **48**, 9, pp. 1580–1589.

[101] Morelli, M. and Mengali, U. (2001). A comparison of pilot-aided channel estimation methods for OFDM systems, *IEEE Trans. Signal Proc.* **49**, pp. 3065–3073.

[102] Morelli, M. and Sanguinetti, L. (2005). Estimation of channel statistics for iterative detection of OFDM signals, *IEEE Trans. Wireless Commun.* **4**, pp. 1360–1365.

[103] Muquet, B., de Courville, M. and Duhamel, P. (2001). Subspace-based blind and semi-blind channel estimation for OFDM systems, *IEEE Trans. Signal Proc.* **50**, pp. 1699–1712.

[104] Nee, R. V. and Prasad, R. (2000). *OFDM for wireless multimedia communications* (Artech House Publishers).

[105] Neeser, F. and Massey, J. (1993). Proper complex random processes with applications to information theory, *IEEE Trans. Info. Theory* **39**, pp. 1292–1302.

[106] Negi, R. and Cioffi, J. (1998). Pilot tone selection for channel estimation in a mobile OFDM system, *IEEE Trans. Consumer Electronics* **44**, pp. 1122–1128.

[107] Nogami, H. and Nagashima, T. (1995). A frequency and timing period acquisition technique for OFDM systems, *In Proc. Personal, Indoor and Mobile Radio Communications (PIMRC) 1995, Toronto, Canada* **3**, pp. 1010–1015.

[108] Ochiai, H. and Imai, H. (2000). Performance of the deliberate clipping with adaptive symbol selection for strictly band-limited OFDM systems, *IEEE Journal Select. Areas Commun.* **18**, pp. 2270–2277.

[109] Ochiai, H. and Imai, H. (2001). On the distribution of the peak-to-average power ratio in OFDM signals, *IEEE Trans. Commun.* **49**, 2, pp. 282–289.

[110] Ochiai, H. and Imai, H. (2002). Performance analysis of deliberately clipped OFDM signals, *IEEE Trans. Commun.* **50**, 1, pp. 89–101.

[111] O'Neill, R. and Lopes, L. (1995). Envelope variations and spectral splatter in clipped multicarrier signals, *In Proc. IEEE PIMRC 1995, Toronto, Canada* , pp. 71–75.

[112] P802.16a/D3-2001, I. (2002). *Draft Amendment to IEEE Standard for Local and metropolitan area networks, Part 16: Air Interface for Fixed Broadband Wireless Access Systems-Amendment 2: Medium Access Control Modifications and Additional Physical Layer Specifications for 2-11 GHz.*

[113] P802.16e/D4-2004, I. (2004). *Draft IEEE Standard for Local and Metropolitan Area Networks - Part 16: Air Interface for Fixed Broadband Wireless Access Systems.*

[114] Papandreou, N. and Antonakopoulos, T. (2005). A new computationally efficient discrete bit-loading algorithm for DMT applications, *IEEE Trans. Commun.* **53**, pp. 785–789.

[115] Papoulis, A. (1991). *Probability, Random Variables, and Stochastic Processes*, 3rd edn. (McGraw-Hill).

[116] Park, S., Kim, Y. and Kang, C. (2004). Iterative receiver for joint detection and channel estimation in OFDM systems under mobile radio channels, *IEEE Trans. Vehicular Technology* **53**, 5, pp. 1316–1326.

[117] Paterson, K. (2000). Generalized Reed–Muller codes and power control in OFDM modulation, *IEEE Trans. Info. Theory* **46**, 1, pp. 104–120.

[118] Pollet, T., Spruyt, P. and Moeneclaey, M. (1994). The BER performance of OFDM systems using non-synchronized sampling, *In Proc. Globecom 1994, San Francisco, California* **1**, pp. 253–257.

[119] Popovic, B. (1991). Synthesis of power efficient multitone signals with flat amplitude spectrum, *IEEE Trans. Commun.* **39**, pp. 1031–1033.

[120] Porter, G. C. (1968). Error distribution and diversity performance of a frequency differential PSK HF modem, *IEEE Trans. Commun.* **COM-16**, pp. 567–575.

[121] Powers, E. and Zimmerman, M. (1968). A digital implementation of a multichannel data modem, *Proc. IEEE ICC, Philadelphia, Pennsylvania* .

[122] Press, W., Teukolsky, S., Vetterling, W. and Flannery, B. (1992). *Numerical Recipes in C - The Art of Scientific Computing*, 2nd edn. (Cambridge Univ. Press, New York).

[123] Proakis, J. (2001). *Digital Communications*, 4th edn. (McGraw Hill).

[124] Pun, M., Morelli, M. and Kuo, C.-C. J. (2005). Joint synchronization and channel estimation in uplink OFDMA systems, *in Proc. ICASSP2005, Philadelphia, Pennsylvania* , pp. 857–860.

[125] Pun, M., Morelli, M. and Kuo, C.-C. J. (2006). Maximum-likelihood synchronization and channel estimation for OFDMA uplink transmissions, *IEEE Trans. Commun.* **54**, 4, pp. 726–736.

[126] Pun, M., Morelli, M. and Kuo, C.-C. J. (2007). Iterative detection and frequency synchronization for OFDMA uplink transmissions, *IEEE Trans. Wireless Commun.* **6**, 2, pp. 629–639.

[127] Pun, M., Tsai, S. and Kuo, C.-C. J. (2004). Joint maximum likelihood estimation of carrier frequency offset and channel for uplink OFDMA systems, *in Proc. Globecom 2004, Dallas, Texas* , pp. 3748 – 3752.

[128] Raheli, R., Polydoros, A. and Tzou, C.-K. (1995). Per-survivor processing: A general approach to MLSE in uncertain environments, *IEEE Trans. Commun.* **43**, pp. 354–364.

[129] RCT, E. D. (March 2001). *Interaction Channel for Digital Terrestrial Television (RCT) Incorporating Multiple Access OFDM.*

[130] Rhee, W. and Cioffi, J. (2000). Increase in capacity of multiuser OFDM system using dynamic subchannel allocation, *In Proc. IEEE VTC-2000*

Spring, Tokyo, Japan **2**, pp. 1085–1089.

[131] Ring, D. (1947). *Mobile Telephony - wide area coverage* (Technical Report, Bell Laboratories).

[132] Rinne, J. and Renfors, M. (1996). Pilot spacing in orthogonal frequency division multiplexing systems on practical channels, *IEEE Trans. Consum. Electron.* **42**, pp. 959–962.

[133] Rohling, H. and Grunheid, R. (1996). Performance of an OFDM-TDMA mobile communication system, *in Proc. IEEE VTC Spring 1996, Melbourne, Australia* , pp. 1589–1593.

[134] Rowe, H. (1982). Memoryless nonlinearities with gaussian inputs: Elementary results, *Bell Syst. Tech. J.* **61**, pp. 1519–1525.

[135] Roy, R. and Kailath, T. (1989). ESPRIT-estimation of signal parameters via rotational invariant techniques, *IEEE Trans. Acoustic, Speech and Signal Proc.* **37**, pp. 984–995.

[136] R.W.Shafer and L.R.Rabiner (1973). A digital signal processing approach to interpolation, *Proceedings of IEEE* **61**, pp. 692–702.

[137] Saeedi, H., Sharif, M. and Marvasti, F. (2002). Clipping noise cancellation in OFDM systems using oversampled signal reconstruction, *IEEE Commun. Letters* **6**, 2, pp. 73–75.

[138] Salkintzis, A. (2004). Interworking techniques and architectures for WLAN/3G integration toward 4G mobile data networks, *IEEE Personal Comm.* **11**, pp. 50–61.

[139] Saltzberg, B. (1967). Performance of an efficient parallel data transmission system, *IEEE Trans. Commun. Technology* **COM-15**, pp. 805–811.

[140] Sari, H. and Karam, G. (1998). Orthogonal frequency-division multiple access and its application to CATV networks, *European Trans. Commun.* **45**, pp. 507–516.

[141] Sari, H., Levy, Y. and Karam, G. (1996). OFDMA: A new multiple access technique and its application to interactive CATV networks, *in Proc. European Conference on Multimedia Applications, Services and Techniques* , pp. 117–127.

[142] Schmidl, T. and Cox, D. (1997). Robust frequency and timing synchronization for OFDM, *IEEE Trans. Commun.* **45**, 12, pp. 1613–1621.

[143] Schmidt, R. (1986). Multiple emitter location and signal parameter estimation, *In Proc. RADC Spectral Estimation Workshop* **34**, 3, pp. 243–258.

[144] Sharif, M., Gharavi-Alkhansari, M. and Khalaj, B. (2003). On the peak-to-average power of OFDM signals based on oversampling, *IEEE Trans. Commun.* **51**, 1, pp. 72–78.

[145] Sharif, M. and Khalaj, B. (2001). Peak to mean envelope power ratio of oversampled OFDM signals: An analytical approach, *In Proc. IEEE Int. Conf. Comm. (ICC), St. Petersburg, Russia* **5**, pp. 1476–1480.

[146] Shi, K. and Serpedin, E. (2004). Coarse frame and carrier synchronization of OFDM systems: a new metric and comparison, *IEEE Trans. on Wireless Commun.* **3**, 4, pp. 1271–1284.

[147] Song, G. and Li, Y. (2005). Cross-layer optimization for OFDM wireless networks-part I: theoretical framework, *IEEE Trans. Wireless Commun.* **4**,

pp. 614–624.

[148] Speth, M., Fechtel, S., Fock, G. and Meyr, H. (1999). Optimum receiver design for wireless broadband systems using OFDM, Part I, *IEEE Trans. Commun.* **47**, pp. 1668–1677.

[149] Stuber, G., Barry, J., McLaughlin, S., Li, Y., Ingram, M. and Pratt, T. (2004). Broadband MIMO-OFDM wireless communications, *Proceedings of the IEEE* **92**, 2, pp. 271–294.

[150] Tarokh, V. and Jafarkhani, H. (2000). On the computation and reduction of the peak-to-average power ratio in multicarrier communications, *IEEE Trans. Commun.* **48**, pp. 37–44.

[151] Tarokh, V., Jafarkhani, H. and Calderbank, A. R. (1999). Space-time block codes from orthogonal designs, *IEEE Trans. Inform. Theory* **45**, pp. 1456–1467.

[152] Telatar, I. (1999). Capacity of multi-antenna gaussian channels, *European Trans. Telecommun. (ETT)* **10**, pp. 585–595.

[153] Tellado, J. (1999). *Peak to average power reduction for muticarrier modulation* (Ph.D. dissertation, Stanford University).

[154] Tellado, J. and Cioffi, J. (1997). PAR reduction in multicarrier transmission systems, *ANSI Document, T1E1.4 Technical Subcommittee* **97-367**, pp. 97–367.

[155] Tellambura, C. (2001a). Computation of the continuous-time PAR of an OFDM signal with BPSK subcarriers, *IEEE Commun. Letters* **5**, 5, pp. 185–187.

[156] Tellambura, C. (2001b). Improved phase factor computation for the PAR reduction of an OFDM signal using PTS, *IEEE Commun. Letters* **5**, 4, pp. 135–137.

[157] Thoen, S., der Perre, L. V., Engels, M. and Man, H. D. (2002). Adaptive loading for OFDM/SDMA-based wireless networks, *IEEE Trans. Commun.* **50**, pp. 1798–1810.

[158] Tonello, A. (2002). Multiuser detection and turbo multiuser decoding for asynchronous multitone multiple access systems, *in Proc. IEEE Vehicular Technology Conf., Vancouver, BC, Canada* , pp. 970–974.

[159] Tuttlebee, W. H. W. (1999). Software-defined radio: facets of a developing technology, *IEEE Personal Comm.* **6**, pp. 38–44.

[160] UMTS, E. (1998). *Universal Mobile Telecommunication System (UMTS) (TR 101 112),*.

[161] UTRA (1998). *Submission of Proposed Radio Transmission Technologies, SMG2*.

[162] van de Beek, J., Börjesson, P., Boucheret, M., Landström, D., Arenas, J., Ödling, O., Östberg, C., Wahlqvist, M. and Wilson, S. (1999). A time and frequency synchronization scheme for multiuser OFDM, *IEEE Journal Select. Areas Commun.* **17**, 11, pp. 1900–1914.

[163] van de Beek, J., Sandell, M. and Borjesson, P. (1997). ML estimation of timing and frequency offset in OFDM systems, *IEEE Trans. Signal Proc.* **45**, 7, pp. 1800–1805.

[164] Verdù, S. (1998). *Multiuser Detection* (Cambridge University Press: Cam-

bridge, UK).

[165] Viswanath, P., Tse, D. and Laroia, R. (2002). Opportunistic beamforming using dumb antennas, *IEEE Trans. Info. theory* **48**, pp. 1277–1294.

[166] Wang, H. and Chen, B. (2004). Asymptotic distributions and peak power analysis for uplink OFDMA signals, *In Proc. IEEE ICASSP, Montreal, Canada* **4**, pp. 17–21.

[167] Wang, X. and Liu, K. (1999). Adaptive channel estimation using cyclic prefix in multicarrier modulation system, *IEEE Commun. Letters* **3**, pp. 291–293.

[168] Warner, W. and Leung, C. (1993). OFDM/FM frame synchronization for mobile radio data communication, *IEEE Trans. on Vehicular Technology* **42**, 3, pp. 302–313.

[169] Wax, M. and Kailath, T. (1985). Detection of signals by information theoretic criteria, *IEEE Trans. Acoustic, Speech and Signal Proc.* **ASSP-33**, pp. 387–392.

[170] Wiesler, A. and Jondral, F. K. (2002). A software radio for second- and third-generation mobile systems, *IEEE Trans. on Vehicular Technology* **51**, pp. 738–748.

[171] Willink, T. and Wittke, P. (1997). Optimization and performance evaluation of multicarrier transmission, *IEEE Trans. Info. Theory* **43**, pp. 426–440.

[172] Wong, C., Cheng, R., Letaief, K. and Murch, R. (1999). Multiuser OFDM with adaptive subcarrier, bit and power allocation, *IEEE Journal Select. Areas Commun.* **17**, pp. 1747–1757.

[173] Wyglinski, A., Labeau, F. and Kabal, P. (2005). Bit loading with BER-constraint for multicarrier systems, *IEEE Trans. Wireless Commun.* **4**, pp. 1383–1387.

[174] Xia, P., Zhou, S. and Giannakis;, G. (2004). Adaptive MIMO-OFDM based on partial channel state information, *IEEE Trans. Signal Processing* **52**, pp. 202–213.

[175] Xia, P., Zhou, S. and Giannakis;, G. (2005). Multiantenna adaptive modulation with beamforming based on bandwidth-constrained feedback, *IEEE Trans. Commun.* **53**, pp. 526–536.

[176] Xie, Y. and Georghiades, C. (2003). Two EM-type channel estimation algorithms for OFDM with transmitter diversity, *IEEE Trans. Commun.* **51**, pp. 106–115.

[177] Yaghoobi, H. (2004). Scalable OFDMA physical layer in IEEE 802.16 WirelessMAN, *Intel Technology Journal* **8**, pp. 201–212.

[178] Yang, B., Letaief, K., Cheng, R. and Cao, Z. (2000). Timing recovery for OFDM transmission, *IEEE Journal Select. Areas Commun.* **18**, 11, pp. 2278–2291.

[179] Yang, B., Letaief, K., Cheng, R. and Cao, Z. (2001). Channel estimation for OFDM transmission in multipath fading channels based on parametric channel modeling, *IEEE Trans. Commun.* **49**, pp. 467–479.

[180] Yu, W. and Cioffi, J. (2001). On constant power water-filling, *In Proc. IEEE GLOBECOM 2001, San Antonio, Texas* , pp. 1665–1669.

[181] Zhang, Y. and Letaief, K. (2005). An efficient resource-allocation scheme for spatial multiuser access in MIMO/OFDM systems, *IEEE Trans. Commun.* **53**, pp. 107–116.

[182] Zhang, Y., Yongacoglu, A. and Chouinard, J.-Y. (2000). Orthogonal frequency division multiple access peak-to-average power ratio reduction using optimized pilot symbols, *In Proc. ICCT 2000, Beijing, China* , pp. 574–577.

[183] Zhou, S. and Giannakis, G. (2001). Finite-alphabet based channel estimation for OFDM and related multicarrier systems, *IEEE Trans. Commun.* **49**, pp. 1402–1414.

[184] Zimmerman, M. and Kirsch, A. (1967). The AN/GSC-10 (KATHRYN) variable rate data modem for HF radio, *IEEE Trans. on Commun.* **COM-15**, pp. 197–205.

Index